SI Base Units

Base Quantity	Name of Unit	Symbol
Length	meter	m
Mass	kilogram	kg
Time	second	s
Electrical current	ampere	A
Temperature	kelvin	K
Amount of substance	mole	mol
Luminous intensity	candela	cd

Derived Units in the SI System

Physical Quantity	Name	Symbol	Units
Energy	joule	J	$\text{kg m}^2\text{ s}^{-2}$
Force	newton	N	kg m s^{-2}
Power	watt	W	$\text{kg m}^2\text{ s}^{-3}$
Electric charge	coulomb	C	A s
Electrical resistance	ohm	Ω	$\text{kg m}^2\text{ s}^{-3}\text{ A}^{-2}$
Electrical potential difference	volt	V	$\text{kg m}^2\text{ s}^{-3}\text{ A}^{-1}$
Electrical capacitance	farad	F	$\text{kg}^{-1}\text{ m}^{-2}\text{ s}^4\text{ A}^2$
Frequency	hertz	Hz	s^{-1}

Prefixes Used with SI Units

Prefix	Symbol	Meaning
Peta-	P	10^{15} or 1,000,000,000,000,000
Tera-	T	10^{12} or 1,000,000,000,000
Giga-	G	10^9 or 1,000,000,000
Mega-	M	10^6 or 1,000,000
Kilo-	k	10^3 or 1,000
Deci-	d	10^{-1} or 1/10
Centi-	c	10^{-2} or 1/100
Milli-	m	10^{-3} or 1/1,000
Micro-	μ	10^{-6} or 1/1,000,000
Nano-	n	10^{-9} or 1/1,000,000,000
Pico-	p	10^{-12} or 1/1,000,000,000,000
Femto-	f	10^{-15} or 1/1,000,000,000,000,000
Atto-	a	10^{-18} or 1/1,000,000,000,000,000,000

Periodic Table of the Elements

1 1A	2 2A	3 3B	4 4B	5 5B	6 6B	7 7B	8 8B	9 8B	10	11 1B	12 2B	13 3A	14 4A	15 5A	16 6A	17 7A	18 8A
1 **H** 1.008																	2 **He** 4.003
3 **Li** 6.941	4 **Be** 9.012											5 **B** 10.81	6 **C** 12.01	7 **N** 14.01	8 **O** 16.00	9 **F** 19.00	10 **Ne** 20.18
11 **Na** 22.99	12 **Mg** 24.31											13 **Al** 26.98	14 **Si** 28.09	15 **P** 30.97	16 **S** 32.07	17 **Cl** 35.45	18 **Ar** 39.95
19 **K** 39.10	20 **Ca** 40.08	21 **Sc** 44.96	22 **Ti** 47.87	23 **V** 50.94	24 **Cr** 52.00	25 **Mn** 54.94	26 **Fe** 55.85	27 **Co** 58.93	28 **Ni** 58.69	29 **Cu** 63.55	30 **Zn** 65.38	31 **Ga** 69.72	32 **Ge** 72.63	33 **As** 74.92	34 **Se** 78.96	35 **Br** 79.90	36 **Kr** 83.80
37 **Rb** 85.47	38 **Sr** 87.62	39 **Y** 88.91	40 **Zr** 91.22	41 **Nb** 92.91	42 **Mo** 95.96	43 **Tc** (98)	44 **Ru** 101.1	45 **Rh** 102.9	46 **Pd** 106.4	47 **Ag** 107.9	48 **Cd** 112.4	49 **In** 114.8	50 **Sn** 118.7	51 **Sb** 121.8	52 **Te** 127.6	53 **I** 126.9	54 **Xe** 131.3
55 **Cs** 132.9	56 **Ba** 137.3	71 **Lu** 175.0	72 **Hf** 178.5	73 **Ta** 180.9	74 **W** 183.8	75 **Re** 186.2	76 **Os** 190.2	77 **Ir** 192.2	78 **Pt** 195.1	79 **Au** 197.0	80 **Hg** 200.6	81 **Tl** 204.4	82 **Pb** 207.2	83 **Bi** 209.0	84 **Po** (209)	85 **At** (210)	86 **Rn** (222)
87 **Fr** (223)	88 **Ra** (226)	103 **Lr** (262)	104 **Rf** (265)	105 **Db** (268)	106 **Sg** (271)	107 **Bh** (272)	108 **Hs** (277)	109 **Mt** (276)	110 **Ds** (281)	111 **Rg** (280)	112 **Cn** (285)	113 (284)	114 **Fl** (289)	115 (288)	116 **Lv** (293)	117 (294)	118 (294)

57 **La** 138.9	58 **Ce** 140.1	59 **Pr** 140.9	60 **Nd** 144.2	61 **Pm** (145)	62 **Sm** 150.4	63 **Eu** 152.0	64 **Gd** 157.3	65 **Tb** 158.9	66 **Dy** 162.5	67 **Ho** 164.9	68 **Er** 167.3	69 **Tm** 168.9	70 **Yb** 173.1
89 **Ac** (227)	90 **Th** 232.0	91 **Pa** 231.0	92 **U** 238.0	93 **Np** (237)	94 **Pu** (244)	95 **Am** (243)	96 **Cm** (247)	97 **Bk** (247)	98 **Cf** (251)	99 **Es** (252)	100 **Fm** (257)	101 **Md** (258)	102 **No** (259)

The 1–18 group designation has been recommended by the International Union of Pure and Applied Chemistry (IUPAC). As of 2013, no names have been assigned for elements 113, 115, 117, and 118.

Problems and Solutions

to accompany

RAYMOND CHANG
JOHN W. THOMAN, JR.

PHYSICAL CHEMISTRY
for the Chemical Sciences

Problems and Solutions

to accompany

RAYMOND CHANG
JOHN W. THOMAN, JR.

PHYSICAL CHEMISTRY
for the Chemical Sciences

Helen O. Leung
AMHERST COLLEGE

Mark D. Marshall

AMHERST COLLEGE

A University Science Book

The MIT Press
Cambridge, Massachusetts
London, England

The MIT Press
Massachusetts Institute of Technology
77 Massachusetts Avenue, Cambridge, MA 02139
mitpress.mit.edu

Print Text ISBN: 978-1-938787-69-0
eText ISBN: 978-1-938787-73-7

Library of Congress Control Number: 2014951009

Printed in the United States of America
10 9 8

EU Authorised Representative: Easy Access System Europe, Mustamäe tee 50,
10621 Tallinn, Estonia I Email: gpsr.requests@easproject.com

Contents

Preface vii

CHAPTER 1 **Introduction and Gas Laws** 1

2 **Kinetic Theory of Gases** 21

3 **The First Law of Thermodynamics** 39

4 **The Second Law of Thermodynamics** 59

5 **Gibbs and Helmholtz Energies and Their Applications** 73

6 **Nonelectrolyte Solutions** 83

7 **Electrolyte Solutions** 101

8 **Chemical Equilibrium** 111

9 **Electrochemistry** 129

10 **Quantum Mechanics** 143

11 **Applications of Quantum Mechanics to Spectroscopy** 165

12 **Electronic Structure of Atoms** 181

13 **Molecular Electronic Structure and the Chemical Bond** 199

14 **Electronic Spectroscopy and Magnetic Resonance Spectroscopy** 215

15 **Chemical Kinetics** 227

16 **Photochemistry** 255

17 **Intermolecular Forces** 263

18 **The Solid State** 271

19 **The Liquid State** 279

20 **Statistical Thermodynamics** 287

Preface

Twenty three years ago the Chemistry Department at Williams College, of which Raymond Chang was the chair at the time, needed a sabbatical replacement for a young physical chemist named John (Jay) W. Thoman, Jr. The department chose to hire Helen, who was looking for her first faculty position. Thus began a very rewarding collaboration that continues to this day. Raymond Chang is a gifted teacher and mentor, and his patient advice was valuable to the professional development of a new colleague just beginning her career. Although away on leave, Jay Thoman was a positive influence as well, providing a "near-peer" example, full of energy and posing exciting questions about the nature of our field. It is a delight to reassemble this dream team and to have Mark officially join it.

The only way to really learn chemistry is to do chemistry. Our intention for this solutions manual is to encourage students of physical chemistry to work many problems—even more than they are assigned! We know that our own mastery of the subject improved by solving all of the problems posed by Raymond Chang and Jay Thoman in their textbook and that new, previously undiscovered connections were revealed to us as we discussed how to approach each one. We hope that students, too, will discuss these solutions because explaining how a question is answered is the best way of being certain that you understand the material yourself.

One thing we all agree on is that there is more to science than simply getting the right answer. Science is a process for recognizing important questions and how to set about seeking answers to those questions. Conversation is an important part of that process, and with our own students we would be certain to discuss the assumptions involved in arriving at a solution and to explain the intermediate steps used to determine an answer. We have tried to approximate that experience in this solutions manual by adopting a similar style; only for the simplest problems are the answers merely quoted.

We have made every effort to ensure consistency in the values of the physical constants used, both internally and with those given in the textbook. Similarly, we strived for a careful and consistent choice of units and symbols. Of course, every measurement has an associated uncertainty that must be considered when comparing results from different experiments or those of experiment with predictions from theory. It is good to practice this aspect of science, too, and we have done so by paying attention to the number of significant figures used in answering each problem. The appropriate number of significant figures is determined by the data given. So, for example, if a pressure is given as 0.729 atm, then the answer to the problem should contain three significant figures. Quantities given as "1 mole" or "5 L" are taken to be exact. Intermediate calculations are done retaining one extra digit, and the final answer is rounded to the correct number of significant figures as a last step. (This is important; rounding too early will cause a loss of significance in your calculations.) If a numerical result from a previous part of a problem is used in a subsequent part, the value with the extra digit is used. In such cases, the answer to the earlier part would look similar to "98.7654 kJ mol^{-1} = 98.765 kJ mol^{-1}," where the first value, 98.7654 kJ mol^{-1}, is used later in the problem and the "correct" answer is the second. Of course, every rule has its exceptions, and we did choose to be more flexible in problems involving exponentiation and logarithms.

Throughout the preparation of this manual, Raymond and Jay provided us with extremely valuable guidance; we are grateful for the opportunity once again to work with them. Jane Ellis of University Science Books organized all the details, smoothed out all the wrinkles that inevitably occur in a project like this, and kept us calm and on task when we would begin to drift. Paul C. Anagnostopoulos of Windfall Software was most kind in dealing with our rudimentary, sometimes regrettable, LaTeX in composing the book. Finally, they say you can best judge executives by the quality of the team they assemble, and by this measure Bruce and Kathy Armbruster are outstanding. They are supportive and encouraging, and it is a true pleasure having them as our publishers.

We don't know if Zephaniah Swift Moore, who left Williams to found Amherst College in 1821, had fond memories of his days in Williamstown, but we do—especially of the colleagues left behind.

Helen O. Leung
Mark D. Marshall
Amherst, Massachusetts

Introduction and Gas Laws

PROBLEMS AND SOLUTIONS

1.2 Some gases, such as NO_2 and NF_2, do not obey Boyle's law at any pressure. Explain.

Boyle's law applies only at constant n and constant T. Both NO_2 and NF_2 undergo association reactions:

$$2NO_2 \rightleftharpoons N_2O_4$$

$$2NF_2 \rightleftharpoons N_2F_4$$

so n is not constant for these gases.

1.4 Some ballpoint pens have a small hole in the main body of the pen. What is the purpose of this hole?

As the pen is used and the ink leaves the pen, the gas inside expands to fill the volume left by the ink. As the volume increases, the pressure inside the pen decreases. The hole is needed to equalize the pressure to allow for easy flow of ink.

1.6 At STP (standard temperature and pressure), 0.280 L of a gas weighs 0.400 g. Calculate the molar mass of the gas.

Following the procedure in Problem 1.5,

$$\mathcal{M} = \frac{mRT}{VP} = \frac{(0.400 \text{ g}) \left(0.08206 \text{ L atm K}^{-1} \text{mol}^{-1}\right) (273.15 \text{ K})}{(0.280 \text{ L}) (1.00 \text{ atm})} = 32.0 \text{ g mol}^{-1}$$

1.8 Calculate the density of HBr in g L^{-1} at 733 mmHg and 46°C. Assume ideal-gas behavior.

Use the expression found in Problem 1.5 for the density of an ideal gas in terms of its pressure, molar mass, and temperature.

$$\rho = \frac{P\mathcal{M}}{RT} = \frac{(733 \text{ mmHg}) \left(\frac{1 \text{ atm}}{760 \text{ mmHg}}\right) (80.91 \text{ g mol}^{-1})}{(0.08206 \text{ L atm K}^{-1} \text{ mol}^{-1}) (273.15 + 46) \text{ K}} = 2.98 \text{ g L}^{-1}$$

1.10 The saturated vapor pressure of mercury is 0.0020 mmHg at 300 K and the density of air at 300 K is 1.18 g L^{-1}. **(a)** Calculate the concentration of mercury vapor in air in mol L^{-1}. **(b)** What is the number of parts per million (ppm) by mass of mercury in air?

(a) An expression for the concentration of mercury vapor can be obtained by rearranging the ideal gas equation, $PV = nRT$.

$$\text{Concentration} = \frac{n}{V} = \frac{P}{RT}$$

$$= \frac{(0.0020 \text{ mmHg}) \left(\frac{1 \text{ atm}}{760 \text{ mmHg}}\right)}{(0.08206 \text{ L atm K}^{-1} \text{ mol}^{-1}) (300 \text{ K})}$$

$$= 1.07 \times 10^{-7} \text{ mol L}^{-1}$$

$$= 1.1 \times 10^{-7} \text{ mol L}^{-1}$$

(b) In 1 L,

$$n_{\text{Hg}} = 1.07 \times 10^{-7} \text{ mol}$$

$$m_{\text{Hg}} = \left(1.07 \times 10^{-7} \text{ mol}\right) \left(200.6 \text{ g mol}^{-1}\right) = 2.15 \times 10^{-5} \text{ g}$$

$$m_{\text{air}} = 1.18 \text{ g}$$

$$m_{\text{total}} = m_{\text{air}} + m_{\text{Hg}} \approx m_{\text{air}} = 1.18 \text{ g}$$

$$\text{ppm of Hg in air} = \frac{m_{\text{Hg}}}{m_{\text{total}}} \times 10^6 = \frac{2.15 \times 10^{-5} \text{ g}}{1.18 \text{ g}} \times 10^6 = 18$$

1.12 Sodium bicarbonate (NaHCO$_3$) is called baking soda because when heated, it releases carbon dioxide gas, which causes cookies, doughnuts, and bread to rise during baking. **(a)** Calculate the volume (in liters) of CO$_2$ produced by heating 5.0 g of NaHCO$_3$ at 180°C and 1.3 atm. **(b)** Ammonium bicarbonate (NH$_4$HCO$_3$) has also been used as a leavening agent. Suggest one advantage and one disadvantage of using NH$_4$HCO$_3$ instead of NaHCO$_3$ for baking.

(a) The following reaction takes place when sodium bicarbonate is heated:

$$2\text{NaHCO}_3(s) \longrightarrow \text{Na}_2\text{CO}_3(s) + \text{H}_2\text{O}(g) + \text{CO}_2(g)$$

$$n_{\text{NaHCO}_3} = \frac{5.0 \text{ g}}{84.0 \text{ g mol}^{-1}} = 5.95 \times 10^{-2} \text{ mol}$$

$$n_{\text{CO}_2} = n_{\text{NaHCO}_3} \left(\frac{1 \text{ mol CO}_2}{2 \text{ mol NaHCO}_3}\right)$$

$$= \left(5.95 \times 10^{-2} \text{ mol NaHCO}_3\right) \left(\frac{1 \text{ mol CO}_2}{2 \text{ mol NaHCO}_3}\right) = 2.98 \times 10^{-2} \text{ mol}$$

$$V_{CO_2} = \frac{n_{CO_2}RT}{P}$$

$$= \frac{(2.98 \times 10^{-2}\text{ mol})\,(0.08206\text{ L atm K}^{-1}\text{ mol}^{-1})\,(273.15 + 180)\text{ K}}{1.3\text{ atm}} = 0.85\text{ L}$$

(b) Ammonium bicarbonate decomposes upon heating according to the following equation:

$$NH_4HCO_3(s) \longrightarrow NH_3(g) + H_2O(g) + CO_2(g)$$

The advantage in using the ammonium salt is that more gas is produced per gram of reactant. (The molar mass of NH_4HCO_3 is 79.1 g mol^{-1}, smaller than that of $NaHCO_3$.) The disadvantage is that one of the gases is ammonia. The strong odor of ammonia would not make the ammonium salt a good choice for baking.

1.14 **(a)** What volume of air at 1.0 atm and 22°C is needed to fill a 0.98-L bicycle tire to a pressure of 5.0 atm at the same temperature? (Note that 5.0 atm is the gauge pressure, which is the difference between the pressure in the tire and atmospheric pressure. Initially, the gauge pressure in the tire was 0 atm.) **(b)** What is the total pressure in the tire when the gauge reads 5.0 atm? **(c)** The tire is pumped with a hand pump full of air at 1.0 atm; compressing the gas in the cylinder adds all the air in the pump to the air in the tire. If the volume of the pump is 33% of the tire's volume, what is the gauge pressure in the tire after 3 full strokes of the pump?

(a) Enough gas (air) must be added to increase the pressure in the tire by 5.0 atm, since it starts at a gauge pressure of 0.0 atm. This is the same amount of gas whose volume is desired at ambient conditions. At constant n and T,

$$P_1V_1 = P_2V_2$$

Letting the pressure and volume in the tire be P_1 and V_1, respectively, and denoting the same quantities at ambient conditions as P_2 and V_2,

$$V_2 = \frac{P_1V_1}{P_2} = \frac{(5.0\text{ atm})(0.98\text{ L})}{1.0\text{ atm}} = 4.9\text{ L}$$

(b) When the gauge pressure reads 5.0 atm, the total pressure in the tire is 6.0 atm.

(c) Since the tire and the pump are at the same temperature, the amount of gas each contains at 1.0 atm is proportional to their volumes. Thus three strokes of the pump will add to the tire 99% of the amount of gas that the tire originally contains at 1.0 atm. At constant V and T,

$$\frac{P_1}{n_1} = \frac{P_2}{n_2}$$

Here, $n_2 = 1.99n_1$,

$$P_2 = \frac{(1.0\text{ atm})}{n_1}(1.99n_1) = 1.99\text{ atm}$$

When the total pressure in the tire is 1.99 atm, the gauge pressure reads 0.99 atm.

1.16 Nitrogen forms several gaseous oxides. One of them has a density of 1.27 g L^{-1} measured at 764 mmHg and 150°C. Write the formula of the compound.

The density of a gas is related to its molar mass as shown in Problem 1.5:

$$\mathcal{M} = \frac{\rho RT}{P} = \frac{\left(1.27 \text{ g L}^{-1}\right)\left(0.08206 \text{ L atm K}^{-1}\text{ mol}^{-1}\right)(273.15 + 150)\text{ K}}{(764 \text{ mmHg})\left(\frac{1 \text{ atm}}{760 \text{ mmHg}}\right)} = 43.9 \text{ g mol}^{-1}$$

Some nitrogen oxides and their molar masses are NO: 30.0 g mol^{-1}; N_2O: 44.0 g mol^{-1}; NO_2: 46.0 g mol^{-1}. The nitrogen oxide is N_2O.

1.18 An ultra-high-vacuum pump can reduce the pressure of air from 1.0 atm to 1.0×10^{-12} mmHg. Calculate the number of air molecules in a liter at this pressure and 298 K. Compare your results with the number of molecules in 1.0 L at 1.0 atm and 298 K. Assume ideal-gas behavior.

When $P = 1.0 \times 10^{-12}$ mmHg:

$$n = \frac{PV}{RT} = \frac{\left(1.0 \times 10^{-12} \text{ mmHg}\right)\left(\frac{1 \text{ atm}}{760 \text{ mmHg}}\right)(1.0 \text{ L})}{\left(0.08206 \text{ L atm K}^{-1}\text{ mol}^{-1}\right)(298 \text{ K})} = 5.38 \times 10^{-17} \text{ mol}$$

$$\text{Number of molecules} = \left(5.38 \times 10^{-17} \text{ mol}\right)\left(\frac{6.022 \times 10^{23} \text{ molecules}}{1 \text{ mol}}\right)$$

$$= 3.24 \times 10^{7} \text{ molecules} = 3.2 \times 10^{7} \text{ molecules}$$

When $P = 1.0$ atm:

$$n = \frac{PV}{RT} = \frac{(1.0 \text{ atm})(1.0 \text{ L})}{\left(0.08206 \text{ L atm K}^{-1}\text{ mol}^{-1}\right)(298 \text{ K})} = 0.0409 \text{ mol}$$

$$\text{Number of molecules} = (0.0409 \text{ mol})\left(\frac{6.022 \times 10^{23} \text{ molecules}}{1 \text{ mol}}\right)$$

$$= 2.46 \times 10^{22} \text{ molecules} = 2.5 \times 10^{22} \text{ molecules}$$

The number of molecules present when $P = 1.0$ atm is 7.6×10^{14} times greater than when $P = 1.0 \times 10^{-12}$ mmHg.

1.20 The density of dry air at 1.00 atm and 34.4°C is 1.15 g L^{-1}. Calculate the composition of air (percent by mass) assuming that it contains only nitrogen and oxygen and behaves like an ideal gas. (*Hint*: First calculate the "molar mass" of air, then the mole fractions, and then the mass fractions of O_2 and N_2.)

This problem is similar to Problem 1.17. From the density of air, calculate its molar mass, $\mathcal{M}_{\text{air}}$ (see Problem 1.5), which in turn yields the mole fraction of oxygen, x_{O_2}, and the mole fraction of nitrogen, x_{N_2}. Once the mole fractions are obtained, the composition of air can be calculated.

$$\mathcal{M}_{air} = \frac{\rho_{air}RT}{P_{air}} = \frac{\left(1.15 \text{ g L}^{-1}\right)\left(0.08206 \text{ L atm K}^{-1}\text{mol}^{-1}\right)(273.15 + 34.4)\text{ K}}{1.00 \text{ atm}}$$

$$= 29.02 \text{ g mol}^{-1}$$

$$x_{O_2}\mathcal{M}_{O_2} + x_{N_2}\mathcal{M}_{N_2} = \mathcal{M}_{air} = 29.02 \text{ g mol}^{-1}$$

The sum of all mole fractions is unity, that is, $x_{O_2} + x_{N_2} = 1$, or $x_{N_2} = 1 - x_{O_2}$. Use this relation in the above equation,

$$x_{O_2}\mathcal{M}_{O_2} + \left(1 - x_{O_2}\right)\mathcal{M}_{N_2} = 29.02 \text{ g mol}^{-1}$$

$$x_{O_2}\left(32.00 \text{ g mol}^{-1}\right) + \left(1 - x_{O_2}\right)\left(28.02 \text{ g mol}^{-1}\right) = 29.02 \text{ g mol}^{-1}$$

$$32.00x_{O_2} + 28.02 - 28.02x_{O_2} = 29.02$$

$$3.98x_{O_2} = 1.00$$

$$x_{O_2} = 0.251$$

Therefore, $x_{N_2} = 1 - x_{O_2} = 1 - 0.251 = 0.749$.

In 1 mol of air, there are 0.251 mol of O_2 and 0.749 mol of N_2. The corresponding masses are therefore:

$$\text{mass of } O_2 = (0.251 \text{ mol})\left(32.00 \text{ g mol}^{-1}\right) = 8.03 \text{ g}$$

$$\text{mass of } N_2 = (0.749 \text{ mol})\left(28.02 \text{ g mol}^{-1}\right) = 21.0 \text{ g}$$

Therefore,

$$\% \text{ } O_2 \text{ by mass} = \frac{8.03 \text{ g}}{8.03 \text{ g} + 21.0 \text{ g}} \times 100\% = 28\%$$

$$\% \text{ } N_2 \text{ by mass} = 1 - \% \text{ } O_2 \text{ by mass} = 1 - 28\% = 72\%$$

1.22 Two bulbs of volumes V_A and V_B are connected by a stopcock. The number of moles of gases in the bulbs are n_A and n_B, respectively, and initially the gases are at the same pressure, P, and temperature, T. Show that the final pressure of the system, after the stopcock has been opened, is equal to P. Assume ideal-gas behavior.

Gas A and gas B both obey the ideal gas equation, that is, before the stopcock is open,

$$PV_A = n_A RT$$
$$PV_B = n_B RT$$

When the stopcock is open,

$$V_{total} = V_A + V_B$$
$$n_{total} = n_A + n_B$$

The total pressure is

$$P_{total} = \frac{n_{total} RT}{V_{total}} = \frac{(n_A + n_B)\,RT}{V_A + V_B} = \frac{n_A RT + n_B RT}{V_A + V_B}$$

From above, $n_A RT = PV_A$ and $n_B RT = PV_B$. Upon substitution into the expression for P_{total},

$$P_{total} = \frac{PV_A + PV_B}{V_A + V_B} = \frac{P\,(V_A + V_B)}{V_A + V_B} = P$$

1.24 A mixture containing nitrogen and hydrogen weighs 3.50 g and occupies a volume of 7.46 L at 300 K and 1.00 atm. Calculate the mass percent of these two gases. Assume ideal-gas behavior.

First calculate the total number of moles of the mixture, n_{mix}, which, together with the mass of the mixture, m_{mix}, is used to determine the number of moles of N_2 and the number of moles of H_2, and consequently, the mass percent of these gases.

$$n_{mix} = \frac{PV}{RT} = \frac{(1.00\ \text{atm})\,(7.46\ \text{L})}{(0.08206\ \text{L atm K}^{-1}\,\text{mol}^{-1})\,(300\ \text{K})} = 0.3030\ \text{mol}$$

The mass of the mixture is

$$m_{mix} = n_{N_2}\mathcal{M}_{N_2} + n_{H_2}\mathcal{M}_{H_2} = 3.50\ \text{g}$$

Because $n_{N_2} + n_{H_2} = n_{mix} = 0.3030$ mol,

$$n_{H_2} = 0.3030\ \text{mol} - n_{N_2}$$

Therefore,

$$m_{mix} = n_{N_2}\mathcal{M}_{N_2} + \left(0.3030\ \text{mol} - n_{N_2}\right)\mathcal{M}_{H_2} = 3.50\ \text{g}$$

$$n_{N_2}\left(28.02\ \text{g mol}^{-1}\right) + \left(0.3030\ \text{mol} - n_{N_2}\right)\left(2.016\ \text{g mol}^{-1}\right) = 3.50\ \text{g}$$

$$28.02 n_{N_2}\ \text{mol}^{-1} + 0.6108 - 2.016 n_{N_2}\ \text{mol}^{-1} = 3.50$$

$$26.00 n_{N_2}\ \text{mol}^{-1} = 2.889$$

$$n_{N_2} = 0.1111\ \text{mol}$$

$$\text{mass of N}_2 = n_{N_2}\left(28.02\ \text{g mol}^{-1}\right) = (0.1111\ \text{mol})\left(28.02\ \text{g mol}^{-1}\right) = 3.113\ \text{g}$$

$$\text{mass of H}_2 = 3.50\ \text{g} - 3.113\ \text{g} = 0.387\ \text{g}$$

$$\text{mass \% of N}_2 = \frac{3.113\ \text{g}}{3.50\ \text{g}} \times 100\% = 88.9\%$$

$$\text{mass \% of H}_2 = \frac{0.387\ \text{g}}{3.50\ \text{g}} \times 100\% = 11.1\%$$

1.26 Death by suffocation in a sealed container is normally caused not by oxygen deficiency but by CO_2 poisoning, which occurs at about 7% CO_2 by volume. For what length of time would it be safe to be in a sealed room $10 \times 10 \times 20$ ft? [Source: "Eco-Chem," J. A. Campbell, *J. Chem. Educ.* **49**, 538 (1972).]

The source of the excess CO_2 is that which is exhaled and which had as its source the O_2 that was inhaled and metabolized. Thus, to calculate how much CO_2 is added to the room, calculate how much O_2 is depleted. (This assumes a 1:1 molar ratio between CO_2 formed and O_2 used with little hydrogen oxidized to H_2O by inhaled oxygen. The ratio is actually about 1.2 O_2:1 CO_2.)

The air becomes lethal (due to CO_2) after 7% of the gas (to 1 sig. fig.) becomes CO_2, or

$$(10 \text{ ft})(10 \text{ ft})(20 \text{ ft}) \left(\frac{28.3 \text{ L}}{1 \text{ ft}^3} \right) (0.07) = 4.0 \times 10^3 \text{ L of } CO_2.$$

A person breathes about 0.5 L of air 12 times per minute, and the air is about 20% O_2. Thus,

$$(12 \text{ min}^{-1})(0.5 \text{ L})(0.20) = 1.2 \text{ L } O_2 \text{ min}^{-1}.$$

About 30% of this inhaled O_2 is absorbed in the lungs, so that a person typically uses

$$(0.30)(1.2 \text{ L } O_2 \text{ min}^{-1}) = 0.36 \text{ L } O_2 \text{ min}^{-1}.$$

For a calm, quiet person about half this amount, or 0.2 L min^{-1} would be enough, where we have rounded to 1 sig. fig. Thus one person could last

$$\left(\frac{4.0 \times 10^3 \text{ L of } O_2}{0.2 \text{ L min}^{-1}} \right) \left(\frac{1 \text{ h}}{60 \text{ min}} \right) \left(\frac{1 \text{ day}}{24 \text{ h}} \right) = 14 \text{ days}$$

1.28 A mixture of helium and neon gases is collected over water at 28.0°C and 745 mmHg. If the partial pressure of helium is 368 mmHg, what is the partial pressure of neon? (*Note:* Vapor pressure of water at 28°C is 28.3 mmHg.)

P_{Ne} can be determined by rearranging the equation $P_{total} = P_{He} + P_{Ne} + P_{H_2O}$:

$$P_{Ne} = P_{total} - P_{He} - P_{H_2O} = 745 \text{ mmHg} - 368 \text{ mmHg} - 28.3 \text{ mmHg} = 349 \text{ mmHg}$$

1.30 A piece of sodium metal reacts completely with water as follows:

$$2Na(s) + 2H_2O(l) \longrightarrow 2NaOH(aq) + H_2(g)$$

The hydrogen gas generated is collected over water at 25.0°C. The volume of the gas is 246 mL measured at 1.00 atm. Calculate the number of grams of sodium used in the reaction. (*Note:* Vapor pressure of water at 25°C is 0.0313 atm.)

First calculate the number of moles of H_2 from the ideal gas law, from which the number of moles of Na, and therefore, the mass of Na used in the reaction can be determined.

$$P_{H_2} + P_{H_2O} = P_{total}$$

$$P_{H_2} = P_{total} - P_{H_2O} = 1.00 \text{ atm} - 0.0313 \text{ atm} = 0.969 \text{ atm}$$

$$n_{H_2} = \frac{P_{H_2} V}{RT} = \frac{(0.969 \text{ atm}) (0.246 \text{ L})}{\left(0.08206 \text{ L atm K}^{-1} \text{ mol}^{-1}\right) (273.15 + 25.0) \text{ K}} = 9.74 \times 10^{-3} \text{ mol}$$

According to the chemical equation,

$$n_{Na} = n_{H_2} \left(\frac{2 \text{ mol Na}}{1 \text{ mol H}_2}\right) = \left(9.74 \times 10^{-3} \text{ mol H}_2\right) \left(\frac{2 \text{ mol Na}}{1 \text{ mol H}_2}\right) = 1.95 \times 10^{-2} \text{ mol}$$

The mass of Na used is

$$m_{Na} = \left(1.95 \times 10^{-2} \text{ mol}\right) \left(22.99 \text{ g mol}^{-1}\right) = 0.45 \text{ g}$$

1.32   Helium is mixed with oxygen gas for deep sea divers. Calculate the percent by volume of oxygen gas in the mixture if the diver has to submerge to a depth where the total pressure is 4.2 atm. The partial pressure of oxygen is maintained at 0.20 atm at this depth.

At constant P and T, $n \propto V$. Thus, % by volume = mol %, and mole fraction is directly related to mol %.

$$P_{O_2} = x_{O_2} P_{total}$$

$$x_{O_2} = \frac{P_{O_2}}{P_{total}} = \frac{0.20 \text{ atm}}{4.2 \text{ atm}} = 0.048$$

The mole fraction of O_2 is 0.048, or the percent by volume of O_2 = 4.8%.

1.34   The partial pressure of carbon dioxide in air varies with the seasons. Would you expect the partial pressure in the Northern Hemisphere to be higher in the summer or winter? Explain.

Plant photosynthesis is a major contributor to the seasonal variation of the amount of carbon dioxide in the atmosphere. Thus, in the Northern Hemisphere the partial pressure of CO_2 is higher in the winter when less CO_2 is being utilized in photosynthesis.

1.36   Describe how you would measure, by either chemical or physical means (other than mass spectrometry), the partial pressures of a mixture of gases: **(a)** CO_2 and H_2, **(b)** He and N_2.

(a) A measurement of the total pressure of the mixture can be made at known temperature and volume. A chemical separation may then be used to measure the amount of a single component. A good choice is the reaction between CO_2 and sodium hydroxide

$$CO_2(g) + 2NaOH(aq) \longrightarrow Na_2CO_3(aq) + H_2O(l)$$

This leaves only the H_2 gas and water vapor. The partial pressure of H_2 can now be determined under the same conditions of temperature and volume after correcting for the known vapor pressure of water. Finally, the partial pressure of CO_2 is calculated from

$$P_{CO_2} = P_{total} - P_{H_2}$$

(b) In this case there is no convenient chemical means of separation, but there is a significant difference in boiling points that can be utilized. As in part **(a)** the total pressure of the mixture is first measured. The temperature is then lowered until the nitrogen liquefies (b.p. N_2: 77 K). At this point the He is still gaseous (b.p. He: 4 K), and its pressure can be measured. Charles' Law is then used to calculate the pressure of He at the temperature of the original total pressure measurement (assuming a constant volume container). The partial pressure of N_2 is then the difference between the total pressure and helium pressure at this temperature.

1.38 A 1.00-L bulb and a 1.50-L bulb, connected by a stopcock, are filled, respectively, with argon at 0.75 atm and helium at 1.20 atm at the same temperature. Calculate the total pressure, the partial pressure of each gas, and the mole fraction of each gas after the stopcock has been opened. Assume ideal-gas behavior.

At constant n and T, $P_1 V_1 = P_2 V_2$, where 1 and 2 denote the state before and after the stopcock is opened, respectively.

For Ar,

$$P_2 = \frac{P_1 V_1}{V_2} = \frac{(0.75 \text{ atm}) (1.00 \text{ L})}{2.50 \text{ L}} = 0.30 \text{ atm} = P_{Ar}$$

For He,

$$P_2 = \frac{P_1 V_1}{V_2} = \frac{(1.20 \text{ atm}) (1.50 \text{ L})}{2.50 \text{ L}} = 0.720 \text{ atm} = P_{He}$$

The total pressure is

$$P = P_{Ar} + P_{He} = 0.30 \text{ atm} + 0.720 \text{ atm} = 1.02 \text{ atm}$$

The mole fractions are

$$x_{Ar} = \frac{P_{Ar}}{P} = \frac{0.30 \text{ atm}}{1.02 \text{ atm}} = 0.29$$

$$x_{He} = 1 - x_{Ar} = 1 - 0.29 = 0.71$$

1.40 Suggest two demonstrations to show that gases do not behave ideally.

One demonstration is quite common. Namely, the condensation of a gas at low temperatures and/or high pressures to form a liquid, such as was used in Problem 1.36(b). The condensation demonstrates the existence of attractive forces between molecules. A second demonstration would be to plot the compressibility factor, $Z = P\overline{V}/RT$ vs. P. Deviations from unity show that the gas does not behave ideally.

1.42 The van der Waals constants a and b for benzene are 18.00 atm L^2 mol^{-2} and 0.115 L mol^{-1}, respectively. Calculate the critical constants for benzene.

Calculate $\overline{V}_c$ from Eq. 1.19, T_c from Eq. 1.20, and P_c from Eq. 1.22.

$$\overline{V}_c = 3b = 3\left(0.115\,\text{L mol}^{-1}\right) = 0.345\,\text{L mol}^{-1}$$

$$T_c = \frac{8a}{27Rb} = \frac{8\left(18.00\,\text{atm L}^2\,\text{mol}^{-2}\right)}{27\left(0.08206\,\text{L atm K}^{-1}\,\text{mol}^{-1}\right)\left(0.115\,\text{L mol}^{-1}\right)} = 565\,\text{K}$$

$$P_c = \frac{a}{27b^2} = \frac{18.00\,\text{atm L}^2\,\text{mol}^{-2}}{27\left(0.115\,\text{L mol}^{-1}\right)^2} = 50.4\,\text{atm}$$

1.44   Without referring to a table, select from the following list the gas that has the largest value of b in the van der Waals equation: CH_4, O_2, H_2O, CCl_4, Ne.

The van der Waals constant b is related to the size of the molecule, so look for the largest molecule, which is CCl_4 in this case.

1.46   At 300 K, the second virial coefficients (B) of N_2 and CH_4 are $-4.2\,\text{cm}^3\,\text{mol}^{-1}$ and -15 $\text{cm}^3\,\text{mol}^{-1}$, respectively. Which gas behaves more ideally at this temperature?

According to the equation

$$Z = 1 + \frac{B}{V} + \cdots$$

the closer to zero the value of B, the closer is Z to unity, that is, the more ideal is the gas. According to the data given, N_2 behaves more ideally than CH_4.

1.48   Consider the virial equation $Z = 1 + B'P + C'P^2$, which describes the behavior of a gas at a certain temperature. From the following plot of Z versus P, deduce the signs of B' and C' (< 0, $= 0$, > 0).

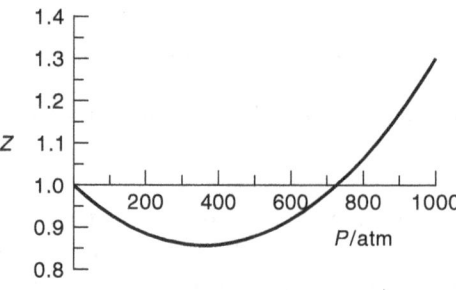

We see that the slope of the Z versus P plot is given by

$$\frac{dZ}{dP} = B' + 2C'P$$

As $P \to 0$, $\frac{dZ}{dP} \to B'$.

Additionally, the curvature of the graph is given by

$$\frac{d^2 Z}{d P^2} = 2C'$$

Near $P = 0$ the graph has a negative slope, starting at $Z = 1$ and dipping below 1. It does display positive curvature, with the graph turning around to rise above 1. Thus $B' < 0$ and $C' > 0$.

1.50 Derive the van der Waals constants a and b in terms of the critical constants by recognizing that at the critical point, $\left(\partial P / \partial \overline{V}\right)_T = 0$ and $\left(\partial^2 P / \partial \overline{V}^2\right)_T = 0$. (This problem requires a knowledge of partial differentiation.)

Rearrange the van der Waals equation, $\left(P + \frac{an^2}{V^2}\right)(V - nb) = nRT$, to give

$$P = \frac{nRT}{V - nb} - \frac{an^2}{V^2}$$

Divide the numerator and denominator of the first term by n, and those of the second term by n^2,

$$P = \frac{RT}{\overline{V} - b} - \frac{a}{\overline{V}^2} \qquad (1.50.1)$$

Take the first and second derivatives of P with respect to $\overline{V}$,

$$\left(\frac{\partial P}{\partial \overline{V}}\right)_T = -\frac{RT}{\left(\overline{V} - b\right)^2} + \frac{2a}{\overline{V}^3} \qquad (1.50.2)$$

$$\left(\frac{\partial^2 P}{\partial \overline{V}^2}\right)_T = \frac{2RT}{\left(\overline{V} - b\right)^3} - \frac{6a}{\overline{V}^4} \qquad (1.50.3)$$

At the critical point, $P = P_c$, $\overline{V} = \overline{V}_c$, and $T = T_c$. In addition, $\left(\frac{\partial P}{\partial \overline{V}}\right)_T = 0$ and $\left(\frac{\partial^2 P}{\partial \overline{V}^2}\right)_T = 0$. Therefore, Eqs. 1.50.2 and 1.50.3 become

$$-\frac{RT_c}{\left(\overline{V}_c - b\right)^2} + \frac{2a}{\overline{V}_c^3} = 0 \qquad (1.50.4)$$

$$\frac{2RT_c}{\left(\overline{V}_c - b\right)^3} - \frac{6a}{\overline{V}_c^4} = 0 \qquad (1.50.5)$$

Rearrange Eqs. 1.50.4 and 1.50.5:

$$\frac{2a}{\overline{V}_c^3} = \frac{RT_c}{\left(\overline{V}_c - b\right)^2} \tag{1.50.6}$$

$$\frac{6a}{\overline{V}_c^4} = \frac{2RT_c}{\left(\overline{V}_c - b\right)^3} \tag{1.50.7}$$

Now divide Eq. 1.50.6 by Eq. 1.50.7:

$$\frac{\frac{2a}{\overline{V}_c^3}}{\frac{6a}{\overline{V}_c^4}} = \frac{\frac{RT_c}{\left(\overline{V}_c - b\right)^2}}{\frac{2RT_c}{\left(\overline{V}_c - b\right)^3}}$$

$$\frac{\overline{V}_c}{3} = \frac{\overline{V}_c - b}{2}$$

$$2\overline{V}_c = 3\overline{V}_c - 3b$$

$$\overline{V}_c = 3b \tag{1.50.8}$$

Substitute Eq. 1.50.8 into 1.50.6:

$$\frac{2a}{(3b)^3} = \frac{RT_c}{(3b - b)^2}$$

$$T_c = \frac{2a}{27b^3}\frac{4b^2}{R}$$

$$= \frac{8a}{27Rb} \tag{1.50.9}$$

At the critical point, Eq. 1.50.1 becomes

$$P_c = \frac{RT_c}{\overline{V}_c - b} - \frac{a}{\overline{V}_c^2} \tag{1.50.10}$$

P_c can be solved for by substituting Eqs. 1.50.8 and 1.50.9 into Eq. 1.50.10:

$$P_c = \frac{R\left(\frac{8a}{27Rb}\right)}{3b - b} - \frac{a}{(3b)^2}$$

$$= \frac{4a}{27b^2} - \frac{a}{9b^2}$$

$$= \frac{a}{27b^2}$$

1.52 A CO_2 fire extinguisher is located on the outside of a building in Massachusetts. During the winter months, one can hear a sloshing sound when the extinguisher is gently shaken. In the summertime, there is often no sound when it is shaken. Explain. Assume that the extinguisher has no leaks and that it has not been used.

The critical temperature for CO_2 is $T_c = 304.2$ K $= 31.1°$C $= 88°$F. On a hot summer day when $T > 88°$F, CO_2 is a supercritical fluid. During more temperate seasons, the CO_2 in the extinguisher is present as liquid with vapor above it and can slosh around.

1.54 It has been said that every breath we take, on average, contains molecules once exhaled by Wolfgang Amadeus Mozart (1756–1791). The following calculations demonstrate the validity of this statement. **(a)** Calculate the total number of molecules in the atmosphere. (*Hint*: Use the result in Problem 1.53 and 29.0 g mol^{-1} as the molar mass of air.) **(b)** Assuming the volume of every breath (inhaled or exhaled) is 500 mL, calculate the number of molecules exhaled in each breath at 37°C, which is the body temperature. **(c)** If Mozart's life span was exactly 35 years, how many molecules did he exhale in that period (given that an average person breathes 12 times per minute)? **(d)** Calculate the fraction of molecules in the atmosphere that were exhaled by Mozart. How many of Mozart's molecules do we inhale with each breath of air? Round your answer to one significant figure. **(e)** List three important assumptions in these calculations.

(a) The number of air molecules in the atmosphere can be calculated from the number of moles of air in the atmosphere.

$$\text{Number of moles of air} = \frac{\text{mass of air}}{\text{molar mass of air}}$$

$$= \frac{5.27 \times 10^{21} \text{ g}}{29.0 \text{ g mol}^{-1}} = 1.817 \times 10^{20} \text{ mol}$$

$$\text{Number of air molecules} = \left(1.817 \times 10^{20} \text{ mol}\right) \left(\frac{6.022 \times 10^{23} \text{ molecules}}{1 \text{ mol}}\right)$$

$$= 1.09 \times 10^{44} \text{ molecules}$$

(b) For each breath,

$$\text{Moles of molecules inhaled or exhaled} = \frac{PV}{RT}$$

$$= \frac{(1.00 \text{ atm}) (500 \text{ mL}) \left(\frac{1 \text{ L}}{1000 \text{ mL}}\right)}{(0.08206 \text{ L atm K}^{-1} \text{ mol}^{-1}) (273.15 + 37) \text{ K}}$$

$$= 1.965 \times 10^{-2} \text{ mol}$$

$$\text{Number of molecules inhaled or exhaled} = \left(1.965 \times 10^{-2} \text{ mol}\right) \left(\frac{6.022 \times 10^{23} \text{ molecules}}{1 \text{ mol}}\right)$$

$$= 1.18 \times 10^{22} \text{ molecules}$$

(c) The number of molecules Mozart exhaled in his life span can be determined by his life span and the number of molecules exhaled/breath calculated in part **(b)**.

$$\text{Number of minutes in Mozart's life span} = (35 \text{ yr}) \left(\frac{365 \text{ day}}{1 \text{ yr}}\right) \left(\frac{24 \text{ h}}{1 \text{ day}}\right) \left(\frac{60 \text{ min}}{1 \text{ h}}\right)$$

$$= 1.84 \times 10^7 \text{ min}$$

$$\text{Number of breaths in Mozart's life span} = (12 \text{ breaths/min}) \left(1.84 \times 10^7 \text{ min}\right)$$

$$= 2.21 \times 10^8 \text{ breaths}$$

Number of molecules exhaled by Mozart

$$= \left(2.21 \times 10^8 \text{ breaths}\right) \left(1.18 \times 10^{22} \text{ molecules/breath}\right)$$

$$= 2.6 \times 10^{30} \text{ molecules}$$

(d) The fraction of molecules in the atmosphere that was exhaled by Mozart can be determined by data in parts (a) and (c):

Fraction of molecules in the atmosphere that was exhaled by Mozart

$$= \frac{\text{number of molecules exhaled by Mozart}}{\text{number of air molecules in the atmosphere}} = \frac{2.6 \times 10^{30} \text{ molecules}}{1.09 \times 10^{44} \text{ molecules}}$$

$$= 2.4 \times 10^{-14}$$

In a single breath, 1.18×10^{22} molecules are inhaled. Therefore,

Number of Mozart's molecules we inhale/breath

$$= \left(1.18 \times 10^{22} \text{ molecules}\right) \left(2.4 \times 10^{-14}\right) = 3 \times 10^8 \text{ molecules}$$

(e) Aside from the estimates of typical breathing rates and volumes, which do not count as major assumptions here, there are some serious assumptions that have been made. These include (1) that the molecules in Mozart's breath have been homogeneously distributed throughout the atmosphere, (2) that Mozart did not exhale the same molecules in different breaths, that is, each breath involved a different 1.18×10^{22} molecules, and (3) that all of the exhaled molecules are still in the atmosphere and have not been removed by incorporation into living matter.

1.56 A relation known as the barometric formula is useful for estimating the change in atmospheric pressure with altitude. **(a)** Starting with the knowledge that atmospheric pressure decreases with altitude, we have $dP = -\rho g dh$, where ρ is the density of air, g is the acceleration due to gravity (9.81 m s^{-2}), and P and h are the pressure and height, respectively. Assuming ideal-gas behavior and constant temperature, show that the pressure P at height h is related to the pressure at sea level P_0 ($h = 0$) by $P = P_0 e^{-g\mathcal{M}h/RT}$. (*Hint:* For an ideal gas, $\rho = P\mathcal{M}/RT$, where $\mathcal{M}$ is the molar mass.) **(b)** Calculate the atmospheric pressure at a height of 5.0 km, assuming the temperature is constant at 5.0°C, given that the average molar mass of air is 29.0 g mol^{-1}.)

(a) To understand the origin of the statement $dP = -\rho g \, dh$, consider the difference in pressure between the two heights h and $h + dh$. This difference is due to the weight (force due to gravity) of the section of air of volume $A \, dh$, where A is the area over which this weight is distributed. The pressure difference will be the weight of this section divided by area. The density, ρ is the connection between volume and mass, $m = \rho V$. Thus, the section of air, which has volume, $V = A \, dh$, has mass $m = \rho A \, dh$. The force is directed downwards while the altitude is measured upwards. Thus, force and dh have opposite directions so that the force is $F = -\rho g A \, dh$. Finally, we see that the pressure difference $dP = F/A = -\rho g A \, dh/A = -\rho g \, dh$.

To continue on with the solution, substitute $\rho = \frac{P\mathcal{M}}{RT}$ into the dP expression:

$$dP = -\frac{P\mathcal{M}}{RT}g\,dh$$

Arrange variables containing P on the left hand side of the equation, and constants and the differential dh on the right hand side of the equation:

$$\frac{dP}{P} = -\frac{\mathcal{M}}{RT}\,g\,dh$$

This equation is then integrated. The limits for h are 0 and h while the corresponding limits for P are P_0 and P.

$$\int_{P_0}^{P}\frac{dP}{P} = -\int_{0}^{h}\frac{\mathcal{M}}{RT}\,g\,dh$$

$$\ln\frac{P}{P_0} = -\frac{\mathcal{M}}{RT}\,g\,h$$

$$\frac{P}{P_0} = e^{-g\mathcal{M}h/RT}$$

$$P = P_0 e^{-g\mathcal{M}h/RT}$$

(b) First calculate $g\mathcal{M}h/RT$:

$$\frac{g\mathcal{M}h}{RT} = \frac{\left(9.81\text{ m s}^{-2}\right)\left(29.0\times10^{-3}\text{ kg mol}^{-1}\right)\left(5.0\times10^{3}\text{ m}\right)}{\left(8.314\text{ J K}^{-1}\text{ mol}^{-1}\right)(273.15+5.0)\text{ K}}$$

$$= 0.615$$

Thus,

$$P = P_0 e^{-g\mathcal{M}h/RT} = (1.0\text{ atm})\,e^{-0.615} = 0.54\text{ atm}$$

1.58 One way to gain a physical understanding of b in the van der Waals equation is to calculate the "excluded volume." Assume that the distance of closest approach between two similar spherical molecules is the sum of their radii ($2r$). **(a)** Calculate the volume around each molecule into which the center of another molecule cannot penetrate. **(b)** From your result in **(a)**, calculate the excluded volume for one mole of molecules, which is the constant b. How does this compare with the sum of the volumes of 1 mole of the same molecules?

(a) We see from the figure that two hard spheres of radius r cannot approach each other more closely than $2r$ (measured from the centers). Thus there is a sphere of radius $2r$ surrounding each hard sphere from which other hard spheres are excluded. The excluded volume/pair of molecules $= \frac{4}{3}\pi\,(2r)^3 = \frac{32}{3}\pi r^3 = 8\left(\frac{4}{3}\pi r^3\right)$, or eight times the volume of an individual molecule.

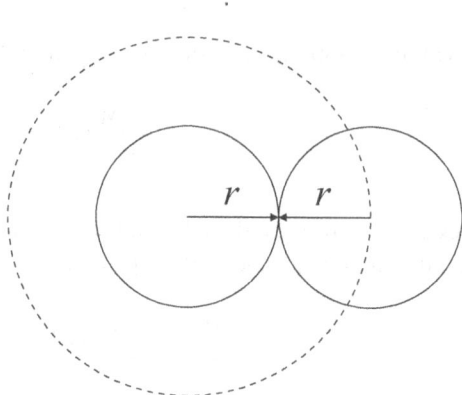

(b) The result in part **(a)** is for a pair of molecules, so the excluded volume/molecule $= \frac{1}{2}\left(\frac{32}{3}\pi r^3\right) = \frac{16}{3}\pi r^3$; Therefore, the excluded volume/mol $= \frac{16}{3}N_A\pi r^3$ where N_A is Avogadro's number. The sum of the volumes of a mole of molecules (treated as hard spheres of radius r) is $\frac{4}{3}N_A\pi r^3$. The excluded volume is four times the volume of the spheres themselves.

1.60   Express the van der Waals equation in the form of Equation 1.10. Derive relationships between the van der Waals constants (a and b) and the virial coefficients (B, C, and D) given that

$$\frac{1}{1-x} = 1 + x + x^2 + x^3 + \cdots \quad |x| < 1$$

The van der Waals equation can be rearranged to give (see Problem 1.50)

$$P = \frac{RT}{\overline{V}-b} - \frac{a}{\overline{V}^2}$$

Substitute this expression into $Z = \frac{P\overline{V}}{RT}$:

$$Z = \left(\frac{RT}{\overline{V}-b} - \frac{a}{\overline{V}^2}\right)\frac{\overline{V}}{RT}$$

$$= \frac{\overline{V}}{\overline{V}-b} - \frac{a}{\overline{V}RT}$$

$$= \frac{1}{1-\frac{b}{\overline{V}}} - \frac{a}{\overline{V}RT}$$

Because $b < \overline{V}$, $\frac{b}{\overline{V}} < 1$, the following expression applies:

$$\frac{1}{1-x} = 1 + x + x^2 + x^3 + \cdots$$

and can be used to approximate $\frac{1}{1-\frac{b}{\overline{V}}}$ in the expression for Z:

$$Z = 1 + \frac{b}{\overline{V}} + \frac{b^2}{\overline{V}^2} + \frac{b^3}{\overline{V}^3} + \cdots - \frac{a}{\overline{V}RT}$$

$$= 1 + \left(b - \frac{a}{RT}\right)\frac{1}{\overline{V}} + \frac{b^2}{\overline{V}^2} + \frac{b^3}{\overline{V}^3} + \cdots$$

In terms of the virial coefficients, Z can be written as

$$Z = 1 + \frac{B}{\overline{V}} + \frac{C}{\overline{V}^2} + \frac{D}{\overline{V}^3} + \cdots$$

The following expressions are obtained when the coefficients for $\frac{1}{\overline{V}}$, $\frac{1}{\overline{V}^2}$, and $\frac{1}{\overline{V}^3}$ are compared in these expressions for Z.

$$B = b - \frac{a}{RT}$$

$$C = b^2$$

$$D = b^3$$

1.62 From Equations 1.10 and 1.11, show that $B' = B/RT$ and $C' = \left(C - B^2\right)/(RT)^2$. (*Hint*: From Equation 1.10, first obtain expressions for P and P^2. Next, substitute these expressions into Equation 1.11.]

Substitute $Z = \frac{P\overline{V}}{RT}$ into Eq. 1.10:

$$\frac{P\overline{V}}{RT} = 1 + \frac{B}{\overline{V}} + \frac{C}{\overline{V}^2} + \frac{D}{\overline{V}^3} + \cdots$$

Rearranging this equation to yield P and keeping terms up to $\frac{1}{\overline{V}^3}$, the following expression is obtained:

$$P = \left(\frac{RT}{\overline{V}}\right)\left(1 + \frac{B}{\overline{V}} + \frac{C}{\overline{V}^2} + \frac{D}{\overline{V}^3} + \cdots\right) = \frac{RT}{\overline{V}} + \frac{BRT}{\overline{V}^2} + \frac{CRT}{\overline{V}^3} + \cdots$$

Square the expression for P and keep terms up to $\frac{1}{\overline{V}^3}$:

$$P^2 = \frac{R^2 T^2}{\overline{V}^2} + \frac{2BR^2T^2}{\overline{V}^3} + \cdots$$

Substitute the expressions for P and P^2 into Eq. 1.11:

$$Z = 1 + B'P + C'P^2$$

$$= 1 + B'\left(\frac{RT}{\overline{V}} + \frac{BRT}{\overline{V}^2} + \frac{CRT}{\overline{V}^3} + \cdots\right) + C'\left(\frac{R^2T^2}{\overline{V}^2} + \frac{2BR^2T^2}{\overline{V}^3} + \cdots\right)$$

$$= 1 + \frac{B'RT}{\overline{V}} + \frac{B'BRT + C'R^2T^2}{\overline{V}^2} + \frac{B'CRT + 2BC'R^2T^2}{\overline{V}^3} + \cdots$$

The following expression is obtained when the coefficients for $\frac{1}{\tilde{V}}$ in the above expression for Z and Eq. 1.10 are compared:

$$B = B'RT$$

Thus,

$$B' = \frac{B}{RT}$$

The following expression is obtained when the coefficients for $\frac{1}{\tilde{V}^2}$ in the above expression for Z and Eq. 1.10 are compared:

$$C = B'BRT + C'R^2T^2$$

Thus,

$$C' = \frac{C - B'BRT}{R^2T^2}$$

Substitute $B' = \frac{B}{RT}$ into the C' expression:

$$C' = \frac{C - \frac{B}{RT}BRT}{R^2T^2} = \frac{C - B^2}{(RT)^2}$$

1.64   A mixture of methane (CH_4) and ethane (C_2H_6) is stored in a container at 294 torr. The gases are burned in air to form CO_2 and H_2O. If the pressure of CO_2 is 356 torr measured at the same temperature and volume as the original mixture, then calculate the mole fractions of the gases.

Let x be the mole fraction of CH_4. The mole fraction of C_2H_6 is then $1 - x$. It follows that the partial pressures of CH_4 and C_2H_6 in the mixture are

$$P_{CH_4} = (294 \text{ torr}) \, x$$

$$P_{C_2H_6} = (294 \text{ torr}) \, (1 - x)$$

The following reactions take place when CH_4 and C_2H_6 are burned in air:

$$CH_4 + 2O_2 \rightarrow CO_2 + 2H_2O$$

$$C_2H_6 + \frac{7}{2}O_2 \rightarrow 2CO_2 + 3H_2O$$

At constant T and V, the pressure of a gas is proportional to the number of moles of gas present. Thus, the pressure of CO_2 produced from CH_4 is P_{CH_4} and the pressure of CO_2 produced from C_2H_6 is $2P_{C_2H_6}$, which sum to 356 torr:

$$P_{CH_4} + 2P_{C_2H_6} = 356 \text{ torr}$$

$$(294 \text{ torr}) \, x + 2 \, (294 \text{ torr}) \, (1 - x) = 356 \text{ torr}$$

$$294x = 232$$

$$x = 0.789$$

The mole fraction of CH_4 in the mixture is 0.789 while that of C_2H_6 is $1 - x = 0.211$.

1.66 A gaseous hydrocarbon in a container of volume 20.2 L at 350 K and 6.63 atm reacts with an excess of oxygen to form 205.1 g of CO_2 and 168.0 g of H_2O. What is the molecular formula of the hydrocarbon?

The amount of hydrocarbon is

$$n_{\text{hydrocarbon}} = \frac{PV}{RT} = \frac{(6.63 \text{ atm}) (20.2 \text{ L})}{(0.08206 \text{ L atm K}^{-1} \text{ mol}^{-1}) (350 \text{ K})} = 4.663 \text{ mol}$$

The C and H in the products come directly from the hydrocarbon:

$$n_C = n_{CO_2} = \frac{205.1 \text{ g}}{44.01 \text{ g mol}^{-1}} = 4.660 \text{ mol}$$

$$n_H = n_{H_2O} \left(\frac{2 \text{ mol H}}{1 \text{ mol H}_2O}\right) = \frac{168.0 \text{ g}}{18.02 \text{ g mol}^{-1}} \left(\frac{2 \text{ mol H}}{1 \text{ mol H}_2O}\right) = 18.65 \text{ mol}$$

Since the number of moles of hydrocarbon is the same as the number of moles of C, there must be only one carbon in the molecular formula. Additionally, the mole ratio C:H = 4.660:18.65 = 1:4, giving a molecular formula of CH_4 for the hydrocarbon.

1.68 **(a)** Show that the pressure P (in pascals) exerted by a fluid is given by $P = hdg$, where h is the column of the fluid in meters, d is the density in kg m^{-3}, and g is the acceleration due to gravity (9.81 m s^{-2}). **(b)** The volume of an air bubble that starts at the bottom of a lake at 5.24°C increases by a factor of 6 as it rises to the surface of water where the temperature is 18.73°C and the air pressure is 0.973 atm. The density of the lake water is 1.02 g cm^{-3}. Use the equation in **(a)** to determine the depth of the lake in meters.

(a) The pressure exerted by a fluid is the weight of the fluid divided by the cross sectional area (A) of the fluid. The weight of the fluid is a product of its mass and gravity, mg, which can also be expressed using the density of the fluid as Vdg. Therefore,

$$P = \frac{Vdg}{A} = hdg$$

(b) Calculate the pressure experienced by the bubble at the bottom of the lake:

$$\frac{P_{\text{bottom}} V_{\text{bottom}}}{T_{\text{bottom}}} = \frac{P_{\text{surface}} V_{\text{surface}}}{T_{\text{surface}}}$$

$$P_{\text{bottom}} = \left(\frac{P_{\text{surface}} V_{\text{surface}}}{T_{\text{surface}}}\right) \left(\frac{T_{\text{bottom}}}{V_{\text{bottom}}}\right)$$

$$= \left[\frac{(0.973 \text{ atm}) (6 V_{\text{bottom}})}{(273.15 + 18.73) \text{ K}}\right] \left[\frac{(273.15 + 5.24) \text{ K}}{V_{\text{bottom}}}\right]$$

$$= 5.568 \text{ atm}$$

The pressure at the bottom of the lake is due to both the water and the atmosphere. Therefore, the pressure exerted by the water is

$$P_{\text{water}} = P_{\text{bottom}} - P_{\text{surface}} = 5.568 \text{ atm} - 0.973 \text{ atm} = 4.595 \text{ atm}$$

The depth of the lake, h, can now be evaluated by rearranging $P = hdg$:

$$h = \frac{P}{dg} = \frac{(4.595 \text{ atm}) \left(\frac{101.3 \times 10^3 \text{ Pa}}{1 \text{ atm}} \right)}{\left(1.02 \text{ g cm}^{-3} \right) \left(\frac{1 \text{ kg}}{1000 \text{ g}} \right) \left(\frac{100 \text{ cm}}{1 \text{ m}} \right)^3 \left(9.81 \text{ m s}^{-2} \right)}$$

$$= 46.5 \text{ m}$$

1.70 A closed 7.8-L flask contains 1.0 g of water. At what temperature will half of the water be in the vapor phase? (*Hint*: Look up the vapor pressure of water in the inside back matter.)

When half the water (0.50 g, or 2.77×10^{-2} mol) is in the vapor phase,

$$\frac{P}{T} = \frac{nR}{V} = \frac{\left(2.77 \times 10^{-2} \text{ mol} \right) \left(0.08206 \text{ L atm K}^{-1} \text{ mol}^{-1} \right)}{7.8 \text{ L}}$$

$$= 2.91 \times 10^{-4} \text{ atm K}^{-1} \left(\frac{760 \text{ mmHg}}{1 \text{ atm}} \right)$$

$$= 0.22 \text{ mmHg K}^{-1}$$

The table of vapor pressures suggests that this value is obtained when the temperature is approximately 45°C. (The vapor pressure of water is 71.88 mmHg at this temperature, giving a P/T value of 0.226 mmHg K^{-1}).

CHAPTER 2

Kinetic Theory of Gases

PROBLEMS AND SOLUTIONS

2.2 Is temperature a microscopic or macroscopic concept? Explain.

Temperature is a macroscopic concept, because one of the postulates of the kinetic theory of gases is that it deals with a very large number of molecules. Temperature is proportional to the average kinetic energy of the molecules in the system, and for this average to be meaningful, it must be taken over a large number of molecules.

2.4 If 2.0×10^{23} argon (Ar) atoms strike 4.0 cm^2 of wall per second at a $90°$ angle to the wall when moving with a speed of $45{,}000 \text{ cm s}^{-1}$, what pressure (in atm) do they exert on the wall?

$F = $ Force exerted by Ar atoms

$\quad = \left(\text{Force exerted by 1 Ar atom}\right) \left(\text{Number of Ar atoms}\right)$

$\quad = \left(\text{Change in momentum for 1 Ar atom/time}\right) \left(\text{Number of Ar atoms}\right)$

$\quad = \left(\dfrac{2mv}{1 \text{ s}}\right) \left(2.0 \times 10^{23}\right)$

$\quad = \dfrac{2\left[(39.95 \text{ amu}) \left(1.6605 \times 10^{-27} \text{ kg amu}^{-1}\right)\right] \left[\left(45{,}000 \text{ cm s}^{-1}\right) \left(\frac{1 \text{ m}}{100 \text{ cm}}\right)\right]}{1 \text{ s}} \left(2.0 \times 10^{23}\right)$

$\quad = 11.9 \text{ kg m s}^{-2} = 11.9 \text{ N}$

$\dfrac{F}{A} = $ Pressure exerted on the wall by Ar atoms

$\quad = \dfrac{11.9 \text{ N}}{\left(4.0 \text{ cm}^2\right) \left(\frac{1 \text{ m}}{100 \text{ cm}}\right)^2} \left(\dfrac{1 \text{ Pa}}{1 \text{ N m}^{-2}}\right)$

$\quad = \left(2.98 \times 10^4 \text{ Pa}\right) \left(\dfrac{1 \text{ atm}}{101.3 \times 10^3 \text{ Pa}}\right) = 0.29 \text{ atm}$

2.6   Calculate the average translational kinetic energy for a N_2 molecule and for 1 mole of N_2 at 20°C.

For one molecule, $\overline{E}_{trans} = \frac{3}{2}k_B T$.

For one mole of molecules, $\overline{E}_{trans} = \frac{3}{2}RT$.

Therefore, for a N_2 molecule,

$$\overline{E}_{trans} = \frac{3}{2}\left(1.381 \times 10^{-23}\,\text{J K}^{-1}\right)(273.15 + 20)\ \text{K} = 6.07 \times 10^{-21}\,\text{J}$$

whereas for a mole of N_2 molecules,

$$\overline{E}_{trans} = \frac{3}{2}\left(8.314\,\text{J K}^{-1}\,\text{mol}^{-1}\right)(273.15 + 20)\ \text{K}$$

$$= 3.65 \times 10^{3}\,\text{J mol}^{-1} = 3.65\,\text{kJ mol}^{-1}$$

2.8   The c_{rms} of CH_4 is 846 m s^{-1}. What is the temperature of the gas?

The temperature of the gas can be calculated by rearranging $c_{rms} = \sqrt{\frac{3RT}{\mathcal{M}}}$

$$\frac{3RT}{\mathcal{M}} = c_{rms}^2$$

$$T = \frac{c_{rms}^2 \mathcal{M}}{3R} = \frac{\left(846\text{ m s}^{-1}\right)^2 \left(16.04 \times 10^{-3}\text{ kg mol}^{-1}\right)}{3\left(8.314\text{ J K}^{-1}\text{ mol}^{-1}\right)} = 460\text{ K}$$

2.10   At what temperature will He atoms have the same c_{rms} value as N_2 molecules at 25°C? Solve this problem without calculating the value of c_{rms} for N_2.

This problem is similar to Problem 2.7.

$$T_{He} = \frac{\mathcal{M}_{He}}{\mathcal{M}_{N_2}}T_{N_2} = \left(\frac{4.003\text{ g mol}^{-1}}{28.02\text{ g mol}^{-1}}\right)(273.15 + 25)\text{ K} = 42.6\text{ K}$$

2.12   Plot the speed distribution function for **(a)** He, O_2, and UF_6 at the same temperature, and **(b)** CO_2 at 300 K and 1000 K.

   (a) In the plot, note that the heavier the molecules, the narrower the speed distribution and the smaller the most probable speed; whereas the lighter the molecules, the wider the speed distribution and the greater the most probable speed.

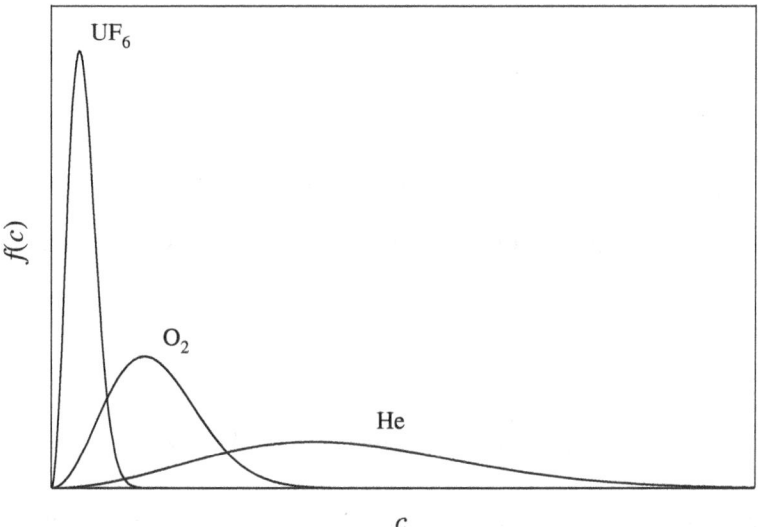

(b) In the plot, note that the lower the temperature, the narrower the speed distribution and the smaller the most probable speed; whereas the higher the temperature, the wider the speed distribution and the greater the most probable speed.

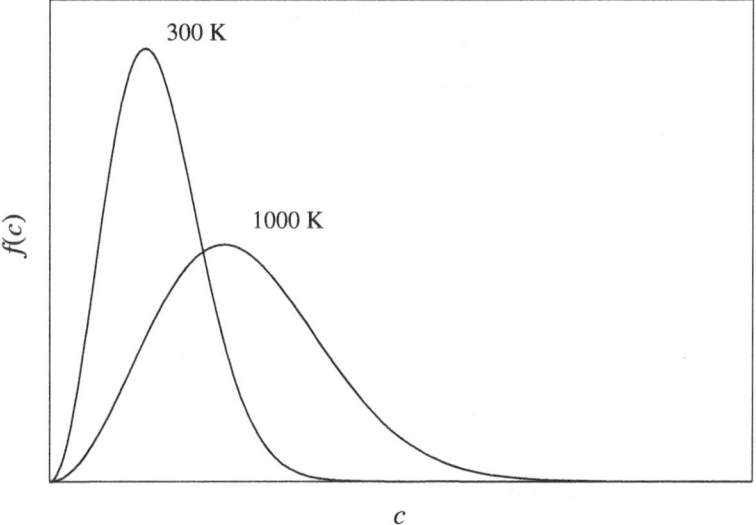

2.14 A N_2 molecule at 20°C is released at sea level to travel upward. Assuming that the temperature is constant and that the molecule does not collide with other molecules, how far would it travel (in meters) before coming to rest? Do the same calculation for a He atom. [*Hint*: To calculate the altitude, h, the molecule will travel, equate its kinetic energy with the potential energy, mgh, where m is the mass and g the acceleration due to gravity (9.81 m s^{-2}).]

This problem is an application of the law of conservation of energy. In this case, the total energy of the molecule remains the same, but it is changed from one form (kinetic energy) to another (potential energy). Therefore,

$$\text{kinetic energy} = \text{potential energy}$$

$$\frac{3}{2}k_B T = mgh$$

$$h = \frac{3k_B T}{2mg}$$

For a N_2 molecule,

$$h = \frac{3\left(1.381 \times 10^{-23} \text{ J K}^{-1}\right)(273.15 + 20) \text{ K}}{2\,(28.02 \text{ amu})\left(1.6605 \times 10^{-27} \text{ kg amu}^{-1}\right)(9.81 \text{ m s}^{-2})} = 1.33 \times 10^4 \text{ m}$$

For a He atom,

$$h = \frac{3\left(1.381 \times 10^{-23} \text{ J K}^{-1}\right)(273.15 + 20) \text{ K}}{2\,(4.003 \text{ amu})\left(1.6605 \times 10^{-27} \text{ kg amu}^{-1}\right)(9.81 \text{ m s}^{-2})} = 9.31 \times 10^4 \text{ m}$$

A He atom, being 1/7 the mass of a N_2 molecule, would travel 7 times higher than a N_2 molecule.

2.16   At a certain temperature, the speeds of six gaseous molecules in a container are 2.0 m s^{-1}, 2.2 m s^{-1}, 2.6 m s^{-1}, 2.7 m s^{-1}, 3.3 m s^{-1}, and 3.5 m s^{-1}. Calculate the root-mean-square speed and the average speed of the molecules. These two average values are close to each other, but the root-mean-square value is always the larger of the two. Why?

The average speed of the molecules is

$$\bar{c} = \frac{\sum\limits_{i=1}^{6} c_i}{N}$$

$$= \frac{(2.0 + 2.2 + 2.6 + 2.7 + 3.3 + 3.5) \text{ m s}^{-1}}{6}$$

$$= 2.7 \text{ m s}^{-1}$$

The mean-square speed for the molecules is

$$\overline{c^2} = \frac{\sum\limits_{i=1}^{6} c_i^2}{N}$$

$$= \frac{\left(2.0^2 + 2.2^2 + 2.6^2 + 2.7^2 + 3.3^2 + 3.5^2\right) \text{ m}^2 \text{ s}^{-2}}{6}$$

$$= 7.67 \text{ m}^2 \text{ s}^{-2}$$

The root-mean-square speed for the molecules is

$$c_{rms} = \sqrt{\overline{c^2}} = 2.8 \text{ m s}^{-1}$$

The rms average is always larger than the straight average because squaring favors (more heavily weights) the larger values and biases the result towards the larger values.

2.18 Following the same procedure used in the chapter to find the value of $\bar{c}$, derive an expression for c_{rms}. (*Hint*: You need to consult the *Handbook of Chemistry and Physics* to evaluate definite integrals.)

By definition,

$$\overline{c^2} = \int_0^\infty c^2 f(c)\,dc$$

$$= \int_0^\infty c^2 \left[4\pi c^2 \left(\frac{m}{2\pi k_{\text{B}}T} \right)^{\frac{3}{2}} e^{-\frac{mc^2}{2k_{\text{B}}T}} \right] dc$$

$$= 4\pi \left(\frac{m}{2\pi k_{\text{B}}T} \right)^{\frac{3}{2}} \int_0^\infty c^4 e^{-\frac{mc^2}{2k_{\text{B}}T}}\,dc$$

From a table of integrals,

$$\int_0^\infty x^{2n} e^{-ax^2}\,dx = \frac{(1)\,(3)\,(5)\cdots(2n-1)}{2^{n+1}a^n} \sqrt{\frac{\pi}{a}}$$

This is applied to the $\overline{c^2}$ integral by identifying

$$x = c$$
$$n = 2$$
$$a = \frac{m}{2k_{\text{B}}T}$$

and resulting in,

$$\overline{c^2} = 4\pi \left(\frac{m}{2\pi k_{\text{B}}T} \right)^{\frac{3}{2}} \frac{(1)\,(3)}{8 \left(\frac{m}{2k_{\text{B}}T} \right)^2} \sqrt{\frac{\pi}{\frac{m}{2k_{\text{B}}T}}}$$

$$= 4\pi \left(\frac{m}{2\pi k_{\text{B}}T} \right)^{\frac{3}{2}} \frac{3\sqrt{\pi}}{8} \left(\frac{2k_{\text{B}}T}{m} \right)^{\frac{5}{2}}$$

$$= \frac{3k_{\text{B}}T}{m} \left(\frac{N_A}{N_A} \right)$$

$$= \frac{3RT}{\mathcal{M}}$$

Consequently,

$$c_{\text{rms}} = \sqrt{\overline{c^2}} = \sqrt{\frac{3RT}{\mathcal{M}}}$$

2.20 Calculate the values of c_{rms}, c_{mp}, and $\bar{c}$ for argon at 298 K.

$$c_{rms} = \sqrt{\frac{3RT}{\mathcal{M}}} = \sqrt{\frac{3\left(8.314 \text{ J K}^{-1} \text{ mol}^{-1}\right)(298 \text{ K})}{39.95 \times 10^{-3} \text{ kg mol}^{-1}}} = 431 \text{ m s}^{-1}$$

$$c_{mp} = \sqrt{\frac{2RT}{\mathcal{M}}} = \sqrt{\frac{2\left(8.314 \text{ J K}^{-1} \text{ mol}^{-1}\right)(298 \text{ K})}{39.95 \times 10^{-3} \text{ kg mol}^{-1}}} = 352 \text{ m s}^{-1}$$

$$\bar{c} = \sqrt{\frac{8RT}{\pi\mathcal{M}}} = \sqrt{\frac{8\left(8.314 \text{ J K}^{-1} \text{ mol}^{-1}\right)(298 \text{ K})}{\pi\left(39.95 \times 10^{-3} \text{ kg mol}^{-1}\right)}} = 397 \text{ m s}^{-1}$$

2.22   Derive an expression for the most probable translational energy for an ideal gas. Compare your result with the average translational energy for the same gas.

The problem is similar to Problem 2.19. The most probable translational (or kinetic) energy for an ideal gas can be calculated by setting $\frac{df(E)}{dE} = 0$ and then solving for E_{mp}.

$$f(E) = 2\pi E^{\frac{1}{2}}\left(\frac{1}{\pi k_B T}\right)^{\frac{3}{2}} e^{-\frac{E}{k_B T}}$$

$$= 2\pi \left(\frac{1}{\pi k_B T}\right)^{\frac{3}{2}} E^{\frac{1}{2}} e^{-\frac{E}{k_B T}}$$

In differentiating $f(E)$ with respect to E, care must be taken to apply the chain rule to the two terms $E^{\frac{1}{2}}$ and $e^{-\frac{E}{k_B T}}$.

$$\frac{df(E)}{dE} = 2\pi \left(\frac{1}{\pi k_B T}\right)^{\frac{3}{2}}\left[\frac{1}{2} E^{-\frac{1}{2}} e^{-\frac{E}{k_B T}} + E^{\frac{1}{2}} e^{-\frac{E}{k_B T}}\left(-\frac{1}{k_B T}\right)\right]$$

$$= 2\pi \left(\frac{1}{\pi k_B T}\right)^{\frac{3}{2}} e^{-\frac{E}{k_B T}}\left(\frac{1}{2E^{\frac{1}{2}}} - \frac{E^{\frac{1}{2}}}{k_B T}\right)$$

When $\frac{df(E)}{dE} = 0$, $E = E_{mp}$. The above equation becomes

$$2\pi \left(\frac{1}{\pi k_B T}\right)^{\frac{3}{2}} e^{-\frac{E_{mp}}{k_B T}}\left(\frac{1}{2E_{mp}^{\frac{1}{2}}} - \frac{E_{mp}^{\frac{1}{2}}}{k_B T}\right) = 0$$

$$\frac{1}{2E_{mp}^{\frac{1}{2}}} - \frac{E_{mp}^{\frac{1}{2}}}{k_B T} = 0$$

Multiply the above equation by $2E_{mp}^{\frac{1}{2}}$:

$$1 - \frac{2E_{mp}}{k_B T} = 0$$

$$E_{mp} = \frac{k_B T}{2} = \frac{\bar{E}_{trans}}{3}$$

2.24 How does the mean free path of a gas depend on **(a)** the temperature at constant volume, **(b)** the density, **(c)** the pressure at constant temperature, **(d)** the volume at constant temperature, and **(e)** the size of molecules.

The mean free path is given by Equation 2.18 in the text

$$\lambda = \frac{1}{\sqrt{2}\pi d^2 \left(\frac{N}{V}\right)}$$

Although it is possible to answer this question solely by reference to the equation, it is useful to have an understanding of the physical basis for the effects observed. The key physical quantity is the density of the gas.

(a) The mean free path is independent of temperature at constant volume. T does not appear in the equation. [See particularly the discussion following Equation 2.19 in the text.] As the temperature is increased the molecules are moving faster, but the average distance between them is not affected. The mean time between collisions decreases, but the mean distance traveled between collisions remains the same.

(b) As the density increases, the mean free path decreases, since $\frac{N}{V}$ appears in the denominator. In a more dense gas, the molecules are more closely spaced.

(c) As the pressure increases at constant temperature, the mean free path decreases. These conditions lead to a decrease in volume, hence an increase in density. The molecules are being squeezed closer together.

(d) As the volume increases at constant temperature, the mean free path increases. As the molecules move into the expanded volume, they move further apart from each other.

(e) As the size of the molecules increases, the mean free path decreases. The collision diameter, d, appears in the denominator of the equation. Larger molecules do not have to travel as far before they run into each other.

2.26 Calculate the mean free path and the binary number of collisions per liter per second between HI molecules at 300 K and 1.00 atm. The collision diameter of the HI molecules may be taken to be 5.10 Å. Assume ideal-gas behavior.

The ideal gas law is used to calculate $\frac{N}{V}$, which is then used to calculate the mean free path.

$$PV = nRT = \frac{N}{N_A}RT$$

$$\frac{N}{V} = \frac{PN_A}{RT} = \frac{(1.00 \text{ atm}) \left(6.022 \times 10^{23} \text{ mol}^{-1}\right)}{\left(0.08206 \text{ L atm K}^{-1} \text{mol}^{-1}\right) (300 \text{ K})} = 2.446 \times 10^{22} \text{ L}^{-1} \left(\frac{1000 \text{ L}}{1 \text{ m}^3}\right)$$

$$= 2.446 \times 10^{25} \text{ m}^{-3}$$

$$\lambda = \frac{1}{\sqrt{2}\pi d^2 \left(\frac{N}{V}\right)} = \frac{1}{\sqrt{2}\pi \left(5.10 \times 10^{-10} \text{ m}\right)^2 \left(2.446 \times 10^{25} \text{ m}^{-3}\right)}$$

$$= 3.54 \times 10^{-8} \text{ m} = 35.4 \text{ nm}$$

The binary number of collisions depends on the average molecular speed, which is

$$\bar{c} = \sqrt{\frac{8RT}{\pi \mathcal{M}}} = \sqrt{\frac{8\left(8.314 \text{ J K}^{-1} \text{ mol}^{-1}\right)(300 \text{ K})}{\pi\left(127.9 \times 10^{-3} \text{ kg mol}^{-1}\right)}} = 222.8 \text{ m s}^{-1}$$

$$Z_{11} = \frac{\sqrt{2}}{2}\pi d^2 \bar{c} \left(\frac{N}{V}\right)^2 = \frac{\sqrt{2}}{2}\pi \left(5.10 \times 10^{-10} \text{ m}\right)^2 \left(222.8 \text{ m s}^{-1}\right)\left(2.446 \times 10^{25} \text{ m}^{-3}\right)^2$$

$$= \left(7.702 \times 10^{34} \text{ m}^{-3}\text{ s}^{-1}\right)\left(\frac{1 \text{ m}^3}{1000 \text{ L}}\right) = 7.70 \times 10^{31} \text{ collisions L}^{-1}\text{s}^{-1}$$

2.28 Suppose that helium atoms in a sealed container all start with the same speed, 2.74×10^4 cm s^{-1}. The atoms are then allowed to collide with one another until the Maxwell distribution is established. What is the temperature of the gas at equilibrium? Assume that there is no heat exchange between the gas and its surroundings.

The total translational energy of the helium atoms can be determined from the initial speed of the atoms. Because energy is conserved, this is also the total translational energy of the atoms after equilibrium is reached. Translational energy is a function of temperature, thus, the latter can be calculated once the former is known.

Suppose there are N helium atoms. Because all the atoms have the same speed, the total translational energy is

$$E_{\text{trans}} = N\left(\frac{1}{2}mv^2\right) = N\left(\overline{E}_{\text{trans}}\right)$$

$$N\left(\frac{1}{2}mv^2\right) = N\left(\frac{3}{2}k_{\text{B}}T\right)$$

$$T = \frac{mv^2}{3k_{\text{B}}}$$

$$= \frac{(4.003 \text{ amu})\left(1.6605 \times 10^{-27} \text{ kg amu}^{-1}\right)\left(2.74 \times 10^4 \text{ cm s}^{-1}\right)^2 \left(\frac{1\text{ m}}{100\text{ cm}}\right)^2}{3\left(1.381 \times 10^{-23} \text{ J K}^{-1}\right)} = 12.0 \text{ K}$$

2.30 Calculate the value of Z_1 and Z_{11} for mercury (Hg) vapor at 40°C, both at $P = 1.0$ atm and at $P = 0.10$ atm. How do these two quantities depend on pressure?

The number density and average molecular speed need to be determined before calculating Z_1 and Z_{11}. The collision diameter of Hg is 4.26 Å (Table 2.1).

At $P = 1.0$ atm,

$$\frac{N}{V} = \frac{PN_A}{RT} \quad \text{(See Problem 2.26)}$$

$$= \frac{(1.0 \text{ atm})\left(6.022 \times 10^{23} \text{ mol}^{-1}\right)}{\left(0.08206 \text{ L atm K}^{-1} \text{ mol}^{-1}\right)(273.15 + 40) \text{ K}} = 2.34 \times 10^{22} \text{ L}^{-1} \left(\frac{1000 \text{ L}}{1 \text{ m}^3}\right)$$

$$= 2.34 \times 10^{25} \text{ m}^{-3}$$

$$\bar{c} = \sqrt{\frac{8RT}{\pi \mathcal{M}}} = \sqrt{\frac{8 \left(8.314 \text{ J K}^{-1} \text{ mol}^{-1}\right)(273.15 + 40) \text{ K}}{\pi \left(200.6 \times 10^{-3} \text{ kg}\right)}} = 181.8 \text{ m s}^{-1}$$

$$Z_1 = \sqrt{2}\pi d^2 \bar{c}\frac{N}{V} = \sqrt{2}\pi \left(4.26 \times 10^{-10} \text{ m}\right)^2 \left(181.8 \text{ m s}^{-1}\right)\left(2.34 \times 10^{25} \text{ m}^{-3}\right)$$

$$= 3.43 \times 10^9 \text{ collisions s}^{-1} = 3.4 \times 10^9 \text{ collisions s}^{-1}$$

$$Z_{11} = \frac{\sqrt{2}}{2}\pi d^2 \bar{c}\left(\frac{N}{V}\right)^2 = \frac{1}{2}Z_1\left(\frac{N}{V}\right)$$

$$= \frac{1}{2}\left(3.43 \times 10^9 \text{ collisions s}^{-1}\right)\left(2.34 \times 10^{25} \text{ m}^{-3}\right)$$

$$= 4.0 \times 10^{34} \text{ collisions m}^{-3} \text{ s}^{-1}$$

For an ideal gas, $\frac{N}{V} = \frac{PN_A}{RT}$. The results above show $Z_1 \propto P$, whereas $Z_{11} \propto P^2$. A reduction in P to one tenth its original value (from 1.0 atm to 0.10 atm) will likewise reduce Z_1 to one tenth its value at $P = 1.0$ atm, but Z_{11} will decrease to $\frac{1}{10^2} = \frac{1}{100}$ of its value at $P = 1.0$ atm. That is, at $P = 0.10$ atm,

$$Z_1 = 3.4 \times 10^8 \text{ collisions s}^{-1}$$

$$Z_{11} = 4.0 \times 10^{32} \text{ collisions m}^{-3} \text{ s}^{-1}$$

2.32 Calculate the values of the average speed and collision diameter for ethylene at 288 K. The viscosity of ethylene is $99.8 \times 10^{-7} \text{ N s m}^{-2}$ at the same temperature.

$$\bar{c} = \sqrt{\frac{8RT}{\pi \mathcal{M}}} = \sqrt{\frac{8 \left(8.314 \text{ J K}^{-1} \text{ mol}^{-1}\right)(288 \text{ K})}{\pi \left(28.05 \times 10^{-3} \text{ kg mol}^{-1}\right)}} = 466.2 \text{ m s}^{-1}$$

$$\eta = \frac{m\bar{c}}{3\sqrt{2}\pi d^2}$$

$$d = \left(\frac{m\bar{c}}{3\sqrt{2}\pi \eta}\right)^{\frac{1}{2}} = \left[\frac{(28.05 \text{ amu})\left(1.6605 \times 10^{-27} \text{ kg amu}^{-1}\right)\left(466.2 \text{ m s}^{-1}\right)}{3\sqrt{2}\pi \left(99.8 \times 10^{-7} \text{ N s m}^{-2}\right)}\right]^{\frac{1}{2}}$$

$$= 4.04 \times 10^{-10} \text{ m} = 4.04 \text{ Å}$$

2.34 Derive Equation 2.28 from Equation 2.14.

The rate of diffusion (effusion) is proportional to average molecular speed,

$$r \propto \bar{c} = \sqrt{\frac{8RT}{\pi \mathcal{M}}}$$

Thus, for two gases of different molar masses, but at the same temperature,

$$\frac{r_1}{r_2} = \frac{\sqrt{\frac{8RT}{\pi \mathcal{M}_1}}}{\sqrt{\frac{8RT}{\pi \mathcal{M}_2}}} = \sqrt{\frac{\mathcal{M}_2}{\mathcal{M}_1}}$$

2.36   Nickel forms a gaseous compound of the formula $Ni(CO)_x$. What is the value of x given the fact that under the same conditions of temperature and pressure, methane (CH_4) effuses 3.3 times faster than the compound?

$$\frac{r_{CH_4}}{r_{Ni(CO)_x}} = 3.3 = \sqrt{\frac{\mathcal{M}_{Ni(CO)_x}}{\mathcal{M}_{CH_4}}}$$

$$\mathcal{M}_{Ni(CO)_x} = (3.3)^2 \left(\mathcal{M}_{CH_4}\right) = (3.3)^2 \left(16.04 \text{ g mol}^{-1}\right) = 1.75 \times 10^2 \text{ g mol}^{-1}$$

Calculate the expected molar mass of $Ni(CO)_x$ from its chemical formula, and equate the expression with the quantity obtained above.

$$\mathcal{M}_{Ni(CO)_x} = 58.69 \text{ g mol}^{-1} + x \left(12.01 \text{ g mol}^{-1} + 16.00 \text{ g mol}^{-1}\right) = 1.75 \times 10^2 \text{ g mol}^{-1}$$

$$28.01x = 1.16 \times 10^2$$

$$x = 4.1 \approx 4$$

2.38   Uranium-235 can be separated from uranium-238 by the effusion process using UF_6. Assuming a 50:50 mixture at the start, what is the percentage of enrichment after a single stage of separation?

This problem is the same as the example given in the text.

$$\text{Separation factor} = s = \sqrt{\frac{^{235}UF_6}{^{238}UF_6}} = \sqrt{\frac{\mathcal{M}_{^{238}UF_6}}{\mathcal{M}_{^{235}UF_6}}} = \sqrt{\frac{238 + (6)(19)}{235 + (6)(19)}} = 1.0043$$

According to this result, $^{235}UF_6$ effuses 1.0043 times faster than $^{235}UF_6$. Therefore, $^{235}UF_6$ is enriched by factor of 0.43% after a single stage of separation.

2.40   The rate (r_{eff}) at which molecules confined to a volume V effuse through an orifice of area A is given by $\left(\frac{1}{4}\right) nN_A \bar{c} A / V$, where n is the number of moles of the gas. An automobile tire of volume 30.0 L and pressure 1,500 torr is punctured as it runs over a sharp nail. **(a)** Calculate the effusion rate if the diameter of the hole is 1.0 mm. **(b)** How long would it take to lose half of the air in the tire through effusion? Assume a constant effusion rate and constant volume. The molar mass of air is 29.0 g and the temperature is 32.0°C.

(a) The effusion rate can be calculated once n, $\bar{c}$, and A are determined. The ideal gas law is used to calculate n.

$$n = \frac{PV}{RT} = \frac{(1500 \text{ torr}) \left(\frac{1 \text{ atm}}{760 \text{ torr}}\right) (30.0 \text{ L})}{(0.08206 \text{ L atm K}^{-1} \text{ mol}^{-1})(273.15 + 32.0) \text{ K}} = 2.365 \text{ mol}$$

$$\bar{c} = \sqrt{\frac{8RT}{\pi \mathcal{M}}} = \sqrt{\frac{8\left(8.314 \text{ J K}^{-1}\text{ mol}^{-1}\right)(273.15 + 32.0) \text{ K}}{\pi\left(29.0 \times 10^{-3} \text{ kg mol}^{-1}\right)}} = 472.0 \text{ m s}^{-1}$$

$$A = \pi\left(\frac{d}{2}\right)^2 = \pi\left(\frac{1.0 \text{ mm}}{2}\right)^2\left(\frac{1 \text{ m}}{1000 \text{ mm}}\right)^2 = 7.854 \times 10^{-7} \text{ m}^2$$

The effusion rate is

$$r = \frac{nN_A\bar{c}A}{4V}$$

$$= \frac{(2.365 \text{ mol})\left(6.022 \times 10^{23} \text{ molecules mol}^{-1}\right)\left(472.0 \text{ m s}^{-1}\right)\left(7.854 \times 10^{-7} \text{ m}^2\right)}{4\,(30.0 \text{ L})\left(\frac{1 \text{ m}^3}{1000 \text{ L}}\right)}$$

$$= 4.400 \times 10^{21} \text{ molecules s}^{-1} = 4.40 \times 10^{21} \text{ molecules s}^{-1}$$

(b) The time it would take to lose half of the molecules is

$$t = \frac{\text{\# of molecules effused}}{\text{effusion rate}}$$

$$= \frac{\left(\frac{2.365 \text{ mol}}{2}\right)\left(6.022 \times 10^{23} \text{ molecules mol}^{-1}\right)}{4.400 \times 10^{21} \text{ molecules s}^{-1}}$$

$$= 161.8 \text{ s} = 2.70 \text{ min}$$

2.42 Explain the equipartition of energy theorem. Why does it fail for diatomic and polyatomic molecules?

The equipartition of energy theorem states that the energy of a molecule is equally divided among all of its degrees of freedom. Based on classical mechanics, the theorem fails for diatomic and polyatomic molecules because it does not take the quantization of energy into account. Specifically, the theorem fails for the electronic and vibrational degrees of freedom of molecules where the spacings between quantized energy levels are much greater than k_BT. The theorem may be successfully applied to the translational and rotational degrees of freedom, although for low temperatures, where k_BT becomes smaller than the spacing between quantized rotational energy levels, it will fail for this degree of freedom as well.

2.44 Calculate the mean kinetic energy ($\bar{E}_{\text{trans}}$) in joules of the following molecules at 350 K: **(a)** He, **(b)** CO_2, and **(c)** UF_6. Explain your results.

Regardless of the identity of the gas, the mean kinetic energy of the molecules is

$$\bar{E}_{\text{trans}} = \frac{3}{2}k_BT = \frac{3}{2}(1.381 \times 10^{-23} \text{ J K}^{-1})(350 \text{ K}) = 7.25 \times 10^{-21} \text{ J}$$

2.46 Calculate the value of $\overline{C}_V$ for H_2, CO_2, and SO_2, assuming that only translational and rotational motions contribute to the heat capacities. Compare your results with the values listed in Table 2.3. Explain the differences.

For molecules, only the translational and rotational degrees of freedom fully contribute to the heat capacity. The electronic and vibrational degrees of freedom, due to the large spacing between quantized energy levels, are inaccessible at room temperature (except for partial contributions from low frequency vibrational motions).

H_2 has 3 translational degrees of freedom and 2 rotational degrees of freedom, therefore, the total energy for 1 mole of H_2 is

$$\overline{U} = \frac{3}{2}RT + \frac{2}{2}RT = \frac{5}{2}RT$$

Because $\overline{U}$ is molar total energy, the molar heat capacity is

$$\overline{C}_V = \left(\frac{\partial \overline{U}}{\partial T}\right)_V = \frac{5}{2}R = 20.79 \text{ J K}^{-1}\text{mol}^{-1}$$

The agreement with the measured value of $20.50 \text{ J K}^{-1}\text{mol}^{-1}$ is quite good, indicating little contribution from the single vibrational degree of freedom.

CO_2 is linear, so it has 3 translational degrees of freedom and 2 rotational degrees of freedom, therefore, the total energy for 1 mole of CO_2 is the same as that for 1 mole of H_2. Consequently, $\overline{C}_V$ for CO_2 is also calculated to be $20.79 \text{ J K}^{-1}\text{mol}^{-1}$. This is significantly lower than the measured value of $28.82 \text{ J K}^{-1}\text{mol}^{-1}$ and is indicative of some vibrational contribution. In fact, CO_2 has two, low-energy bending vibrations into which some thermal energy may be distributed, although not to the extent necessary for the equipartition of energy theorem to be valid.

SO_2 is nonlinear, and has 3 translational degrees of freedom and 3 rotational degrees of freedom, therefore, the total energy for 1 mole of SO_2 is

$$\overline{U} = \frac{3}{2}RT + \frac{3}{2}RT = 3RT$$

Thus, the calculated molar heat capacity for SO_2 is

$$\overline{C}_V = 3R = 24.94 \text{ J K}^{-1}\text{mol}^{-1}$$

Again the calculated value is significantly lower than that measured ($31.51 \text{ J K}^{-1}\text{mol}^{-1}$) for the same reason as for CO_2. In both cases, however, the measured value is lower than that predicted by the equipartition theorem when all degrees of freedom are included.

2.48 The typical energy differences between successive rotational, vibrational, and electronic energy levels are 5.0×10^{-22} J, 5.0×10^{-20} J, and 1.0×10^{-18} J respectively. Calculate the ratios of the numbers of molecules in the two adjacent energy levels (higher to lower) in each case at 298 K.

Let the numbers of molecules in the lower and higher level be N_1 and N_2, respectively. The ratio between the numbers of molecules in these two levels is

$$\frac{N_2}{N_1} = e^{-\frac{\Delta E}{k_B T}} = e^{-\frac{\Delta E}{(1.381\times10^{-23}\,\text{J K}^{-1})(298\,\text{K})}} = e^{-\frac{\Delta E}{4.115\times10^{-21}\,\text{J}}}$$

Rotational energy levels

$$\frac{N_2}{N_1} = e^{-\frac{5.0\times10^{-22}\,\text{J}}{4.115\times10^{-21}\,\text{J}}} = 0.89$$

Vibrational energy levels

$$\frac{N_2}{N_1} = e^{-\frac{5.0\times10^{-20}\,\text{J}}{4.115\times10^{-21}\,\text{J}}} = 5.3 \times 10^{-6}$$

Electronic energy levels

$$\frac{N_2}{N_1} = e^{-\frac{1.0\times10^{-18}\,\text{J}}{4.115\times10^{-21}\,\text{J}}} = e^{-243} \approx 0$$

2.50   Consider 1 mole each of gaseous He and N_2 at the same temperature and pressure. State which gas (if any) has the greater value for: **(a)** $\bar{c}$, **(b)** c_{rms}, **(c)** $\overline{E}_{\text{trans}}$, **(d)** Z_1, **(e)** Z_{11}, **(f)** density, **(g)** mean free path, **(h)** viscosity.

Both gases are at the same pressure and temperature, hence the same volume and density as well. The two important differences are of size and mass: He is both smaller and less massive than N_2. N_2 has 7 times the mass of He, and although it is not listed in Table 2.1, we can find on-line or in the literature, that the collision diameter of N_2 is about 1.45 times that of He.

(a) Being less massive, He will have the greater $\bar{c}$. From Equation 2.14, $\bar{c} \propto \mathcal{M}^{-1/2}$.

(b) For the same reason as part **(a)**, He will have the greater c_{rms}. Note $c_{\text{rms}} = \sqrt{\frac{3RT}{\mathcal{M}}} \propto \mathcal{M}^{-1/2}$.

(c) Both gases have the same $\overline{E}_{\text{trans}}$, as both are at the same temperature. Note $\overline{E}_{\text{trans}} = \frac{3}{2}k_B T$.

(d) $Z_1 \propto \frac{d^2}{\sqrt{\mathcal{M}}}$, where the mass dependence arises through the $\bar{c}$ term. He is moving $\sqrt{7} = 2.6$ times faster than N_2, while the d^2 term represents an increase in "target" size of a factor of about two for the N_2. The increased speed of He outweighs the larger size of N_2 to give He the greater Z_1.

(e) $Z_{11} = \frac{1}{2}Z_1\left(\frac{N}{V}\right)$. Since the densities of the two gases are the same, He will have the greater Z_{11} as well.

(f) The densities are the same.

(g) With the greater size, N_2 has the smaller mean free path. From Equation 2.18, $\lambda \propto d^{-2}$.

(h) From Equation 2.27, $\eta \propto \frac{m\bar{c}}{d^2} \propto \frac{\sqrt{m}}{d^2}$. Thus although He is lighter, its smaller size and consequent larger mean free path leads to He having the greater viscosity.

2.52   At 298 K, the $\overline{C}_V$ of SO_2 is greater than that of CO_2. At very high temperatures (> 1000 K), the $\overline{C}_V$ of CO_2 is greater than that of SO_2. Explain.

At high temperatures, $k_B T$ becomes large enough that the vibrational degrees of freedom of a molecule can fully contribute to the heat capacity. CO_2 is a linear molecule with 3 translational, 2 rotational, and $3(3) - 5 = 4$ vibrational degrees of freedom. Recalling that translational and rotational degrees of freedom each contributes $\frac{R}{2}$ to $\overline{C}_V$ while each vibrational degree of freedom contributes R,

$$\overline{C}_{V,CO_2} = 3\frac{R}{2} + 2\frac{R}{2} + 4R = \frac{13}{2}R$$

SO_2, on the other hand, is not linear. It has 3 translational and 3 rotational degrees of freedom and $3(3) - 6 = 3$ vibrational degrees of freedom. Thus,

$$\overline{C}_{V,SO_2} = 3\frac{R}{2} + 3\frac{R}{2} + 3R = 6R$$

We see that in the linear molecule the substitution of an extra vibrational degree of freedom for the absent rotational degree of freedom leads to a larger $\overline{C}_V$, since vibrational degrees of freedom contribute twice as much to the heat capacity.

At room temperature, however, the vibrational degrees of freedom contribute only slightly, and the extra rotational degree of freedom for the non-linear SO_2 molecule gives it the larger $\overline{C}_V$.

2.54   The following apparatus can be used to measure atomic and molecular speed. A beam of metal atoms is directed at a rotating cylinder in a vacuum. A small opening in the cylinder allows the atoms to strike a target area. Because the cylinder is rotating, atoms traveling at different speeds will strike the target at different position. In time, a layer of the metal will deposit on the target area, and the variation in its thickness is found to correspond to Maxwell's speed distribution. In one experiment it is found that at 850°C, some bismuth (Bi) atoms struck the target at a point 2.80 cm from the spot directly opposite the slit. The diameter of the cylinder is 15.0 cm and it is rotating at 130 revolutions per second. **(a)** Calculate the speed (m s^{-1}) at which the target is moving (*Hint*: The circumference of a circle is given by $2\pi r$, where r is the radius). **(b)** Calculate the time (in seconds) it takes for the target to travel 2.80 cm. **(c)** Determine the speed of the Bi atoms. Compare your result in **(c)** with the c_{rms} value for Bi at 850°C. Comment on the difference.

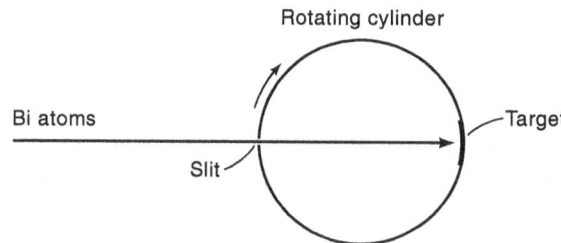

(a)   The speed at which a point on the target moves is equal to the product of the circumference of the cylinder and the number of revolutions per second. Calling the speed of the target s,

$$s = \left(\frac{130 \text{ revolutions}}{1 \text{ s}}\right)\left(\frac{2\pi\left(\frac{15.0 \text{ cm}}{2}\right)}{1 \text{ revolution}}\right)\left(\frac{1 \text{ m}}{100 \text{ cm}}\right) = 61.26 \text{ m s}^{-1} = 61.3 \text{ m s}^{-1}$$

(b)   The time, t, it takes for a point on the target to move 2.80 cm is given by

$$t = \left(\frac{2.80 \text{ cm}}{61.26 \text{ m s}^{-1}}\right)\left(\frac{1 \text{ m}}{100 \text{ cm}}\right) = 4.571 \times 10^{-4} \text{ s} = 4.57 \times 10^{-4} \text{ s}$$

(c)   Since the Bi atoms struck the target at a point 2.80 cm from the spot directly opposite the slit, it took the same amount of time for the atoms to travel the 15.0 cm diameter of the cylinder as it took for the target to move 2.80 cm, as calculated in part (b). Thus the speed, c, of these atoms is

$$c = \left(\frac{15.0 \text{ cm}}{4.571 \times 10^{-4} \text{ s}}\right)\left(\frac{1 \text{ m}}{100 \text{ cm}}\right) = 328 \text{ m s}^{-1}$$

This can be compared with

$$c_{\text{rms}} = \sqrt{\frac{3RT}{\mathcal{M}}} = \sqrt{\frac{3\left(8.314 \text{ J K}^{-1} \text{mol}^{-1}\right)(273.15 + 850) \text{ K}}{209.0 \times 10^{-3} \text{ kg mol}^{-1}}} = 366 \text{ m s}^{-1}$$

The difference is not unexpected, since c_{rms} is an average speed of the sample of atoms (assuming it is thermally equilibrated) while the speed calculated in part (c) is the speed of a particular group of atoms. Nevertheless, the magnitudes of the two speeds are comparable. Furthermore, the measured speed is quite close to

$$c_{\text{mp}} = \sqrt{\frac{8RT}{\pi\mathcal{M}}} = \sqrt{\frac{8\left(8.314 \text{ J K}^{-1} \text{mol}^{-1}\right)(273.15 + 850) \text{ K}}{\pi(209.0 \times 10^{-3} \text{ kg mol}^{-1})}} = 337 \text{ m s}^{-1}$$

which would be expected to give the most prominent spot in the measurement.

2.56 The escape velocity, v, from Earth's gravitational field is given by $(2GM/r)^{1/2}$, where G is the universal gravitational constant $(6.67 \times 10^{-11} \text{ m}^3 \text{ kg}^{-1} \text{ s}^{-2})$, M is the mass of Earth $(6.0 \times 10^{24} \text{ kg})$, and r is the distance from the center of Earth to the object, in meters. Compare the average speeds of He and N_2 molecules in the thermosphere (altitude about 100 km, $T = 250$ K). Which of the two molecules will have a greater tendency to escape? The radius of Earth is 6.4×10^6 m.

The distance between molecules in the thermosphere and the center of Earth (r) is 6.4×10^6 m plus 100 km, which is 6.5×10^6 m. The escape velocity is

$$v = \sqrt{\frac{2G\mathcal{M}}{r}} = \sqrt{\frac{2\left(6.67 \times 10^{-11} \text{ m}^3 \text{ kg}^{-1} \text{ s}^{-2}\right)\left(6.0 \times 10^{24} \text{ kg}\right)}{6.5 \times 10^6 \text{ m}}} = 1.1 \times 10^4 \text{ m s}^{-1}$$

The average speed of He molecules is

$$\bar{c} = \sqrt{\frac{8RT}{\pi\mathcal{M}}} = \sqrt{\frac{8\left(8.314 \text{ J K}^{-1} \text{mol}^{-1}\right)(250 \text{ K})}{\pi\left(4.003 \times 10^{-3} \text{ kg mol}^{-1}\right)}} = 1.15 \times 10^3 \text{ m s}^{-1}$$

The average speed of N_2 molecules is

$$\bar{c} = \sqrt{\frac{8RT}{\pi\mathcal{M}}} = \sqrt{\frac{8\left(8.314 \text{ J K}^{-1} \text{mol}^{-1}\right)(250 \text{ K})}{\pi\left(28.02 \times 10^{-3} \text{ kg mol}^{-1}\right)}} = 435 \text{ m s}^{-1}$$

Recalling the form of the Maxwell distribution of molecular speeds, there are indeed molecules moving faster than the average, and although the high speed "tail" of the distribution extends out to very large speeds, it does approach the x-axis fairly rapidly. Thus, there will be many more He atoms with sufficient speed to escape the Earth's gravitational field than there are N_2 molecules. Indeed, it is for this reason that there are no appreciable amounts of He or H_2 in the Earth's atmosphere.

2.58 Compare the ratio of the number of O_3 molecules with a speed of 1300 m s^{-1} at 360 K to the number with that speed at 293 K.

This problem is very similar to Problem 2.21, except that problem wanted a ratio of the numbers of molecules with two different speeds at the same temperature. Here, the ratio of the numbers of molecules with the same speed, but at two different temperatures is desired.

$$\frac{dN_2/N}{dN_1/N} = \frac{4\pi c^2 \left(\frac{m}{2\pi k_B T_2}\right)^{\frac{3}{2}} e^{-\frac{mc^2}{2k_B T_2}} dc}{4\pi c^2 \left(\frac{m}{2\pi k_B T_1}\right)^{\frac{3}{2}} e^{-\frac{mc^2}{2k_B T_1}} dc}$$

$$= \left(\frac{T_1}{T_2}\right)^{3/2} e^{-\frac{mc^2}{2k_B}\left(\frac{1}{T_2}-\frac{1}{T_1}\right)}$$

$$= \left(\frac{293\ \text{K}}{360\ \text{K}}\right)^{\frac{3}{2}} e^{-\frac{(48.00\ \text{amu})(1.6605\times10^{-27}\ \text{kg amu}^{-1})(1300\ \text{m s}^{-1})^2}{2(1.381\times10^{-23}\ \text{J K}^{-1})}\left(\frac{1}{360\ \text{K}}-\frac{1}{293\ \text{K}}\right)}$$

$$= 16.3$$

2.60 Apply your knowledge of the kinetic theory of gases to the following situations. **(a)** Two flasks of volumes V_1 and V_2 (where $V_2 > V_1$) contain the same number of helium atoms at the same temperature. **(i)** Compare the root-mean-square (rms) speeds and average kinetic energies of the helium (He) atoms in the flasks. **(ii)** Compare the frequency and the force with which the He atoms collide with the walls of their containers. **(b)** Equal numbers of He atoms are placed in two flasks of the same volume at temperatures T_1 and T_2 (where $T_2 > T_1$). **(i)** Compare the rms speeds of the atoms in the two flasks. **(ii)** Compare the frequency and the force with which the He atoms collide with the walls of their containers. **(c)** Equal numbers of He and neon (Ne) atoms are placed in two flasks of the same volume. The temperature of both gases is 74°C. Comment on the validity of the following statements: **(i)** The rms speed of He is equal to that of Ne. **(ii)** The average kinetic energies of the two gases are equal. **(iii)** The rms speed of each He atom is 1.47×10^3 m s^{-1}.

(a) With the two samples at the same temperature, **(i)** c_{rms} and average kinetic energies of the atoms in the two flasks are the same. Likewise, **(ii)** the force with which the He atoms strike the walls of the containers is the same in each flask, but the frequency of collision is greater in the smaller flask, V_1. It will have the higher pressure.

(b) Now at the same volume, but different temperatures, **(i)** c_{rms} is greater in the flask with the higher temperature, T_2, and **(ii)** in the flask with the higher temperature, T_2, the atoms collide with the walls both with greater force and with greater frequency than in the lower temperature sample.

(c) (i) The statement is false, because the lighter He atoms have a greater rms speed.

(ii) True, because the samples are at the same temperature, the average kinetic energies of their atoms are the same.

(iii) True and False. The rms speed of the sample is

$$c_{rms} = \sqrt{\frac{3RT}{\mathcal{M}}}$$

$$= \sqrt{\frac{3\left(8.314\ \text{J K}^{-1}\ \text{mol}^{-1}\right)(273.15+74)\ \text{K}}{4.003\times 10^{-3}\ \text{kg mol}^{-1}}}$$

$$= 1.47\times 10^3\ \text{m s}^{-1}$$

but there is a wide distribution of speeds of individual He atoms.

2.62 Use the kinetic theory of gases to explain why hot air rises.

The molecules in a sample of hot air have greater average kinetic energy and are thus able to achieve a greater gravitational potential energy and rise to a higher altitude. This is, of course, a microscopic, molecular view. A macroscopic explanation would be that hot air has a smaller density (see Problem 1.8) than does cooler air and is more buoyant. Thus, it will "float" on the cooler air.

2.64 Two ideal gases A and B are heated to different temperatures. If their pressures and densities are the same at these temperatures, show that their average speeds must also be the same.

Because $P_A = P_B$ and $\dfrac{m_A}{V_A} = \dfrac{m_B}{V_B}$,

$$\frac{P_A V_A}{m_A} = \frac{P_B V_B}{m_B}$$

$$\frac{n_A R T_A}{m_A} = \frac{n_B R T_B}{m_B}$$

$$\frac{T_A}{\mathcal{M}_A} = \frac{T_B}{\mathcal{M}_B}$$

The average speed of gas A is

$$\bar{c}_A = \sqrt{\frac{8RT_A}{\pi\mathcal{M}_A}}$$

which, according to the relation derived between the temperatures of molar masses of the two gases, gives

$$\bar{c}_A = \sqrt{\frac{8RT_B}{\pi\mathcal{M}_B}} = \bar{c}_B$$

2.66 Identify the gas whose c_{rms} is 2.82 times that of HI at the same temperature.

The desired relationship is $c_{rms, gas} = 2.82 c_{rms, HI}$, or $\frac{c_{rms, gas}}{c_{rms, O_2}} = 2.82$. Since $c_{rms} = \sqrt{\frac{3RT}{\mathcal{M}}}$,

$$\frac{c_{rms, gas}}{c_{rms, HI}} = \frac{\sqrt{\frac{3RT}{\mathcal{M}_{gas}}}}{\sqrt{\frac{3RT}{\mathcal{M}_{HI}}}} = \sqrt{\frac{\mathcal{M}_{HI}}{\mathcal{M}_{gas}}} = 2.82$$

$$\mathcal{M}_{gas} = \frac{\mathcal{M}_{HI}}{2.82^2} = \frac{127.9 \text{ g mol}^{-1}}{7.952} = 16.1 \text{ g mol}^{-1}$$

The gas is likely CH_4.

CHAPTER 3

The First Law of Thermodynamics

PROBLEMS AND SOLUTIONS

3.2 What is heat? How does heat differ from thermal energy? Under what condition is heat transferred from one system to another?

Heat is the transfer of energy from one object to another as a result of a temperature difference between the two objects. Heat is only transferred between systems when they are at different temperatures.

Thermal energy is that part of the energy of a system that is associated with the random motion (translational, vibrational, and rotational) of atoms and molecules.

3.4 A 7.24-g sample of ethane occupies 4.65 L at 294 K. **(a)** Calculate the work done when the gas expands isothermally against a constant external pressure of 0.500 atm until its volume is 6.87 L. **(b)** Calculate the work done if the same expansion occurs reversibly.

(a) $w = -P_{ex}\Delta V = -(0.500 \text{ atm})(6.87 \text{ L} - 4.65 \text{ L})\left(\dfrac{101.3 \text{ J}}{1 \text{ L atm}}\right) = -112 \text{ J}$

(b) $w = -nRT \ln \dfrac{V_2}{V_1} = -\dfrac{7.24 \text{ g}}{30.07 \text{ g mol}^{-1}}\left(8.314 \text{ J K}^{-1} \text{ mol}^{-1}\right)(294 \text{ K}) \ln \dfrac{6.87 \text{ L}}{4.65 \text{ L}} = -230 \text{ J}$

Note that work done in the reversible process in **(b)** is greater in magnitude than that in the irreversible process in **(a)**.

3.6 Calculate the work done by the reaction

$$Zn(s) + H_2SO_4(aq) \rightarrow ZnSO_4(aq) + H_2(g)$$

when 1.0 mole of hydrogen gas is collected at 273 K and 1.0 atm. (Neglect volume changes other than the change in gas volume.)

The gas expands until its pressure is the same as the external pressure of 1.0 atm. Furthermore, since volume changes other than gas are neglected, the change in volume of the system is

$$\Delta V = V_{H_2} = \frac{nRT}{P_{H_2}} = \frac{nRT}{P_{ex}}$$

The work done is

$$w = -P_{ex}\Delta V = -P_{ex}V_f = -P_{ex}\left(\frac{nRT}{P_{ex}}\right) = -nRT$$

$$= -(1.0 \text{ mol})\left(8.314 \text{ J K}^{-1}\text{mol}^{-1}\right)(273 \text{ K}) = -2.3 \times 10^3 \text{ J}$$

3.8 Some driver's test manuals state that the stopping distance quadruples as the velocity doubles. Justify this statement by using mechanics and thermodynamic arguments.

The kinetic energy of a moving vehicle is given by $E_{kin} = \frac{1}{2}mv^2$. If the velocity doubles, the kinetic energy quadruples. If $v_2 = 2v_1$, then $E_2 = \frac{1}{2}m\left(2v_1\right)^2 = 4E_1$. Assuming kinetic energy is dissipated at a constant rate in the brakes and between the tires and the road as heat through friction, then the stopping distance is proportional to the energy. The doubling in velocity quadruples the kinetic energy and thus quadruples the stopping distance.

3.10 An ideal gas is compressed isothermally by a force of 85 newtons acting through 0.24 meter. Calculate the values of ΔU and q.

$\Delta U = 0$ for an ideal gas undergoing an isothermal process. Using this information, q can be calculated from w, the work done on the system.

$$w = (\text{force})\,(\text{distance}) = (85 \text{ N})\,(0.24 \text{ m}) = 20 \text{ J}$$

Because $\Delta U = 0 = q + w$, $q = -w = -20$ J.

3.12 A thermos bottle containing milk is shaken vigorously. Consider the milk as the system. **(a)** Will the temperature rise as a result of the shaking? **(b)** Has heat been added to the system? **(c)** Has work been done on the system? **(d)** Has the system's internal energy changed?

The thermos bottle ensures that $q = 0$, since it prevents the transfer of heat to or from the surroundings.

(a) Energy (mechanical) has been added to the system by shaking. The random motion of the molecules in the milk is increased, and the temperature rises. $\Delta T > 0$.

(b) The insulation of the thermos bottle prevents the transfer of heat between the system and surroundings. $q = 0$.

(c) Work has been done on the system through shaking. $w > 0$.

(d) The internal energy of the system has increased. $\Delta U > 0$.

3.14 An ideal gas is compressed isothermally from 2.0 atm and 2.0 L to 4.0 atm and 1.0 L. Calculate the values of ΔU and ΔH if the process is carried out **(a)** reversibly and **(b)** irreversibly.

ΔU and ΔH of an ideal gas depend only on T. Therefore, for any isothermal process [either process **(a)** or **(b)**], $\Delta U = 0$ and $\Delta H = 0$.

3.16 A piece of potassium metal is added to water in a beaker. The reaction that takes place is

$$2K(s) + 2H_2O(l) \rightarrow 2KOH(aq) + H_2(g)$$

Predict the signs of w, q, ΔU, and ΔH.

---

w is negative because the hydrogen gas produced expands and does work on the surroundings.

q is negative because thermal energy is generated in this reaction.

Since $\Delta U = q + w$, it must be negative also.

The process takes place at constant pressure, therefore $q = q_P = \Delta H$. Thus, ΔH is negative.

3.18 Consider a cyclic process involving a gas. If the pressure of the gas varies during the process but returns to the original value at the end, is it correct to write $\Delta H = q_P$?

No, it is not correct. For a cyclic process, $\Delta H = 0$, since H is a state function, but the heat, or q, associated with a process is not a state function and depends on the path. Depending on the path used to achieve the cyclic process q may take on a variety of values.

3.20 One mole of an ideal gas undergoes an isothermal expansion at 300 K from 1.00 atm to a final pressure while performing 200 J of expansion work. Calculate the final pressure of the gas if the external pressure is 0.20 atm.

$$w = -P_{ex}\left(V_2 - V_1\right) = -P_{ex}\left(\frac{nRT}{P_2} - \frac{nRT}{P_1}\right)$$

$$= -P_{ex}nRT\left(\frac{1}{P_2} - \frac{1}{P_1}\right)$$

$$\frac{1}{P_2} = -\frac{w}{P_{ex}nRT} + \frac{1}{P_1}$$

$$= -\frac{-200\text{ J}}{(0.20\text{ atm})\,(1\text{ mol})\left(8.314\text{ J K}^{-1}\text{ mol}^{-1}\right)(300\text{ K})} + \frac{1}{1.00\text{ atm}} = 1.401\text{ atm}^{-1}$$

$$P_2 = 0.71\text{ atm}$$

3.22 A 10.0-g sheet of gold with a temperature of 18.0°C is laid flat on a sheet of iron that weighs 20.0 g and has a temperature of 55.6°C. Given that the specific heats of Au and Fe are 0.129 $\text{J g}^{-1}\,{}^{\circ}\text{C}^{-1}$ and 0.444 $\text{J g}^{-1}\,{}^{\circ}\text{C}^{-1}$, respectively, what is the final temperature of the combined metals? Assume that no heat is lost to the surroundings. (*Hint*: The heat gained by the gold must be equal to the heat lost by the iron.)

The final temperature of the sheet of gold is the same as that of the sheet of iron when thermal equilibrium is reached, and is denoted by T_f. Furthermore, the amount of heat gained by gold is the same as that lost by iron, that is,

$$q_{Au} = -q_{Fe}$$

The "−" sign is used to indicate that q_{Au} and q_{Fe} are of opposite sign.

The above relation gives

$$m_{Au}s_{Au}\Delta T_{Au} = -m_{Fe}s_{Fe}\Delta T_{Fe}$$

$$(10.0 \text{ g})\left(0.129 \text{ J g}^{-1}{}^{\circ}\text{C}^{-1}\right)(T_f - 18.0^{\circ}\text{C}) = -(20.0 \text{ g})\left(0.444 \text{ J g}^{-1}{}^{\circ}\text{C}^{-1}\right)(T_f - 55.6^{\circ}\text{C})$$

$$\left(1.29 \text{ J}{}^{\circ}\text{C}^{-1}\right)(T_f - 18.0^{\circ}\text{C}) = -\left(8.88 \text{ J}{}^{\circ}\text{C}^{-1}\right)(T_f - 55.6^{\circ}\text{C})$$

$$T_f - 18.0^{\circ}\text{C} = -\frac{8.88 \text{ J}{}^{\circ}\text{C}^{-1}}{1.29 \text{ J}{}^{\circ}\text{C}^{-1}}(T_f - 55.6^{\circ}\text{C})$$

$$= -6.884\left(T_f - 55.6^{\circ}\text{C}\right)$$

$$= -6.884T_f + 382.8^{\circ}\text{C}$$

$$7.884T_f = 400.8^{\circ}\text{C}$$

$$T_f = 50.8^{\circ}\text{C}$$

3.24 The molar heat of vaporization for water is 44.01 kJ mol^{-1} at 298 K and 40.79 kJ mol^{-1} at 373 K. Give a qualitative explanation of the difference in these two values.

At the higher temperature, the water molecules have more kinetic energy and are moving faster. Thus, there are on average fewer hydrogen bonds holding the individual water molecules together. Consequently at the higher temperature, less thermal energy must be supplied to break the remaining intermolecular attractions and allow the molecules to enter the gas phase.

3.26 The heat capacity ratio (γ) for a gas with the molecular formula X_2Y is 1.38. What can you deduce about the structure of the molecule?

The molecule X_2Y is either linear or bent (*e.g.* CO_2 or H_2O). Assuming ideal behavior, $\overline{C}_P = \overline{C}_V + R$, and the heat capacity ratio may be rewritten as

$$\gamma = \frac{\overline{C}_P}{\overline{C}_V}$$

$$= \frac{\overline{C}_V + R}{\overline{C}_V}$$

This may be solved for $\overline{C}_V$ to give

$$\overline{C}_V = \frac{R}{\gamma - 1}$$

$$= \frac{R}{0.38}$$

$$= 2.6R$$

At typical temperatures the vibrational contribution to the heat capacity is negligible, and for a linear molecule $\overline{C}_V = 2.5R$, while for a bent molecule, $\overline{C}_V = 3.0R$. Better agreement between prediction and measurement is obtained in the case of the linear molecule, and thus the molecule is most likely linear.

3.28 Which of the following gases has the largest $\overline{C}_V$ value at 298 K? He, N_2, CCl_4, HCl.

A gas composed of non-linear, polyatomic molecules is predicted to have a larger $\overline{C}_V$ than monoatomic gases or those made up of linear molecules. In the list above, only CCl_4 is a non-linear molecule, so it should have the largest $\overline{C}_V$.

3.30 In the nineteenth century, two scientists named Dulong and Petit noticed that the product of the molar mass of a solid element and its specific heat is approximately 25 J $°C^{-1}$. This observation, now called Dulong and Petit's law, was used to estimate the specific heat of metals. Verify the law for aluminum (0.900 J g^{-1} $°C^{-1}$), copper (0.385 J g^{-1} $°C^{-1}$), and iron (0.444 J g^{-1} $°C^{-1}$). The law does not apply to one of the metals. Which one is it? Why?

Al:

$$(26.98 \text{ g}) \left(0.900 \text{ J g}^{-1} °C^{-1}\right) = 24.3 \text{ J} °C^{-1}$$

Cu:

$$(63.55 \text{ g}) \left(0.385 \text{ J g}^{-1} °C^{-1}\right) = 24.5 \text{ J} °C^{-1}$$

Fe:

$$(55.85 \text{ g}) \left(0.444 \text{ J g}^{-1} °C^{-1}\right) = 24.8 \text{ J} °C^{-1}$$

Therefore, Dulong and Petit's law applies to Al, Cu, and Fe. However, since this law applies only to solids, it does not apply to mercury, a liquid. In fact, the product of the molar mass of Hg and its specific heat is

$$(200.6 \text{ g}) \left(0.139 \text{ J g}^{-1} °C^{-1}\right) = 27.9 \text{ J} °C^{-1}$$

3.32 The equation of state for a certain gas is given by $P\left[(V/n) - b\right] = RT$. Obtain an expression for the maximum work done by the gas in a reversible isothermal expansion from V_1 to V_2.

Write P in terms of n, V, and T:

$$P = \frac{RT}{\frac{V}{n} - b} = \frac{nRT}{V - nb}$$

The maximum work done by the gas undergoing an isothermal expansion is

$$w = -\int_{V_1}^{V_2} P \, dV = -\int_{V_1}^{V_2} \frac{nRT}{V - nb} \, dV$$

$$= -nRT \ln (V - nb)\big|_{V_1}^{V_2}$$

$$= -nRT \ln \frac{V_2 - nb}{V_1 - nb}$$

3.34 A quantity of 0.27 mole of neon is confined in a container at 2.50 atm and 298 K and then allowed to expand adiabatically under two different conditions: **(a)** reversibly to 1.00 atm and **(b)** against a constant pressure of 1.00 atm. Calculate the final temperature in each case.

(a) For a reversible adiabatic process, P and V are related as follow:

$$\frac{P_2}{P_1} = \left(\frac{V_1}{V_2}\right)^\gamma$$

Assuming neon to be an ideal gas, write V in terms of n, P, and T:

$$\frac{P_2}{P_1} = \left(\frac{\frac{nRT_1}{P_1}}{\frac{nRT_2}{P_2}}\right)^\gamma = \left(\frac{T_1}{T_2}\right)^\gamma \left(\frac{P_2}{P_1}\right)^\gamma$$

Rearrange the equation to give an expression relating T and P:

$$\left(\frac{T_2}{T_1}\right)^\gamma = \left(\frac{P_2}{P_1}\right)^{\gamma-1}$$

$$\frac{T_2}{T_1} = \left(\frac{P_2}{P_1}\right)^{(\gamma-1)/\gamma}$$

Utilizing the last equation and the fact that γ for an ideal monatomic gas is $\frac{5}{3}$, the final temperature can be calculated:

$$T_2 = T_1 \left(\frac{P_2}{P_1}\right)^{(\gamma-1)/\gamma} = (298 \text{ K}) \left(\frac{1.00 \text{ atm}}{2.50 \text{ atm}}\right)^{(\frac{5}{3}-1)/\frac{5}{3}} = 207 \text{ K}$$

(b) The process is adiabatic with $q = 0$. Thus, $\Delta U = w$. Since $\Delta U = C_V (T_2 - T_1)$, and $w = -P_{ex} (V_2 - V_1)$,

$$C_V (T_2 - T_1) = -P_{ex} (V_2 - V_1)$$

$$\frac{3}{2} nR (T_2 - T_1) = -P_{ex} \left(\frac{nRT_2}{P_2} - \frac{nRT_1}{P_1}\right)$$

$$\frac{3}{2} (T_2 - T_1) = -P_{ex} \left(\frac{T_2}{P_2} - \frac{T_1}{P_1}\right)$$

The expansion occurs until $P_2 = P_{ex}$. Therefore, the above expression simplifies to

$$\frac{3}{2}\left(T_2 - T_1\right) = -T_2 + \frac{P_{ex}T_1}{P_1}$$

$$\frac{5}{2}T_2 = \frac{P_{ex}T_1}{P_1} + \frac{3}{2}T_1 = \left(\frac{P_{ex}}{P_1} + \frac{3}{2}\right)T_1$$

$$T_2 = \frac{2}{5}\left(\frac{P_{ex}}{P_1} + \frac{3}{2}\right)T_1$$

$$= \frac{2}{5}\left(\frac{1.00 \text{ atm}}{2.50 \text{ atm}} + \frac{3}{2}\right)(298 \text{ K}) = 226 \text{ K}$$

3.36 A 0.1375-g sample of magnesium is burned in a constant-volume bomb calorimeter that has a heat capacity of $1769 \text{ J } °\text{C}^{-1}$. The calorimeter contains exactly 300 g of water, and the temperature increases by 1.126°C. Calculate the heat given off by the burning magnesium, in kJ g^{-1} and in kJ mol^{-1}. The specific heat of water is $4.184 \text{ J g}^{-1}°\text{C}^{-1}$.

Heat given off by the burning Mg is absorbed by the calorimeter and water. The heat absorbed is calculated using the heat capacity of the calorimeter, specific heat of water, and the temperature rise.

$$\text{Heat gained by calorimeter and water} = \left(1769 \text{ J }°\text{C}^{-1}\right)(1.126 \text{ }°\text{C})$$

$$+ \left(300 \text{ g}\right)\left(4.184 \text{ J g}^{-1}°\text{C}^{-1}\right)(1.126 \text{ }°\text{C})$$

$$= 3.4052 \times 10^3 \text{ J}$$

Therefore, heat given off by Mg is

$$\frac{3.4052 \times 10^3 \text{ J}}{0.1375 \text{ g}} = 2.4765 \times 10^4 \text{ J g}^{-1} = 24.77 \text{ kJ g}^{-1}$$

or

$$\left(2.4765 \times 10^4 \text{ J g}^{-1}\right)\left(24.31 \text{ g mol}^{-1}\right) = 6.020 \times 10^5 \text{ J mol}^{-1} = 602.0 \text{ kJ mol}^{-1}$$

3.38 A quantity of 2.00×10^2 mL of 0.862 M HCl is mixed with 2.00×10^2 mL of 0.431 M Ba(OH)$_2$ in a constant-pressure calorimeter that has a heat capacity of $453 \text{ J }°\text{C}^{-1}$. The initial temperature of the HCl and Ba(OH)$_2$ solutions is the same at 20.48°C. For the process

$$\text{H}^+(aq) + \text{OH}^-(aq) \rightarrow \text{H}_2\text{O}(l)$$

the heat of neutralization is $-56.2 \text{ kJ mol}^{-1}$. What is the final temperature of the mixed solution?

The final temperature of the solution can be determined once the heat capacity of the calorimeter *plus* solution is known as well as the number of moles of reaction that occurs.

First determine the number of moles of reactants.

$$n_{\text{H}^+} = \left(2.00 \times 10^{-1} \text{ L}\right)\left(0.862 \text{ mol L}^{-1}\right) = 0.1724 \text{ mol}$$

$$n_{\text{OH}^-} = 2\left(2.00 \times 10^{-1} \text{ L}\right)\left(0.431 \text{ mol L}^{-1}\right) = 0.1724 \text{ mol}$$

There is just enough of each reactant to completely react with the other to form 0.1724 mol of the product, H_2O. The thermal energy released by this reaction, under constant pressure, is

$$q_P = \left(-56.2 \text{ kJ mol}^{-1}\right)(0.1724 \text{ mol}) = -9.689 \text{ kJ}$$

This same thermal energy is used to increase the temperature of the calorimeter and its contents, but since these items are *gaining* thermal energy, the sign is switched.

$$9.689 \text{ kJ} = \left(C_{calorimeter} + m_{solution}s_{solution}\right)\Delta T$$

Two assumptions are now made. Namely, that the densities of the solutions are the same as pure water, and likewise that the specific heat of the solution is the same as that of water. (These assumptions are good to 3-5%.) Under these assumptions, $s_{solution} = 4.184 \text{ J g}^{-1}\,^\circ\text{C}^{-1}$, and

$$m_{solution} = m_{HCl} + m_{Ba(OH)_2}$$

$$= \left(2.00 \times 10^2 \text{ mL}\right)\left(1.00 \text{ g mL}^{-1}\right) + \left(2.00 \times 10^2 \text{ mL}\right)\left(1.00 \text{ g mL}^{-1}\right)$$

$$= 4.00 \times 10^2 \text{ g}$$

This allows the determination of ΔT,

$$9.689 \text{ kJ} = \left[453 \text{ J}\,^\circ\text{C}^{-1} + \left(4.00 \times 10^2 \text{ g}\right)\left(4.184 \text{ J g}^{-1}\,^\circ\text{C}^{-1}\right)\right]\left(\frac{1 \text{ kJ}}{1000 \text{ J}}\right)\Delta T$$

$$9.689 = \left(2.127 \,^\circ\text{C}^{-1}\right)\Delta T$$

$$\Delta T = 4.555\,^\circ\text{C}$$

The final temperature of the mixed solution is found from $\Delta T = T_f - T_i$, or $4.555\,^\circ\text{C} = T_f - 20.48\,^\circ\text{C}$ which gives $T_f = 25.04\,^\circ\text{C}$, although the assumptions made limit the accuracy of our answer to $T_f = 25.0\,^\circ\text{C}$.

3.40   Consider the following reaction:

$$2CH_3OH(l) + 3O_2(g) \rightarrow 4H_2O(l) + 2CO_2(g) \qquad \Delta_r H^\circ = -1452.8 \text{ kJ mol}^{-1}$$

What is the value of $\Delta_r H^\circ$ if (a) the equation is multiplied throughout by 2, (b) the direction of the reaction is reversed so that the products become the reactants and vice versa, (c) water vapor instead of liquid water is the product?

(a) $\Delta_r H^\circ = 2\left(-1452.8 \text{ kJ mol}^{-1}\right) = -2905.6 \text{ kJ mol}^{-1}$

(b) $\Delta_r H^\circ = -\left(-1452.8 \text{ kJ mol}^{-1}\right) = 1452.8 \text{ kJ mol}^{-1}$

(c) The reaction

$$2CH_3OH(l) + 3O_2(g) \rightarrow 4H_2O(g) + 2CO_2(g)$$

is a sum of the following equations:

$$2CH_3OH(l) + 3O_2(g) \rightarrow 4H_2O(l) + 2CO_2(g) \quad \Delta_r H_1^\circ = -1452.8 \text{ kJ mol}^{-1}$$

$$4H_2O(l) \rightarrow 4H_2O(g) \qquad\qquad\qquad \Delta_r H_2^\circ$$

The standard enthalpy of reaction for vaporization of H_2O is

$$\Delta_r H_2^{\circ} = 4\Delta_f \overline{H}^{\circ}[H_2O(g)] - 4\Delta_f \overline{H}^{\circ}[H_2O(l)]$$

$$= 4\left(-241.8 \text{ kJ mol}^{-1}\right) - 4\left(-285.8 \text{ kJ mol}^{-1}\right)$$

$$= 176.0 \text{ kJ mol}^{-1}$$

The standard enthalpy of reaction of $2CH_3OH(l) + 3O_2(g) \rightarrow 4H_2O(g) + 2CO_2(g)$ is

$$\Delta_r H^{\circ} = \Delta_r H_1^{\circ} + \Delta_r H_2^{\circ} = -1452.8 \text{ kJ mol}^{-1} + 176.0 \text{ kJ mol}^{-1} = -1276.8 \text{ kJ mol}^{-1}$$

3.42 The standard enthalpies of formation of ions in aqueous solution are obtained by arbitrarily assigning a value of zero to H^+ ions; that is, $\Delta_f \overline{H}^{\circ}[H^+(aq)] = 0$. **(a)** For the following reaction,

$$HCl(g) \rightarrow H^+(aq) + Cl^-(aq) \qquad \Delta_r H^{\circ} = -74.9 \text{ kJ mol}^{-1}$$

calculate the value of $\Delta_f \overline{H}^{\circ}$ for the Cl^- ions. **(b)** The standard enthalpy of neutralization between a HCl solution and a NaOH solution is $-56.2 \text{ kJ mol}^{-1}$. Calculate the standard enthalpy of formation of the hydroxide ion at 25°C.

(a)
$$\Delta_r H^{\circ} = -74.9 \text{ kJ mol}^{-1} = \Delta_f \overline{H}^{\circ}[H^+(aq)] + \Delta_f \overline{H}^{\circ}[Cl^-(aq)] - \Delta_f \overline{H}^{\circ}[HCl(g)]$$

$$\Delta_f \overline{H}^{\circ}[Cl^-(aq)] = -74.9 \text{ kJ mol}^{-1} - \Delta_f \overline{H}^{\circ}[H^+(aq)] + \Delta_f \overline{H}^{\circ}[HCl(g)]$$

$$= -74.9 \text{ kJ mol}^{-1} - 0 \text{ kJ mol}^{-1} + \left(-92.3 \text{ kJ mol}^{-1}\right)$$

$$= -167.2 \text{ kJ mol}^{-1}$$

(b) The neutralization reaction for 1 mole of H_2O is

$$H^+(aq) + OH^-(aq) \rightarrow H_2O(l) \qquad \Delta_r H^{\circ} = -56.2 \text{ kJ mol}^{-1}$$

$$\Delta_r H^{\circ} = -56.2 \text{ kJ mol}^{-1} = \Delta_f \overline{H}^{\circ}[H_2O(l)] - \Delta_f \overline{H}^{\circ}[H^+(aq)] - \Delta_f \overline{H}^{\circ}[OH^-(aq)]$$

$$\Delta_f \overline{H}^{\circ}[OH^-(aq)] = \Delta_f \overline{H}^{\circ}[H_2O(l)] - \Delta_f \overline{H}^{\circ}[H^+(aq)] + 56.2 \text{ kJ mol}^{-1}$$

$$= -285.8 \text{ kJ mol}^{-1} - 0 \text{ kJ mol}^{-1} + 56.2 \text{ kJ mol}^{-1}$$

$$= -229.6 \text{ kJ mol}^{-1}$$

3.44 When 2.00 g of hydrazine decomposed under constant-pressure conditions, 7.00 kJ of heat were transferred to the surroundings:

$$3N_2H_4(l) \rightarrow 4NH_3(g) + N_2(g)$$

What is the $\Delta_r H^{\circ}$ for the reaction?

The reaction describes the decomposition of 3 moles of hydrazine. Therefore, the amount of heat given must be scaled to this amount of reactant.

$$\Delta_r H^\circ = q_P = \left(\frac{-7.00 \text{ kJ}}{\frac{2.00 \text{ g } N_2H_4}{32.05 \text{ g mol}^{-1} N_2H_4}} \right) (3) = -337 \text{ kJ mol}^{-1}$$

3.46 The standard enthalpies of combustion of fumaric acid and maleic acid (to form carbon dioxide and water) are -1336.0 kJ mol^{-1} and -1359.2 kJ mol^{-1}, respectively. Calculate the enthalpy of the following isomerization process:

Maleic acid   Fumaric acid

The chemical equations and the standard enthalpies of combustion of 1 mole of fumaric acid and 1 mole of maleic acid are given below:

$$\text{fumaric} + 3O_2 \rightarrow 4CO_2 + 2H_2O \quad \Delta_r H^\circ = -1336.0 \text{ kJ mol}^{-1}$$

$$\text{maleic} + 3O_2 \rightarrow 4CO_2 + 2H_2O \quad \Delta_r H^\circ = -1359.2 \text{ kJ mol}^{-1}$$

The isomerization reaction (maleic acid $\rightarrow$ fumaric acid) can be obtained as a combination of these two reactions:

$$4CO_2 + 2H_2O \rightarrow \text{fumaric} + 3O_2 \quad \Delta_r H^\circ = 1336.0 \text{ kJ mol}^{-1}$$

$$\text{maleic} + 3O_2 \rightarrow 4CO_2 + 2H_2O \quad \Delta_r H^\circ = -1359.2 \text{ kJ mol}^{-1}$$

Therefore, the enthalpy of the isomerization process is

$$\Delta_r H = 1336.0 \text{ kJ mol}^{-1} - 1359.2 \text{ kJ mol}^{-1} = -23.2 \text{ kJ mol}^{-1}$$

3.48 The standard molar enthalpy of formation of molecular oxygen at 298 K is zero. What is its value at 315 K? (*Hint*: Look up the $\overline{C}_P$ value in Appendix B.)

Let $\Delta_f \overline{H}_1^\circ$ and $\Delta_f \overline{H}_2^\circ$ be the enthalpies of formation of molecular oxygen at 298 K and 315 K, respectively.

$$\Delta_f \overline{H}_2^\circ = \Delta_f \overline{H}_1^\circ + \overline{C}_P \Delta T = 0 \text{ J mol}^{-1} + \left(29.4 \text{ J K}^{-1} \text{ mol}^{-1} \right) (315 \text{ K} - 298 \text{ K}) = 500 \text{ J mol}^{-1}$$

3.50 The hydrogenation of ethylene is

$$C_2H_4(g) + H_2(g) \rightarrow C_2H_6(g)$$

Calculate the change in the enthalpy of hydrogenation from 298 K to 398 K. The $\overline{C}_P^\circ$ values are C_2H_4: 43.6 J K^{-1} mol^{-1} and C_2H_6: 52.7 J K^{-1} mol^{-1}.

$$\Delta_r H_{398} - \Delta_r H_{298} = \Delta \overline{C}_P^0 \, (398 \text{ K} - 298 \text{ K})$$

$$= \left\{ \overline{C}_P^0 [C_2H_6(g)] - \overline{C}_P^0 [C_2H_4(g)] - \overline{C}_P^0 [H_2(g)] \right\} (10 \text{ K})$$

$$= \left(52.7 \text{ J K}^{-1} \text{mol}^{-1} - 43.6 \text{ J K}^{-1} \text{mol}^{-1} - 28.8 \text{ J K}^{-1} \text{mol}^{-1} \right) (10 \text{ K})$$

$$= -197 \text{ kJ mol}^{-1} = -2.0 \times 10^2 \text{ kJ mol}^{-1}$$

3.52   Calculate the standard enthalpy of formation for diamond, given that

$$C(\text{graphite}) + O_2(g) \to CO_2(g) \quad \Delta_r H^\circ = -393.5 \text{ kJ mol}^{-1}$$

$$C(\text{diamond}) + O_2(g) \to CO_2(g) \quad \Delta_r H^\circ = -395.4 \text{ kJ mol}^{-1}$$

The formation reaction of diamond is

$$C(\text{graphite}) \to C(\text{diamond}),$$

which can be thought of as a sum of the reactions:

$$C(\text{graphite}) + O_2(g) \to CO_2(g) \qquad \Delta_r H^\circ = -393.5 \text{ kJ mol}^{-1}$$

$$CO_2(g) \to C(\text{diamond}) + O_2(g) \quad \Delta_r H^\circ = 395.4 \text{ kJ mol}^{-1}$$

Therefore, the standard enthalpy of formation for diamond is the sum of the standard enthalpies of reaction of the two reactions above:

$$\Delta_f \overline{H}^\circ [\text{diamond}] = -393.5 \text{ kJ mol}^{-1} + 395.4 \text{ kJ mol}^{-1} = 1.9 \text{ kJ mol}^{-1}$$

3.54   From the following heats of combustion,

$$CH_3OH(l) + \frac{3}{2}O_2(g) \to CO_2(g) + 2H_2O(l) \quad \Delta_r H^\circ = -726.4 \text{ kJ mol}^{-1}$$

$$C(\text{graphite}) + O_2(g) \to CO_2(g) \qquad\qquad \Delta_r H^\circ = -393.5 \text{ kJ mol}^{-1}$$

$$H_2(g) + \frac{1}{2}O_2(g) \to H_2O(l) \qquad\qquad\quad \Delta_r H^\circ = -285.8 \text{ kJ mol}^{-1}$$

calculate the enthalpy of formation of methanol (CH_3OH) from its elements:

$$C(\text{graphite}) + 2H_2(g) + \frac{1}{2}O_2(g) \to CH_3OH(l)$$

The formation reaction of methanol can be thought of as a sum of the reactions:

$$CO_2(g) + 2H_2O(l) \to CH_3OH(l) + \frac{3}{2}O_2(g) \quad \Delta_r H^\circ = 726.4 \text{ kJ mol}^{-1}$$

$$C(\text{graphite}) + O_2(g) \to CO_2(g) \qquad\qquad \Delta_r H^\circ = -393.5 \text{ kJ mol}^{-1}$$

$$2H_2(g) + O_2(g) \to 2H_2O(l) \qquad\qquad \Delta_r H^\circ = 2\left(-285.8 \text{ kJ mol}^{-1}\right) = -571.6 \text{ kJ mol}^{-1}$$

Therefore, the standard enthalpy of formation of methanol is the sum of the standard enthalpies of reaction of the three reactions above:

$$\Delta_f \overline{H}^\circ[CH_3OH(l)] = 726.4 \text{ kJ mol}^{-1} - 393.5 \text{ kJ mol}^{-1} - 571.6 \text{ kJ mol}^{-1} = -238.7 \text{ kJ mol}^{-1}$$

3.56 Calculate the difference between the values of $\Delta_r H^\circ$ and $\Delta_r U^\circ$ for the oxidation of glucose at 298 K:

$$C_6H_{12}O_6(s) + 6O_2(g) \rightarrow 6CO_2(g) + 6H_2O(l)$$

$\Delta_r H^\circ$ and $\Delta_r U^\circ$ differ from each other if the number of moles of gases after the reaction is not the same as that before the reaction.

$$\Delta_r H^\circ = \Delta_r U^\circ + RT \Delta n$$

Since $\Delta n = 0$, $\Delta_r H^\circ = \Delta_r U^\circ$, or $\Delta_r H^\circ - \Delta_r U^\circ = 0$.

3.58 **(a)** Explain why the bond enthalpy of a molecule is always defined in terms of a gas-phase reaction. **(b)** The bond dissociation enthalpy of F_2 is 158.8 kJ mol^{-1}. Calculate $\Delta_f \overline{H}^\circ$ for F(g).

(a) In the gas phase, molecules are far apart and not affected by intermolecular interactions. The bond dissociation enthalpies so determined thus refer only to the chemical bond between specific atoms and are not influenced by intermolecular interactions.

(b) It is given that $\Delta_r H^\circ = 158.8$ kJ mol^{-1} for the reaction $F_2(g) \rightarrow 2F(g)$. $\Delta_r H^\circ$ is related to the enthalpies of formation of $F_2(g)$ (which is 0) and F(g) in the following manner:

$$\Delta_r H^\circ = 2\Delta_f \overline{H}^\circ[F(g)] - \Delta_f \overline{H}^\circ[F_2(g)] = 2\Delta_f \overline{H}^\circ[F(g)]$$

Therefore,

$$\Delta_f \overline{H}^\circ[F(g)] = \frac{1}{2}\Delta_r H^\circ = \frac{1}{2}\left(158.8 \text{ kJ mol}^{-1}\right) = 79.4 \text{ kJ mol}^{-1}$$

3.60 Use the bond enthalpy values in Table 3.4 to calculate the enthalpy of combustion for ethane,

$$2C_2H_6(g) + 7O_2(g) \rightarrow 4CO_2(g) + 6H_2O(l)$$

Compare your result with that calculated from the enthalpy of formation values of the products and reactants listed in Appendix B.

Calculation of the enthalpy of combustion using bond enthalpies:

Type of bonds broken	Number of bonds broken	Bond enthalpy/kJ mol^{-1}	Enthalpy change/kJ mol^{-1}
C—H	12	414	4968
C—C	2	347	694
O=O	7	498.8	3491.6

Type of bonds formed	Number of bonds formed	Bond enthalpy/kJ mol^{-1}	Enthalpy change/kJ mol^{-1}
C=O	8	799	6392 ·
O—H	12	460	5520

$$\Delta_r H^\circ = (4968 + 694 + 3491.6) \text{ kJ mol}^{-1} - (6392 + 5520) \text{ kJ mol}^{-1} = -2758 \text{ kJ mol}^{-1}$$

Calculation of the enthalpy of combustion using enthalpies of formation:

$$\Delta_r H^\circ = 4\Delta_f \overline{H}^\circ[CO_2(g)] + 6\Delta_f \overline{H}^\circ[H_2O(l)] - 2\Delta_f \overline{H}^\circ[C_2H_6(g)] - 7\Delta_f \overline{H}^\circ[O_2(g)]$$

$$= 4\left(-393.5 \text{ kJ mol}^{-1}\right) + 6\left(-285.8 \text{ kJ mol}^{-1}\right) - 2\left(-84.7 \text{ kJ mol}^{-1}\right) - 7\left(0 \text{ kJ mol}^{-1}\right)$$

$$= -3119.4 \text{ kJ mol}^{-1}$$

The value of $\Delta_r H^\circ$ so calculated is 13% greater than that calculated using bond enthalpies. The value determined using enthalpies of formation is the correct value, since it relies on the first law of thermodynamics. Bond enthalpies are averages determined for similar bonds in many molecules and provide estimates that are typically within 10% of the experimental value for any given, particular reaction. Furthermore, in this reaction, water is produced as a liquid. Bond enthalpies are defined in terms of gas-phase reactions for which intermolecular interaction can be ignored. (See Problem 3.58) Thus, it is not appropriate to use bond enthalpies to estimate this enthalpy of combustion.

3.62 Predict whether the values of q, w, ΔU, ΔH are positive, zero, or negative for each of the following processes: **(a)** melting of ice at 1 atm and 273 K, **(b)** melting of solid cyclohexane at 1 atm and the normal melting point, **(c)** reversible isothermal expansion of an ideal gas, and **(d)** reversible adiabatic expansion of an ideal gas.

For most substances, the volume increases upon melting, but for ice melting to water under the conditions given, the volume decreases. In all cases, the volume change between the solid and liquid phases is small as is the value of the work associated with the melting. For ideal gases, both internal energy U and enthalpy H depend only on temperature. Keeping these points in mind results in the following predictions.

	q	w	ΔU	ΔH
(a)	positive	positive	positive	positive
(b)	positive	negative	positive	positive
(c)	positive	negative	zero	zero
(d)	zero	negative	negative	negative

3.64 The convention of arbitrarily assigning a zero enthalpy value to all the (most stable) elements in the standard state and (usually) 298 K is a convenient way of dealing with the enthalpy changes of chemical processes. This convention does not apply to one kind of process, however. What process is it? Why?

In a nuclear process, there are different elements on both sides of the chemical equation, and this convention would not apply.

3.66   The fuel value of hamburger is about 3.6 kcal g^{-1}. If a person eats 1 pound of hamburger for lunch and if none of the energy is stored in his body, estimate the amount of water that would have to be lost in perspiration in order to keep his body temperature constant. (1 lb = 454 g.)

The fuel value of 1 pound of hamburger is

$$(1 \text{ lb}) \left(\frac{454 \text{ g}}{1 \text{ lb}} \right) \left(3.6 \text{ kcal g}^{-1} \right) \left(\frac{4.184 \text{ kJ}}{1 \text{ kcal}} \right) = 6.84 \times 10^3 \text{ kJ}$$

For the vaporization of water at 298 K, $H_2O(l) \longrightarrow H_2O(g)$, $\Delta_{vap}\overline{H} = 44.0$ kJ mol^{-1}. ($\Delta_{vap}\overline{H}$ is appropriate, since the vaporization is taking place at constant, atmospheric pressure.) Assuming that the entire fuel value of the hamburger is used to vaporize water, it will require that

$$\left(\frac{6.84 \times 10^3 \text{ kJ}}{44.0 \text{ kJ mol}^{-1}} \right) \left(\frac{18.02 \text{ g } H_2O}{1 \text{ mol } H_2O} \right) = 2.8 \times 10^3 \text{ g } H_2O$$

be vaporized, or approximately 2.8 liters of water.

3.68   An oxyacetylene flame is often used in the welding of metals. Estimate the flame temperature produced by the reaction

$$2C_2H_2(g) + 5O_2(g) \rightarrow 4CO_2(g) + 2H_2O(g)$$

Assume the heat generated from this reaction is all used to heat the products. (*Hint*: First calculate the value of $\Delta_r H°$ for the reaction. Next, look up the heat capacities of the products. Assume the heat capacities are temperature independent.)

The enthalpy of reaction for the oxidation of acetylene is

$$\Delta_r H° = 4\Delta_f \overline{H}°[CO_2(g)] + 2\Delta_f \overline{H}°[H_2O(g)] - 2\Delta_f \overline{H}°[C_2H_2(g)] - 5\Delta_f \overline{H}°[O_2(g)]$$

$$= 4 \left(-393.5 \text{ kJ mol}^{-1} \right) + 2 \left(-241.8 \text{ kJ mol}^{-1} \right) - 2 \left(226.6 \text{ kJ mol}^{-1} \right) - 5 \left(0 \text{ kJ mol}^{-1} \right)$$

$$= -2510.8 \text{ kJ mol}^{-1}$$

Because the reaction takes place at constant pressure, the enthalpy of reaction is the same as heat released by the reaction, and this heat is absorbed by the products. The initial temperature, T_i, is assumed to be 298 K, and the final temperature of the products, T_f is

$$q = 2510.8 \text{ kJ mol}^{-1} = \left\{ 4\overline{C}_p°[CO_2(g)] + 2\overline{C}_p°[H_2O(g)] \right\} (T_f - T_i)$$

$$= \left[4 \left(37.1 \text{ J K}^{-1} \text{mol}^{-1} \right) + 2 \left(33.6 \text{ J K}^{-1} \text{mol}^{-1} \right) \right] (T_f - 298 \text{ K}) \left(\frac{1 \text{ kJ}}{1000 \text{ J}} \right)$$

$$T_f - 298 \text{ K} = 1.1646 \times 10^4 \text{ K}$$

$$T_f = 1.194 \times 10^4 \text{ K}$$

3.70 The enthalpies of hydrogenation of ethylene and benzene have been determined at 298 K:

$$C_2H_4(g) + H_2(g) \rightarrow C_2H_6(g) \quad \Delta_r H^\circ = -132 \text{ kJ mol}^{-1}$$

$$C_6H_6(g) + 3H_2(g) \rightarrow C_6H_{12}(g) \quad \Delta_r H^\circ = -246 \text{ kJ mol}^{-1}$$

What would be the enthalpy of hydrogenation for benzene if it contained three isolated, unconjugated double bonds? How would you account for the difference between the calculated value based on this assumption and the measured value?

If benzene contained three isolated, unconjugated double bonds, its enthalpy of hydrogenation could be estimated as three times that for the single double bond in ethylene, or $\Delta_r H^\circ_{\text{calc}} = 3\left(-132 \text{ kJ mol}^{-1}\right) = -396 \text{ kJ mol}^{-1}$. The difference is $-246 \text{ kJ mol}^{-1} - \left(-396 \text{ kJ mol}^{-1}\right) = 150 \text{ kJ mol}^{-1}$. This implies that benzene is 150 kJ mol^{-1} more stable than the hypothetical molecule with three isolated, unconjugated double bonds. This is attributed to the resonance (electron delocalization) energy of benzene.

3.72 The standard enthalpy of formation at 298 K of HF(aq) is -320.1 kJ mol^{-1}; OH$^-$(aq), -229.6 kJ mol^{-1}; F$^-$(aq), -329.11 kJ mol^{-1}; and H$_2$O(l), -285.8 kJ mol^{-1}. **(a)** Calculate the enthalpy of neutralization of HF(aq),

$$HF(aq) + OH^-(aq) \rightarrow F^-(aq) + H_2O(l)$$

(b) Using the value of -55.83 kJ mol^{-1} as the enthalpy change from the reaction

$$H^+(aq) + OH^-(aq) \rightarrow H_2O(l)$$

calculate the enthalpy change for the dissociation

$$HF(aq) \rightarrow H^+(aq) + F^-(aq)$$

(a) $\quad \Delta_r H^\circ = \Delta_f \overline{H}^\circ[F^-(aq)] + \Delta_f \overline{H}^\circ[H_2O(l)] - \Delta_f \overline{H}^\circ[HF(aq)] - \Delta_f \overline{H}^\circ[OH^-(aq)]$

$$= -329.11 \text{ kJ mol}^{-1} + \left(-285.8 \text{ kJ mol}^{-1}\right) - \left(-320.1 \text{ kJ mol}^{-1}\right)$$

$$- \left(-229.6 \text{ kJ mol}^{-1}\right)$$

$$= -65.2 \text{ kJ mol}^{-1}$$

(b) The dissociation of HF can be considered as a sum of the following equations:

$$HF(aq) + OH^-(aq) \rightarrow F^-(aq) + H_2O(l) \quad \Delta_r H^\circ = -65.2 \text{ kJ mol}^{-1}$$

$$H_2O(l) \rightarrow H^+(aq) + OH^-(aq) \quad \Delta_r H^\circ = 55.83 \text{ kJ mol}^{-1}$$

The enthalpy change for the dissociation reaction is therefore

$$\Delta_r H^\circ = -65.2 \text{ kJ mol}^{-1} + 55.83 \text{ kJ mol}^{-1} = -9.4 \text{ kJ mol}^{-1}$$

3.74 Metabolic activity in the human body releases approximately 1.0×10^4 kJ of heat per day. Assuming the body is 50 kg of water, how fast would the body temperature rise if it were an isolated system? How much water must the body eliminate as perspiration to maintain the normal body temperature (98.6°F)? Comment on your results. The heat of vaporization of water may be taken as 2.41 kJ g^{-1}.

If the body absorbs all the heat released and is an isolated system, then the temperature rise, ΔT is related to q in the following fashion:

$$q_{absorbed} = C_P[H_2O(l)]\Delta T$$

$$\Delta T = \frac{q}{C_P[H_2O(l)]} = \frac{(1.0 \times 10^4 \text{ kJ})\left(\frac{1000 \text{ J}}{1 \text{ kJ}}\right)}{\left(\frac{50 \times 10^3 \text{ g}}{18.01 \text{ g mol}^{-1}}\right)\left(75.3 \text{ J K}^{-1} \text{ mol}^{-1}\right)} = 47.8 \text{ K}$$

If the body temperature is to remain constant, then the heat released by metabolic activity must be used for the evaporation of water as perspiration, that is,

$$1.0 \times 10^4 \text{ kJ} = m_{water}\left(2.41 \text{ kJ g}^{-1}\right)$$

$$m_{water} = 4.1 \times 10^3 \text{ g}$$

The actual amount of perspiration is less than this because part of the body heat is lost to the surroundings by convection and radiation.

3.76 Calculate the fraction of the enthalpy of vaporization of water used for the expansion of steam at its normal boiling point.

Treating the steam as an ideal gas, the molar volume of steam at 373.15 K is calculated as

$$\overline{V} = \frac{RT}{P}$$

$$= \frac{\left(0.08206 \text{ L atm K}^{-1} \text{ mol}^{-1}\right)(373.15 \text{ K})}{1 \text{ atm}}$$

$$= 30.621 \text{ L mol}^{-1}$$

The work done in the expansion from liquid water to this volume of steam is then found having made the assumption that the volume of the condensed phase is negligible. (One mole of liquid water has a volume of 18 mL, so the approximation is a good one.)

$$w = -P_{ex}\Delta\overline{V}$$

$$= -(1 \text{ atm})\left(30.621 \text{ L mol}^{-1}\right)\left(\frac{101.3 \text{ J}}{1 \text{ L atm}}\right)\left(\frac{1 \text{ kJ}}{1000 \text{ J}}\right)$$

$$= -3.1019 \text{ kJ mol}^{-1}$$

At 373.15 K the molar enthalpy of vaporization of water is $\Delta_{vap}\overline{H}^\circ = 40.79$ kJ mol^{-1} (see Problem 3.24), so the fraction used for the expansion of steam is

$$\frac{3.1019 \text{ kJ mol}^{-1}}{40.79 \text{ kJ mol}^{-1}} \times 100\% = 7.60\%$$

3.78 Calculate the internal energy of a Goodyear blimp filled with helium gas at 1.2×10^5 Pa (compared to the empty blimp). The volume of the inflated blimp is 5.5×10^3 m^3. If all the internal energy were used to heat 10.0 tons of copper at 21°C, calculate the final temperature of the metal. (*Hint*: 1 ton $= 9.072 \times 10^5$ g.)

The internal energy of a monoatomic (ideal) gas is $\frac{3}{2}nRT$. Likewise for an ideal gas, $PV = nRT$. Thus, for an ideal gas, $U = \frac{3}{2}PV$. For the blimp,

$$U = \frac{3}{2}\left(1.2 \times 10^5 \text{ Pa}\right)\left(5.5 \times 10^3 \text{ m}^3\right)\left(\frac{1\,\text{N}\,\text{m}^{-2}}{1\,\text{Pa}}\right)\left(\frac{1\,\text{J}}{1\,\text{N}\,\text{m}}\right) = 9.90 \times 10^8 \text{ J}$$

If all this energy is used to heat copper, the temperature change is related to the amount of energy used via $q = m_{\text{Cu}}s_{\text{Cu}}\Delta T$, giving

$$9.90 \times 10^8 \text{ J} = (10.0 \text{ ton})\left(\frac{9.072 \times 10^5 \text{ g}}{1 \text{ ton}}\right)\left(24.47 \text{ J}\,°\text{C}^{-1}\,\text{mol}^{-1}\right)\left(\frac{1 \text{ mol Cu}}{63.55 \text{ g Cu}}\right)\Delta T$$

$$9.90 \times 10^8 = 3.49 \times 10^6 \,°\text{C}^{-1}\Delta T$$

$$283.7°\text{C} = \Delta T = T_f - 21°\text{C}$$

Thus, $T_f = 305°$C.

3.80 The combustion of what volume of ethane (C_2H_6), measured at 23.0°C and 752 mmHg, would be required to heat 855 g of water from 25.0°C to 98.0°C?

The combustion reaction is

$$C_2H_6(g) + \frac{7}{2}O_2(g) \longrightarrow 2CO_2(g) + 3H_2O(l)$$

Use ethalpies of formation to determine the enthalpy of combustion.

$$\Delta_r H° = 2\Delta_f \overline{H}°[CO_2(g)] + 3\Delta_f \overline{H}°[H_2O(l)] - \Delta_f \overline{H}°[C_2H_6(g)] - \frac{7}{2}\Delta_f \overline{H}°[O_2(g)]$$

$$= 2\left(-393.5 \text{ kJ mol}^{-1}\right) + 3\left(-285.8 \text{ kJ mol}^{-1}\right) - \left(-84.7 \text{ kJ mol}^{-1}\right) - \frac{7}{2}\left(0 \text{ kJ mol}^{-1}\right)$$

$$= -1559.7 \text{ kJ mol}^{-1}$$

Thus each mole of ethane provides 1559.7 kJ of thermal energy upon combustion at constant pressure.

The thermal energy required to heat the water is

$$q = m_{H_2O}s_{H_2O}\Delta T = (855 \text{ g})\left(4.184 \text{ J g}^{-1}\,°\text{C}^{-1}\right)(98.0 - 25.0)\,°\text{C}\left(\frac{1 \text{ kJ}}{1000 \text{ J}}\right) = 261.1 \text{ kJ}$$

Clearly a fraction of a mole of ethane is required. The actual number of moles is

$$n = (261.1 \text{ kJ})\left(\frac{1 \text{ mol ethane}}{1559.7 \text{ kJ}}\right) = 0.1674 \text{ mol ethane}$$

The volume of the ethane at the state conditions is

$$V = \frac{nRT}{P} = \frac{(0.1674 \text{ mol}) \left(0.08206 \text{ L atm K}^{-1} \text{ mol}^{-1}\right) (273.15 + 23) \text{ K}}{(752 \text{ mmHg}) \left(\frac{1 \text{ atm}}{760 \text{ mmHg}}\right)} = 4.11 \text{ L}$$

3.82 State whether each of the following statements is true or false: **(a)** $\Delta U \approx \Delta H$ except for gases or high-pressure processes. **(b)** In gas compression, a reversible process does maximum work. **(c)** ΔU is a state function. **(d)** $\Delta U = q + w$ for an open system. **(e)** C_V is temperature independent for gases. **(f)** The internal energy of a real gas depends only on temperature.

(a) True

(b) False

(c) False (U is a state function, ΔU is not.)

(d) False (It is true for a *closed* system.)

(e) False

(f) False

3.84 Calculate the work done during the isothermal, reversible expansion of a van der Waals gas. Account physically for the way in which the coefficients a and b appear in the final expression. *Hint*: You need to apply the Taylor series expansion:

$$\ln (1 - x) = -x - \frac{x^2}{2} \cdots \quad \text{for } |x| \ll 1$$

to the expression $\ln (V - nb)$. Recall that the a term represents attraction and the b term repulsion.

$$w = -\int_{V_1}^{V_2} P \, dV$$

$$= -\int_{V_1}^{V_2} \left(\frac{nRT}{V - nb} - \frac{an^2}{V^2} \right) dV$$

$$= -nRT \ln \frac{V_2 - nb}{V_1 - nb} - an^2 \left(\frac{1}{V_2} - \frac{1}{V_1} \right)$$

The ln term can be written as

$$\ln \frac{V_2 - nb}{V_1 - nb} = \ln \frac{V_2 \left(1 - \frac{nb}{V_2}\right)}{V_1 \left(1 - \frac{nb}{V_1}\right)}$$

$$= \ln \frac{V_2}{V_1} + \ln \frac{1 - \frac{nb}{V_2}}{1 - \frac{nb}{V_1}}$$

$$= \ln \frac{V_2}{V_1} + \ln \left(1 - \frac{nb}{V_2}\right) - \ln \left(1 - \frac{nb}{V_1}\right)$$

Assume the volume occupied by the gas molecules is much greater than the volume of the molecules, both before and after the expansion, that is, $V_1 \gg nb$ and $V_2 \gg nb$, then $\frac{nb}{V_1} \ll 1$

and $\frac{nb}{V_2} \ll 1$. Under these condition, the Taylor expansion described in the question can be used to simplify the ln terms.

$$\ln \frac{V_2 - nb}{V_1 - nb} = \ln \frac{V_2}{V_1} + \left(-\frac{nb}{V_2} - \frac{1}{2} \frac{n^2 b^2}{V_2^2} + \cdots \right) - \left(-\frac{nb}{V_1} - \frac{1}{2} \frac{n^2 b^2}{V_1^2} + \cdots \right)$$

$$= \ln \frac{V_2}{V_1} - nb \left(\frac{1}{V_2} - \frac{1}{V_1} \right) - \frac{n^2 b^2}{2} \left(\frac{1}{V_2^2} - \frac{1}{V_1^2} \right) + \cdots$$

Substitute the above expression into w.

$$w = -nRT \left[\ln \frac{V_2}{V_1} - nb \left(\frac{1}{V_2} - \frac{1}{V_1} \right) - \frac{n^2 b^2}{2} \left(\frac{1}{V_2^2} - \frac{1}{V_1^2} \right) + \cdots \right] - an^2 \left(\frac{1}{V_2} - \frac{1}{V_1} \right)$$

$$= -nRT \ln \frac{V_2}{V_1} + (bRT - a) n^2 \left(\frac{1}{V_2} - \frac{1}{V_1} \right) + \frac{n^3 b^2 RT}{2} \left(\frac{1}{V_2^2} - \frac{1}{V_1^2} \right) + \cdots$$

The first term in this last equation is just the result for the ideal gas. It is modified by the succeeding terms to account for intermolecular interactions. The second term shows the balance between attractive and repulsive forces. In an expansion, $V_2 > V_1$ so $\left(\frac{1}{V_2} - \frac{1}{V_1} \right) < 0$. If attractive forces dominate, $a > bRT$ and the entire second term is positive, which cancels some of the (negative) work done in the expansion. Because of the attractive forces, some energy must be used to overcome the intermolecular interactions, and not as much work can be done as in the case of the ideal gas.

On the other hand, if $a < bRT$, the entire second term is negative and enhances the (negative) work done in the expansion. In this case the repulsive forces dominate, and the energy released as the molecules move farther apart from each other is available to do more work than the ideal gas could. The higher order terms reinforce this effect for high densities where the repulsive forces are most significant.

3.86 A 4.0-L ideal gas initially at 2.0 atm and 300 K is isothermally compressed to a final volume of 2.0 L. Calculate the work done if the process is carried out **(a)** reversibly and **(b)** irreversibly. Support your answers with a graphical illustration of the processes.

(a) Since $P_1 V_1 = nRT$,

$$w = -nRT \ln \frac{V_2}{V_1} = -P_1 V_1 \ln \frac{V_2}{V_1}$$

$$= -(2.0 \text{ atm}) (4.0 \text{ L}) \left(\frac{101.3 \text{ J}}{1 \text{ L atm}} \right) \ln \frac{2.0 \text{ L}}{4.0 \text{ L}}$$

$$= 5.6 \times 10^2 \text{ J}$$

(b) The external pressure is the same as the final pressure. Since this is an isothermal process, the final pressure is

$$P_2 = \frac{P_1 V_1}{V_2} = \frac{(2.0 \text{ atm}) (4.0 \text{ L})}{2.0 \text{ L}} = 4.0 \text{ atm}$$

It follows then

$$w = -P_{ex}\Delta V = -P_2\left(V_2 - V_1\right)$$

$$= -(4.0 \text{ atm})(2.0 \text{ L} - 4.0 \text{ L})\left(\frac{101.3 \text{ J}}{1 \text{ L atm}}\right)$$

$$= 8.1 \times 10^2 \text{ J}$$

Graphically, the path for the reversible compression is represented by a solid line while the path for the irreversible compression is shown in two dotted segment. The work done by each process is the area under the path, and it is apparent that the work done by the irreversible process is greater than that by the reversible process.

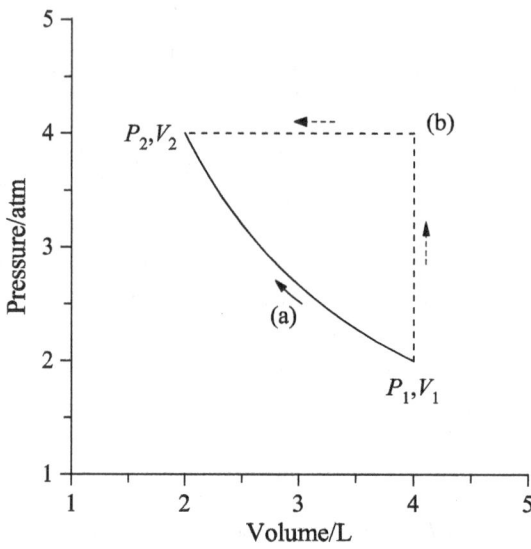

3.88 In acid–base theory, the species $H^+(aq)$ and $H_3O^+(aq)$ are usually treated as the same. This is not the case, however, in thermochemistry. What are the values for $\Delta_f \overline{H}°[H^+(aq)]$ and $\Delta_f \overline{H}°[H_3O^+(aq)]$?

The following two chemical equations describe the autoionization of water and have the same enthalpy of reaction:

$$2H_2O(l) \rightarrow H_3O^+(aq) + OH^-(aq) \quad \Delta_r H°$$

$$H_2O(l) \rightarrow H^+(aq) + OH^-(aq) \quad \Delta_r H°$$

Take the difference between these two equations:

$$H_2O(l) + H^+(aq) \rightarrow H_3O^+(aq) \quad \Delta_r H° = 0$$

The enthalpy of reaction corresponding to the above equation can also be written in terms of enthalpies of formation of the substances involved:

$$\Delta_r H° = 0 = \Delta_f \overline{H}°[H_3O^+(aq)] - \Delta_f \overline{H}°[H^+(aq)] - \Delta_f \overline{H}°[H_2O(l)]$$

$$= \Delta_f \overline{H}°[H_3O^+(aq)] - 0 \text{ kJ mol}^{-1} - \left(-285.8 \text{ kJ mol}^{-1}\right)$$

Thus, $\Delta_f \overline{H}°[H_3O^+(aq)] = -285.8 \text{ kJ mol}^{-1}$.

CHAPTER 4

The Second Law of Thermodynamics

PROBLEMS AND SOLUTIONS

4.2 Suppose that your friend told you of the following extraordinary event. A block of metal weighing 500 g was seen rising spontaneously from the table on which it was resting to a height of 1.00 cm above the table. He stated that the metal had absorbed thermal energy from the table that was then used to raise itself against gravitational pull. **(a)** Does this process violate the first law of thermodynamics? **(b)** How about the second law? Assume that the room temperature was 298 K and that the table was large enough so that its temperature was unaffected by this transfer of energy. (*Hint*: First calculate the decrease in entropy as a result of this process and then estimate the probability for the occurrence of such a process. The acceleration due to gravity is 9.81 m s^{-2}.)

(a) The process would not violate the first law of thermodynamics as long as the amount of thermal energy absorbed from the table was equal to the increase in (gravitational) potential energy.

(b) The amount of energy required is found via

$$E = mgh$$

$$= (500\text{ g}) \left(\frac{1\text{ kg}}{1000\text{ g}}\right) \left(9.81\text{ m s}^{-2}\right) (1.00\text{ cm}) \left(\frac{1\text{ m}}{100\text{ cm}}\right)$$

$$= 4.905 \times 10^{-2}\text{ J}$$

The entropy of the table changed because thermal energy was transferred to the block via heat. Were this heat to be transferred reversibly, the entropy change for the table would be

$$\Delta S = \frac{q_{rev}}{T}$$

$$= \frac{-4.905 \times 10^{-2}\text{ J}}{298\text{ K}}$$

$$= -1.646 \times 10^{-4}\text{ J K}^{-1}$$

since the process occurs at constant temperature. The minus sign is included because thermal energy is leaving the table as heat. The table is assumed large enough so that the change in the state of the table is effectively infinitesimal, and this value is taken to be the entropy change for the process as observed.

Relating the entropy change to the probability that this change happens,

$$\Delta S = k_{\mathrm{B}} \ln \frac{p_2}{p_1}$$

$$-1.646 \times 10^{-4}\,\mathrm{J\,K^{-1}} = 1.381 \times 10^{-23}\,\mathrm{J\,K^{-1}} \ln \frac{p_2}{p_1}$$

$$-1.192 \times 10^{19} = \ln \frac{p_2}{p_1}$$

$$\frac{p_2}{p_1} = e^{-1.192\times 10^{19}} \approx 0$$

The ratio is so small that it is practically zero. That is, the probability of attaining the microstate necessary for this process to occur is effectively zero. The event is practically impossible and would violate the second law of thermodynamics.

4.4 Convert the P–V diagram for the Carnot cycle to a T–S diagram. What is the area of the enclosed portion?

The temperature and entropy change for each step must be determined before plotting a T–S diagram.

Step 1:

The temperature of the system is kept constant at T_2.

$$\Delta S = S_2 - S_1 = \frac{q_2}{T_2} = R \ln \frac{V_2}{V_1}$$

Step 2:

The temperature of the system drops from T_2 to T_1.

Because $q = 0$, $\Delta S = 0$. Thus, the entropy stays at S_2 throughout this step.

Step 3:

The temperature of the system is kept constant at T_1.

$\Delta S = \frac{q_1}{T_1} = R \ln \frac{V_4}{V_3} = -R \ln \frac{V_2}{V_1}$. The initial entropy is S_2 and $\Delta S = -(S_2 - S_1)$ (see Step 1).
Therefore, the final entropy is $S_2 + \Delta S = S_2 - (S_2 - S_1) = S_1$.

Step 4:

The temperature of the system rises from T_1 to T_2.

Because $q = 0$, $\Delta S = 0$. Thus, the entropy stays at S_1 throughout this step.

Using the above information, the T–S diagram is constructed as shown in the figure.

The area of the enclosed portion $= (T_2 - T_1)(S_2 - S_1) = (T_2 - T_1) R \ln \frac{V_2}{V_1}$, which is the same as the magnitude of work done in a Carnot cycle.

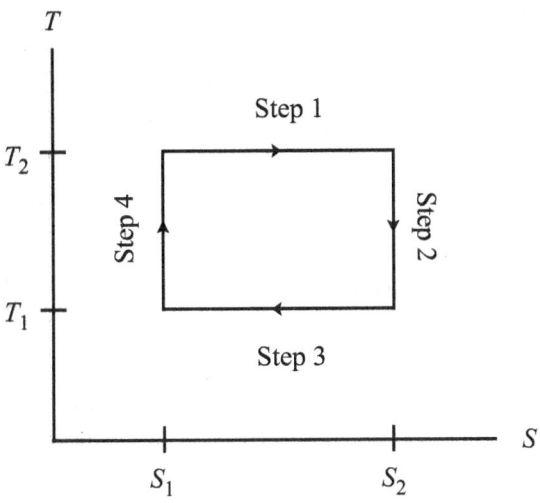

4.6 A heat engine operates between 210°C and 35°C. Calculate the minimum amount of heat that must be withdrawn from the hot source to obtain 2000 J of work.

Let q be the minimum amount of heat that must be withdrawn from the hot source.

$$\text{Efficiency} = \frac{|w|}{q} = 1 - \frac{T_1}{T_2}$$

$$q = \frac{|w|}{1 - \frac{T_1}{T_2}} = \frac{2000 \text{ J}}{1 - \frac{308.15 \text{ K}}{483.15 \text{ K}}} = 5.52 \times 10^3 \text{ J}$$

4.8 One of the many statements of the second law of thermodynamics is: Heat cannot flow from a colder body to a warmer one without external aid. Assume two systems, 1 and 2, at T_1 and T_2 $(T_2 > T_1)$. Show that if a quantity of heat q did flow spontaneously from 1 to 2, the process would result in a decrease in the entropy of the universe. (You may assume that the heat flows very slowly so that the process can be regarded as reversible. Assume also that the loss of heat by system 1 and the gain of heat by system 2 do not affect T_1 and T_2.)

The universe is comprised of the two systems. Since heat flows from system 1 to system 2, $q_1 = -q_{\text{rev}}$ and $q_2 = q_{\text{rev}}$.

$$\Delta S_{\text{univ}} = \Delta S_1 + \Delta S_2 = \frac{q_1}{T_1} + \frac{q_2}{T_2} = \frac{-q_{\text{rev}}}{T_1} + \frac{q_{\text{rev}}}{T_2}$$

$$= q_{\text{rev}} \left(\frac{1}{T_2} - \frac{1}{T_1} \right) < 0 \quad \text{since } T_1 < T_2$$

4.10 Molecules of a gas at any temperature T above the absolute zero are in constant motion. Does this "perpetual motion" violate the laws of thermodynamics?

As long as the gas does no work on the surroundings it violates no law of thermodynamics. Were the gas to do work on the surroundings, then the laws of thermodynamics would require that $w = \Delta U - q$ (the first law) and $\Delta S \geq \frac{q}{T}$ for an isothermal process (the second law). This last requirement means that the state of the system must change when work is done, since for an isothermal process doing work on the surroundings, $\Delta U = 0$ (assuming an ideal gas) and $w < 0$ imply $q > 0$. Thus, $\Delta S > 0$ and the system must have changed state, since S is a state function.

4.12 On a hot summer day, a person tries to cool himself by opening the door of a refrigerator. Is this a wise action, thermodynamically speaking?

No, this is not wise. The refrigerator is not working under ideal conditions, so that it delivers more heat to the room than it extracts. Indeed, even under ideal conditions, the second law of thermodynamics requires an excess of heat delivered to the room over that extracted. The room will get warmer than if the door remained closed.

4.14 Calculate the values of ΔU, ΔH, and ΔS for the following process:

1 mole of liquid water at 25°C and 1 atm → 1 mol of steam at 100°C and 1 atm

The molar heat of vaporization of water at 373 K is 40.79 kJ mol^{-1}, and the molar heat capacity of water is 75.3 J K^{-1} mol^{-1}. Assume the molar heat capacity to be temperature independent and ideal-gas behavior.

The problem can be solved readily by breaking down the process into two steps, each carried out at 1 atm: (1) $H_2O(l)$ is heated from 25°C to 100°C, then (2) $H_2O(l)$ at 100°C is heated to cause the phase transformation to $H_2O(g)$ at 100° C.

Step 1:

$$H_2O(l), 25°C \rightarrow H_2O(l), 100°C$$

$$\Delta H = C_P \Delta T = (1 \text{ mol}) \left(75.3 \text{ J K}^{-1} \text{mol}^{-1}\right)(373.15 \text{ K} - 298.15 \text{ K}) = 5.648 \times 10^3 \text{ J}$$

ΔU is related to ΔH by

$$\Delta U = \Delta H - P\Delta V$$

Since both reactant and product are in the liquid phase, ΔV is negligible. Therefore,

$$\Delta U = \Delta H = 5.648 \times 10^3 \text{ J}$$

$$\Delta S = C_P \ln \frac{T_2}{T_1} = (1 \text{ mol}) \left(75.3 \text{ J K}^{-1} \text{mol}^{-1}\right) \ln \frac{373.15 \text{ K}}{298.15 \text{ K}} = 16.90 \text{ J K}^{-1}$$

Step 2:

$$H_2O(l), 100°C \rightarrow H_2O(g), 100°C$$

$$\Delta H = n\Delta_{vap}\overline{H} = (1 \text{ mol}) \left(40.79 \text{ kJ mol}^{-1}\right) = 40.79 \text{ kJ}$$

To calculate ΔU, the change in volume must first be determined. Since the volume of $H_2O(g)$, V_g is much greater than that of $H_2O(l)$, V_l, the latter is ignored.

$$\Delta U = \Delta H - P\Delta V = \Delta H - P\left(V_g - V_l\right) = \Delta H - PV_g$$

$$= \Delta H - P\frac{nRT}{P} = \Delta H - nRT$$

$$= 40.79 \text{ kJ} - (1 \text{ mol})\left(8.314 \text{ J K}^{-1} \text{ mol}^{-1}\right)(373.15 \text{ K})\left(\frac{1 \text{ kJ}}{1000 \text{ J}}\right)$$

$$= 37.688 \text{ kJ}$$

$$\Delta S = \frac{\Delta H_{vap}}{T_b} = \frac{40.79 \times 10^3 \text{ J}}{373.15 \text{ K}} = 109.31 \text{ J K}^{-1}$$

The values of ΔH, ΔU, and ΔS for the entire process can be obtained by summing the corresponding quantities calculated in the two steps:

$$\Delta H = 5.648 \text{ kJ} + 40.79 \text{ kJ} = 46.44 \text{ kJ}$$

$$\Delta U = 5.648 \text{ kJ} + 37.688 \text{ kJ} = 43.34 \text{ kJ}$$

$$\Delta S = 16.90 \text{ J K}^{-1} + 109.31 \text{ J K}^{-1} = 126.2 \text{ J K}^{-1}$$

4.16 A quantity of 6.0 moles of an ideal gas is reversibly heated at constant volume from 17°C to 35°C. Calculate the entropy change. What would be the value of ΔS if the heating were carried out irreversibly?

At constant volume, $dq_{rev} = C_V dT$.

$$\Delta S = \int \frac{dq_{rev}}{T} = \int \frac{C_V dT}{T} = C_V \ln \frac{T_2}{T_1} = \frac{3}{2}nR \ln \frac{T_2}{T_1}$$

$$= \frac{3}{2}(6.0 \text{ mol})\left(8.314 \text{ J K}^{-1} \text{ mol}^{-1}\right) \ln \frac{308.15 \text{ K}}{290.15 \text{ K}}$$

$$= 4.5 \text{ J K}^{-1}$$

If heating were carried out irreversibly, ΔS is still 4.5 J K^{-1} because S is a state function so that ΔS depends only on the final and initial states. ΔS must be calculated, however, using a reversible pathway.

4.18 A quantity of 35.0 g of water at 25.0°C (called A) is mixed with 160.0 g of water at 86.0°C (called B). **(a)** Calculate the final temperature of the system, assuming that the mixing is carried out adiabatically. **(b)** Calculate the entropy changes of A, B, and the entire system.

(a) Let the final temperature be T_f and the specific heat of water be s. Since the process is carried out adiabatically, the energy entering A as heat, q_A, is equal in magnitude but opposite in sign to that leaving B as heat, q_B.

$$q_A = -q_B$$

$$(35.0 \text{ g}) \, s \, (T_f - 25.0°\text{C}) = - (160.0 \text{ g}) \, s \, (T_f - 86.0°\text{C})$$

$$T_f - 25.0°\text{C} = -4.571 \, (T_f - 86.0°\text{C})$$

$$= -4.571 T_f + 393.1°\text{C}$$

$$5.571 T_f = 418.1°\text{C}$$

$$T_f = 75.05°\text{C} = 75.1°\text{C}$$

(b) Let the entropy change of A be ΔS_A and the entropy change of B be ΔS_B.

$$\Delta S_A = C_P \ln \frac{348.20 \text{ K}}{298.15 \text{ K}} = \left(\frac{35.0 \text{ g}}{18.02 \text{ g mol}^{-1}} \right) \left(75.3 \text{ J K}^{-1} \text{mol}^{-1} \right) \ln \frac{348.20 \text{ K}}{298.15 \text{ K}}$$

$$= 22.70 \text{ J K}^{-1} = 22.7 \text{ J K}^{-1}$$

$$\Delta S_B = C_P \ln \frac{348.20 \text{ K}}{359.15 \text{ K}} = \left(\frac{160.0 \text{ g}}{18.02 \text{ g mol}^{-1}} \right) \left(75.3 \text{ J K}^{-1} \text{mol}^{-1} \right) \ln \frac{348.20 \text{ K}}{359.15 \text{ K}}$$

$$= -20.702 \text{ J K}^{-1} = -20.70 \text{ J K}^{-1}$$

$$\Delta S_{total} = \Delta S_A + \Delta S_B = 22.70 \text{ J K}^{-1} - 20.702 \text{ J K}^{-1} = 2.0 \text{ J K}^{-1}$$

4.20 A sample of helium (He) gas initially at 20°C and 1.0 atm is expanded from 1.2 L to 2.6 L and simultaneously heated to 40°C. Calculate the entropy change for the process.

The number of moles of He can be determined using the initial conditions and the ideal gas law.

$$n = \frac{(1.0 \text{ atm}) \, (1.2 \text{ L})}{(0.08206 \text{ L atm K}^{-1} \text{mol}^{-1}) \, (293.15 \text{ K})} = 4.99 \times 10^{-2} \text{ mol}$$

The problem can be solved by breaking down the process into 2 steps: (1) isothermal expansion from 1.2 L at 1.0 atm to 2.6 L. The temperature is kept at 20°C; (2) heating at constant volume (2.6 L) from 20°C to 40°C. The entropy changes for these two steps, ΔS_1 and ΔS_2 are

$$\Delta S_1 = nR \ln \frac{V_2}{V_1} = \left(4.99 \times 10^{-2} \text{ mol} \right) \left(8.314 \text{ J K}^{-1} \text{mol}^{-1} \right) \ln \frac{2.6 \text{ L}}{1.2 \text{ L}} = 0.321 \text{ J K}^{-1}$$

$$\Delta S_2 = C_V \ln \frac{T_2}{T_1} = \frac{3}{2} nR \ln \frac{T_2}{T_1} = \frac{3}{2} \left(4.99 \times 10^{-2} \text{ mol} \right) \left(8.314 \text{ J K}^{-1} \text{mol}^{-1} \right) \ln \frac{313.15 \text{ K}}{293.15 \text{ K}}$$

$$= 4.11 \times 10^{-2} \text{ J K}^{-1}$$

The entropy for the entire process is

$$\Delta S = \Delta S_1 + \Delta S_2 = 0.321 \text{ J K}^{-1} + 4.11 \times 10^{-2} \text{ J K}^{-1} = 0.36 \text{ J K}^{-1}$$

4.22 One mole of an ideal gas at 298 K expands isothermally from 1.0 L to 2.0 L **(a)** reversibly and **(b)** against a constant external pressure of 12.2 atm. Calculate the values of ΔS_{sys}, ΔS_{surr}, and ΔS_{univ} in both cases. Are your results consist with the nature of the processes?

(a) $\Delta S_{sys} = nR \ln \dfrac{V_2}{V_1} = (1 \text{ mol}) \left(8.314 \text{ J K}^{-1} \text{ mol}^{-1}\right) \ln \dfrac{2.0 \text{ L}}{1.0 \text{ L}} = 5.76 \text{ J K}^{-1} = 5.8 \text{ J K}^{-1}$

$\Delta S_{surr} = -5.8 \text{ J K}^{-1}$

$\Delta S_{univ} = 0 \text{ J K}^{-1}$

(b) ΔS_{sys} is the same above, that is, 5.8 J K^{-1}, since S is a state function, although ΔS has to be calculated using a reversible path.

ΔS_{surr} can be calculated once q_{surr} is determined. The latter quantity is related to q_{sys}, which in turn can be calculated from the work done by the system, w, and the first law of thermodynamics.

$$w = -P_{ex}\Delta V = -(12.2 \text{ atm})(2.0 \text{ L} - 1.0 \text{ L})\left(\dfrac{101.3 \text{ J}}{1 \text{ L atm}}\right) = -1.24 \times 10^3 \text{ J}$$

According to the first law, $\Delta U = q + w$. Since $\Delta U = 0$ for an isothermal process involving an ideal gas, $q_{sys} = q = -w = 1.24 \times 10^3$ J. The entropy change of the surroundings is

$$\Delta S_{surr} = \dfrac{q_{surr}}{T_{surr}} = \dfrac{-q_{sys}}{T_{surr}} = \dfrac{-1.24 \times 10^3 \text{ J}}{298 \text{ K}} = -4.16 \text{ J K}^{-1}$$

Now ΔS_{univ} can be determined:

$$\Delta S_{univ} = 5.76 \text{ J K}^{-1} - 4.16 \text{ J K}^{-1} = 1.6 \text{ J K}^{-1}$$

The results in both parts are consistent with the nature of the processes. Specifically, for a reversible process, $\Delta S_{univ} = 0$, whereas for a spontaneous process, $\Delta S_{univ} > 0$.

4.24 A quantity of 0.54 mole of steam initially at 350°C and 2.4 atm undergoes a cyclic process for which $q = -74$ J. Calculate the value of ΔS for the process.

$\Delta S = 0$ for a cyclic process.

4.26 Use the data in Appendix B to calculate the values of $\Delta_r S^\circ$ of the reactions listed in the previous problem.

(a) $\Delta_r S^\circ = 2\overline{S}^\circ[\text{Fe}_2\text{O}_3(s)] - 4\overline{S}^\circ[\text{Fe}(s)] - 3\overline{S}^\circ[\text{O}_2(g)]$

$= 2\left(90.0 \text{ J K}^{-1} \text{ mol}^{-1}\right) - 4\left(27.2 \text{ J K}^{-1} \text{ mol}^{-1}\right) - 3\left(205.0 \text{ J K}^{-1} \text{ mol}^{-1}\right)$

$= -543.8 \text{ J K}^{-1} \text{ mol}^{-1}$

(b) $\Delta_r S^\circ = \overline{S}^\circ[\text{O}_2(g)] - 2\overline{S}^\circ[\text{O}(g)]$

$= 205.0 \text{ J K}^{-1} \text{ mol}^{-1} - 2\left(161.0 \text{ J K}^{-1} \text{ mol}^{-1}\right)$

$= -117.0 \text{ J K}^{-1} \text{ mol}^{-1}$

(c)
$$\Delta_r S^\circ = \overline{S}^\circ[NH_3(g)] + \overline{S}^\circ[HCl(g)] - \overline{S}^\circ[NH_4Cl(s)]$$
$$= 192.5 \text{ J K}^{-1} \text{mol}^{-1} + 186.5 \text{ J K}^{-1} \text{mol}^{-1} - 94.6 \text{ J K}^{-1} \text{mol}^{-1}$$
$$= 284.4 \text{ J K}^{-1} \text{mol}^{-1}$$

(d)
$$\Delta_r S^\circ = 2\overline{S}^\circ[HCl(g)] - \overline{S}^\circ[H_2(g)] - \overline{S}^\circ[Cl_2(g)]$$
$$= 2\left(186.5 \text{ J K}^{-1} \text{mol}^{-1}\right) - \left(130.6 \text{ J K}^{-1} \text{mol}^{-1}\right) - \left(223.0 \text{ J K}^{-1} \text{mol}^{-1}\right)$$
$$= 19.4 \text{ J K}^{-1} \text{mol}^{-1}$$

These results agree with predictions made in Problem 4.25.

4.28 One mole of an ideal gas is isothermally expanded from 5.0 L to 10 L at 300 K. Compare the entropy changes for the system, surroundings, and the universe if the process is carried out **(a)** reversibly, and **(b)** irreversibly against an external pressure of 2.0 atm.

(a) For the reversible process,

$$\Delta S_{sys} = nR \ln \frac{V_2}{V_1} = (1 \text{ mol}) \left(8.314 \text{ J K}^{-1} \text{mol}^{-1}\right) \ln \frac{10 \text{ L}}{5.0 \text{ L}} = 5.76 \text{ J K}^{-1} = 5.8 \text{ J K}^{-1}$$

$$\Delta S_{surr} = -5.8 \text{ J K}^{-1}$$

$$\Delta S_{univ} = 0 \text{ J K}^{-1}$$

(b) Because S is a state function, ΔS_{sys} is the same as above, that is, 5.8 J K^{-1}.

ΔS_{surr} can be calculated once q_{surr} is determined. The latter quantity is related to q_{sys}, which in turn can be calculated from the work done by the system, w, and the first law of thermodynamics.

$$w = -P_{ex}\Delta V = - (2.0 \text{ atm}) (10 \text{ L} - 5.0 \text{ L}) \left(\frac{101.3 \text{ J}}{1 \text{ L atm}}\right) = -1.01 \times 10^3 \text{ J}$$

According to the first law, $\Delta U = q + w$. For an ideal gas, $\Delta U = 0$ for an isothermal process, and $q_{sys} = q = -w = 1.01 \times 10^3$ J. The entropy change of the surroundings is

$$\Delta S_{surr} = \frac{q_{surr}}{T_{surr}} = \frac{-q_{sys}}{T_{surr}} = \frac{-1.01 \times 10^3 \text{ J}}{300 \text{ K}} = -3.37 \text{ J K}^{-1} = -3.4 \text{ J K}^{-1}$$

Therefore,

$$\Delta S_{univ} = \Delta S_{sys} + \Delta S_{surr} = 5.76 \text{ J K}^{-1} - 3.37 \text{ J K}^{-1} = 2.4 \text{ J K}^{-1}$$

4.30 Consider the reaction

$$N_2(g) + O_2(g) \rightarrow 2NO(g)$$

Calculate the values of $\Delta_r S^\circ$ for the reaction mixture, surroundings, and the universe at 298 K. Why is your result reassuring to Earth's inhabitants?

$$\Delta_r S^\circ = 2\overline{S}^\circ[NO(g)] - \overline{S}^\circ[N_2(g)] - \overline{S}^\circ[O_2(g)]$$

$$= 2\left(210.6 \text{ J K}^{-1} \text{ mol}^{-1}\right) - 191.6 \text{ J K}^{-1} \text{ mol}^{-1} - 205.0 \text{ J K}^{-1} \text{ mol}^{-1}$$

$$= 24.6 \text{ J K}^{-1} \text{ mol}^{-1}$$

ΔS°_{surr} is determined from $\Delta_r H$ and the temperature of the surroundings.

$$\Delta_r H^\circ = 2\Delta_f \overline{H}^\circ[NO(g)] - \Delta_f \overline{H}^\circ[N_2(g)] - \Delta_f \overline{H}^\circ[O_2(g)]$$

$$= 2\left(90.4 \text{ kJ mol}^{-1}\right) - 0 \text{ kJ mol}^{-1} - 0 \text{ kJ mol}^{-1}$$

$$= 180.8 \text{ kJ mol}^{-1}$$

$$\Delta H^\circ_{surr} = -\Delta_r H^\circ = -180.8 \text{ kJ mol}^{-1}$$

$$\Delta S^\circ_{surr} = \frac{\Delta H^\circ_{surr}}{T} = \frac{-180.8 \times 10^3 \text{ J mol}^{-1}}{298 \text{ K}} = -607 \text{ J K}^{-1} \text{ mol}^{-1}$$

Therefore,

$$\Delta S_{univ} = 24.6 \text{ J K}^{-1} \text{ mol}^{-1} - 607 \text{ J K}^{-1} \text{ mol}^{-1} = -582 \text{ J K}^{-1} \text{ mol}^{-1}$$

This is not a spontaneous process at 298 K. Therefore, O_2, which is essential to us, does not react with N_2 to form NO in the atmosphere at 298 K.

4.32 Choose the substance with the greater molar entropy in each of the following pairs: (a) $H_2O(l)$, $H_2O(g)$, (b) NaCl(s), $CaCl_2(s)$, (c) N_2(0.1 atm), N_2(1 atm), (d) C(diamond), C(graphite), (e) $O_2(g)$, $O_3(g)$, (f) ethanol (C_2H_5OH), dimethyl ether (CH_3OCH_3), (g) $N_2O_4(g)$, $2NO_2(g)$, (h) Fe(s) at 298 K, Fe(s) at 398 K. (Unless otherwise stated, assume the temperature is 298 K.)

(a) $H_2O(g)$, a gas has greater entropy than the more ordered liquid.

(b) $CaCl_2(s)$, this is a more complex system than NaCl(s).

(c) N_2 (0.1 atm), at the lower pressure, the gas occupies a larger volume leading to a larger number of microstates for the system.

(d) C(graphite), diamond is a more ordered solid than is graphite.

(e) $O_3(g)$, this is a more complex system than diatomic $O_2(g)$.

(f) Dimethyl ether, ethanol can form hydrogen bonds leading to a more ordered system.

(g) One mole of $N_2O_4(g)$ is a more complex system and has greater entropy than one mole of $NO_2(g)$, although *two* moles of $NO_2(g)$ has greater entropy than one mole of $N_2O_4(g)$.

(h) Fe(s) at 398 K, since it is at a higher temperature.

4.34 Calculate the molar residual entropy of a solid in which the molecules can adopt (a) three, (b) four, and (c) five orientations of equal energy at the absolute zero.

(a) $\overline{S}^{\circ} = R \ln 3 = 9.134 \text{ J K}^{-1} \text{mol}^{-1}$

(b) $\overline{S}^{\circ} = R \ln 4 = 11.53 \text{ J K}^{-1} \text{mol}^{-1}$

(c) $\overline{S}^{\circ} = R \ln 5 = 13.38 \text{ J K}^{-1} \text{mol}^{-1}$

4.36 Explain why $\overline{S}^{\circ}$(graphite) > $\overline{S}^{\circ}$(diamond) at 298 K (see Appendix B). Would this inequality hold at 0 K?

Graphite does not have as highly ordered structure as does diamond, and is expected to have the larger entropy. At 0 K, however, both allotropic forms of carbon are expected to have the same entropy, $\overline{S}^{\circ} = 0$, presuming there are no crystal defects or impurities.

4.38 State the condition(s) under which the following equations can be applied: **(a)** $\Delta S = \Delta H/T$, **(b)** $S_0 = 0$, **(c)** $dS = C_p dT/T$, **(d)** $dS = dq/T$.

(a) Constant pressure and temperature, reversible process.

(b) Absolute zero, no residual entropy and a pure, crystalline substance.

(c) Constant pressure (Note that when this form is integrated, one must either include the explicit temperature dependence of C_P or assume it to be temperature independent.)

(d) Reversible process

4.40 Calculate the entropy change when neon at 25°C and 1.0 atm in a container of volume 0.780 L is allowed to expand to 1.25 L and is simultaneously heated to 85°C. Assume ideal behavior. (*Hint*: Since S is a state function, you can first calculate the value of ΔS for expansion and then calculate the value of ΔS for heating at constant final volume.)

The number of moles of Ne can be determined using the initial conditions and the ideal gas law.

$$n = \frac{(1.0 \text{ atm}) (0.780 \text{ L})}{(0.08206 \text{ L atm K}^{-1} \text{mol}^{-1}) (298.15 \text{ K})} = 3.19 \times 10^{-2} \text{ mol}$$

The problem can be solved by breaking down the process into 2 steps: (1) isothermal expansion from 0.780 L at 1.0 atm to 1.25 L. The temperature is kept at 25°C; (2) heating at constant volume (1.25 L) from 25°C to 85°C. The entropy changes for these two steps, ΔS_1 and ΔS_2 are

$$\Delta S_1 = nR \ln \frac{V_2}{V_1} = \left(3.19 \times 10^{-2} \text{ mol}\right) \left(8.314 \text{ J K}^{-1} \text{mol}^{-1}\right) \ln \frac{1.25 \text{ L}}{0.780 \text{ L}} = 0.125 \text{ J K}^{-1}$$

$$\Delta S_2 = C_V \ln \frac{T_2}{T_1} = \frac{3}{2} nR \ln \frac{T_2}{T_1} = \frac{3}{2} \left(3.19 \times 10^{-2} \text{ mol}\right) \left(8.314 \text{ J K}^{-1} \text{mol}^{-1}\right) \ln \frac{358.15 \text{ K}}{298.15 \text{ K}}$$

$$= 7.29 \times 10^{-2} \text{ J K}^{-1}$$

The entropy for the entire process is

$$\Delta S = \Delta S_1 + \Delta S_2 = 0.125 \text{ J K}^{-1} + 7.29 \times 10^{-2} \text{ J K}^{-1} = 0.20 \text{ J K}^{-1}$$

4.42 One mole of an ideal monatomic gas is compressed from 2.0 atm to 6.0 atm while being cooled from 400 K to 300 K. Calculate the values of ΔU, ΔH, and ΔS for the process.

The thermodynamic quantities can be readily calculated by breaking down the process into 2 steps: (1) isothermal compression at 400 K from 2.0 atm to 6.0 atm; (2) cooling at constant pressure (6.0 atm) from 400 K to 300 K.

Step 1:

For an ideal gas undergoing an isothermal process, $\Delta U = 0$ and $\Delta H = 0$.

$$\Delta S = nR \ln \frac{V_2}{V_1} = nR \ln \frac{P_1}{P_2} = (1 \text{ mol}) \left(8.314 \text{ J K}^{-1} \text{mol}^{-1}\right) \ln \frac{2.0 \text{ atm}}{6.0 \text{ atm}} = -9.13 \text{ J K}^{-1}$$

Step 2:

$$\Delta H = C_P \Delta T = \frac{5}{2} nR \Delta T = \frac{5}{2}(1 \text{ mol}) \left(8.314 \text{ J K}^{-1} \text{mol}^{-1}\right) (300 \text{ K} - 400 \text{ K})$$

$$= -2.079 \times 10^3 \text{ J}$$

$$\Delta U = \Delta H - nR \Delta T = -2.079 \times 10^3 \text{ J} - (1 \text{ mol}) \left(8.314 \text{ J K}^{-1} \text{mol}^{-1}\right)(300 \text{ K} - 400 \text{ K})$$

$$= -1.248 \times 10^3 \text{ J}$$

$$\Delta S = C_P \ln \frac{T_2}{T_1} = \frac{5}{2} nR \ln \frac{T_2}{T_1} = \frac{5}{2} (1 \text{ mol}) \left(8.314 \text{ J K}^{-1} \text{mol}^{-1}\right) \ln \frac{300 \text{ K}}{400 \text{ K}} = -5.980 \text{ J K}^{-1}$$

For the entire process,

$$\Delta H = 0 \text{ J} - 2.079 \times 10^3 \text{ J} = -2.08 \times 10^3 \text{ J}$$

$$\Delta U = 0 \text{ J} - 1.248 \times 10^3 \text{ J} = -1.25 \times 10^3 \text{ J}$$

$$\Delta S = -9.13 \text{ J K}^{-1} - 5.980 \text{ J K}^{-1} = -15.1 \text{ J K}^{-1}$$

4.44 Use the following data to determine the normal boiling point, in kelvins, of mercury. What assumptions must you make in order to do the calculation?

$$\text{Hg}(l): \quad \Delta_f \overline{H}^\circ = 0 \text{ (by definition)}$$

$$\overline{S}^\circ = 75.9 \text{ J K}^{-1} \text{mol}^{-1}$$

$$\text{Hg}(g): \quad \Delta_f \overline{H}^\circ = 60.78 \text{ kJ mol}^{-1}$$

$$\overline{S}^\circ = 175.0 \text{ J K}^{-1} \text{mol}^{-1}$$

The normal boiling point of mercury, T_b is related to $\Delta_{vap} S^\circ$ and $\Delta_{vap} H^\circ$:

$$\Delta_{vap} S^\circ = \frac{\Delta_{vap} H^\circ}{T_b}$$

For 1 mole of mercury, the entropy and enthalpy of vaporization are

$$\Delta_{vap} H^\circ = 60.78 \text{ kJ} - 0 \text{ kJ} = 60.78 \text{ kJ}$$

$$\Delta_{vap} S^\circ = 175.0 \text{ J K}^{-1} - 75.9 \text{ J K}^{-1} = 99.1 \text{ J K}^{-1}$$

Therefore,

$$T_b = \frac{\Delta_{vap}H^\circ}{\Delta_{vap}S^\circ} = \frac{60.78 \times 10^3 \text{ J}}{99.1 \text{ J K}^{-1}} = 613 \text{ K} = 340°C$$

The assumptions made in this calculation are that the values of $\Delta_f\overline{H}^\circ$ and $\overline{S}^\circ$ are temperature independent. These assumptions are quite good because the calculated boiling point of mercury is very close to the actual value of 356.6°C.

4.46   Give a detailed example of each of the following, with an explanation: **(a)** a thermodynamically spontaneous process; **(b)** a process that would violate the first law of thermodynamics; **(c)** a process that would violate the second law of thermodynamics; **(d)** an irreversible process; **(e)** an equilibrium process.

(a) An ice cube melting in a glass of water at 20°C.

(b) A perpetual motion machine of the first kind, such as a rotating flywheel that drives a generator, the output of which is used to keep the flywheel rotating at a constant speed and also to lift a weight.

(c) A perfect air conditioner; it extracts heat from the room and warms the outside without using any energy to do so. (This does not violate the first law, since the energy deposited outside is exactly equal to that removed from inside.)

(d) Same as part **(a)**, an ice cube melting in a glass of water at 20°C.

(e) Water and ice in a closed system at 0°C and 1 atm pressure.

4.48   A refrigerator set at 0°C discharges heat into the kitchen at 20°C. **(a)** How much work would be required to freeze 500 mL of water (about an ice tray's volume)? **(b)** How much heat would be discharged during this process? (The molar enthalpy of fusion of water is 6.01 kJ mol^{-1}, and the refrigerator operates at 35% efficency.)

(a) The work required is calculated from the heat extracted from the cold reservoir and the coefficient of performance of the refrigerator. The heat extracted depends on the amount of water to be frozen, which is calculated by assuming a density of 1 g mL^{-1}.

$$\text{Number of moles of water} = \frac{(500 \text{ mL}) \left(1 \text{ g mL}^{-1}\right)}{18.02 \text{ g mol}^{-1}} = 27.75 \text{ mol}$$

$$\text{Heat extracted from the cold reservoir} = (27.75 \text{ mol}) \left(6.01 \text{ kJ mol}^{-1}\right) = 166.8 \text{ kJ}$$

$$\text{COP} = \frac{T_1}{T_2 - T_1} \text{ (efficiency)} = \frac{273.15 \text{ K}}{293.15 \text{ K} - 273.15 \text{ K}} (0.35) = 4.78$$

$$\text{Work required} = \frac{166.8 \text{ kJ}}{4.78} = 35 \text{ kJ}$$

(b) Heat discharged = 35 kJ + 166.8 kJ = 202 kJ

4.50   Toluene (C_7H_8) has a dipole moment, but benzene (C_6H_6) is nonpolar.

m.pt.	5.5°C	−95°C
b.pt.	80.1°C	110.6°C

Explain why, contrary to our expectation, benzene melts at a much higher temperature than toluene? Why is the boiling point of toluene higher than that of benzene?

Rearranging Equation 4.18 gives $T_f = \frac{\Delta_{fus}H}{\Delta_{fus}S}$. Although toluene as a polar molecule is expected to have a slightly larger $\Delta_{fus}H$ than the non-polar benzene, benzene is a highly symmetric molecule so that it has a much smaller $\Delta_{fus}S$ than toluene. The entropy effect outweighs the enthalpy effect, giving benzene a melting point considerably higher than that of toluene.

For boiling, the situation is different. The entropy of the vapor phase is significantly greater than that of the liquid phase for both molecules. Thus, $\Delta_{vap}S = S_{vap} - S_{liq} \approx S_{vap}$ is roughly the same for both species. Equation 4.19, $T_b = \frac{\Delta_{vap}H}{\Delta_{vap}S}$, then indicates that the boiling point is determined primarily by $\Delta_{vap}H$ which is greater for toluene than benzene so that the polar toluene has the higher boiling point.

4.52 In Section 4.2 we saw that the probability of finding all 100 helium atoms in one half of the cylinder is 8×10^{-31} (see Figure 4.3). Assuming that the age of the universe is 14 billion years, calculate the time in seconds during which this event can be observed.

We need to find the period of time that corresponds to 8×10^{-31} times the age of the universe.

$$\left(14 \times 10^9 \text{ yr}\right) \left(8 \times 10^{-31}\right) \left(\frac{365 \text{ days}}{1 \text{ yr}}\right) \left(\frac{24 \text{ h}}{1 \text{ day}}\right) \left(\frac{3600 \text{ s}}{1 \text{ h}}\right) = 3.5 \times 10^{-13} \text{ s}$$

This is approximately one-third of a picosecond and represents an extremely improbable event!

4.54 Two moles of argon gas initially at 300 K and 2.00 L are heated to 400 K and the volume changed to 6.00 L. Calculate ΔS for the process. Assume ideal behavior and $\overline{C}_V$ to be independent of temperature.

The problem can be solved by breaking down the process into 2 steps: (1) isothermal expansion from 2.00 L to 6.00 L. The temperature is kept at 300 K; (2) heating at constant volume (6.00 L) from 300 K to 400 K. The entropy changes for these two steps, ΔS_1 and ΔS_2 are

$$\Delta S_1 = nR \ln \frac{V_2}{V_1} = (2 \text{ mol}) \left(8.314 \text{ J K}^{-1} \text{mol}^{-1}\right) \ln \frac{6.00 \text{ L}}{2.00 \text{ L}} = 18.27 \text{ J K}^{-1}$$

$$\Delta S_2 = C_V \ln \frac{T_2}{T_1} = \frac{3}{2} nR \ln \frac{T_2}{T_1} = \frac{3}{2} (2 \text{ mol}) \left(8.314 \text{ J K}^{-1} \text{mol}^{-1}\right) \ln \frac{400 \text{ K}}{300 \text{ K}}$$

$$= 7.175 \text{ J K}^{-1}$$

The entropy for the entire process is

$$\Delta S = \Delta S_1 + \Delta S_2 = 18.27 \text{ J K}^{-1} + 7.175 \text{ J K}^{-1} = 25.4 \text{ J K}^{-1}$$

4.56 A certain reaction is spontaneous at 72°C. If the enthalpy change for the reaction is 19 kJmol^{-1}, what is the minimum value of $\Delta \dot{S}$ for the reaction?

For a spontaneous reaction, $\Delta S_{univ} = \Delta S_{sys} + \Delta S_{surr} > 0$. It follows then

$$\Delta S_{sys} > -\Delta S_{surr}$$

$$> -\left(\frac{-\Delta H}{T}\right) = \frac{19 \text{ kJ mol}^{-1}}{(273.15 + 72) \text{ K}}$$

$$> 55 \text{ J K}^{-1} \text{ mol}^{-1}$$

4.58 Calculate the number of microstates (W) when 10 molecules are equally distributed among five energy levels. What is the value of W if one molecule is removed from one state and added to another?

When the molecules are equally distributed among five energy levels,

$$W = \frac{10!}{2!\,2!\,2!\,2!\,2!} = 113400$$

If one molecule is removed from one state and added to another,

$$W = \frac{10!}{1!\,3!\,2!\,2!\,2!} = 75600$$

CHAPTER 5

Gibbs and Helmholtz Energies and Their Applications

PROBLEMS AND SOLUTIONS

5.2 At one time, the domestic gas used for cooking, called "water gas," was prepared as follows:

$$H_2O(g) + C(graphite) \rightarrow CO(g) + H_2(g)$$

From the thermodynamic quantities listed in Appendix B, predict whether this reaction will occur at 298 K. If not, at what temperature will the reaction occur? Assume $\Delta_r H^\circ$ and $\Delta_r S^\circ$ are temperature-independent.

The sign of $\Delta_r G^\circ$ is used to determine if the reaction is spontaneous when the reactants and products are in their standard states. If $\Delta_r G^\circ$ is calculated via $\Delta_r H^\circ$ and $\Delta_r S^\circ$, the dependence of this quantity on temperature can also be assessed.

$$\Delta_r H^\circ = \Delta_f \overline{H}^\circ[CO(g)] + \Delta_f \overline{H}^\circ[H_2(g)] - \Delta_f \overline{H}^\circ[H_2O(g)] - \Delta_f \overline{H}^\circ[C(graphite)]$$

$$= -110.5 \text{ kJ mol}^{-1} + 0 \text{ kJ mol}^{-1} - \left(-241.8 \text{ kJ mol}^{-1}\right) - 0 \text{ kJ mol}^{-1}$$

$$= 131.3 \text{ kJ mol}^{-1}$$

$$\Delta_r S^\circ = \overline{S}^\circ[CO(g)] + \overline{S}^\circ[H_2(g)] - \overline{S}^\circ[H_2O(g)] - \overline{S}^\circ[C(graphite)]$$

$$= 197.9 \text{ J K}^{-1} \text{mol}^{-1} + 130.6 \text{ J K}^{-1} \text{mol}^{-1} - 188.7 \text{ J K}^{-1} \text{mol}^{-1} - 5.7 \text{ J K}^{-1} \text{mol}^{-1}$$

$$= 134.1 \text{ J K}^{-1} \text{mol}^{-1}$$

$$\Delta_r G^\circ = \Delta_r H^\circ - T \Delta_r S^\circ = 131.3 \text{ kJ mol}^{-1} - (298 \text{ K}) \left(134.1 \times 10^{-3} \text{ kJ K}^{-1} \text{mol}^{-1}\right)$$

$$= 91.3 \text{ kJ mol}^{-1}$$

Since $\Delta_r G^\circ > 0$, the reaction is not spontaneous when all the reactants and products are in their standard states and at 298 K. The reaction is spontaneous when $\Delta_r G^\circ < 0$, that is, when

$$\Delta_r H^\circ - T \Delta_r S^\circ < 0$$

$$\Delta_r H^\circ < T \Delta_r S^\circ$$

$$T > \frac{\Delta_r H^\circ}{\Delta_r S^\circ} = \frac{131.3 \text{ kJ mol}^{-1}}{134.1 \times 10^{-3} \text{ kJ K}^{-1} \text{mol}^{-1}} = 979.1 \text{ K}$$

The lower limit of temperature, 979.1 K, is an estimate because $\Delta_r H^\circ$ and $\Delta_r S^\circ$ are not truly temperature independent, as assumed here.

5.4   Without referring to Appendix B, calculate the quantity $(\Delta_r G^\circ - \Delta_r A^\circ)$ for the following reaction at 298 K:

$$C(s) + CO_2(g) \rightarrow 2CO(g)$$

Assume ideal-gas behavior.

The definitions of $\Delta_r G^\circ$ and $\Delta_r A^\circ$ give

$$\Delta_r G^\circ = \Delta_r H^\circ - T\Delta_r S^\circ$$
$$\Delta_r A^\circ = \Delta_r U^\circ - T\Delta_r S^\circ$$

Thus,

$$\begin{aligned}
\Delta_r G^\circ - \Delta_r A^\circ &= \Delta_r H^\circ - \Delta_r U^\circ \\
&= \Delta(PV) \\
&= \Delta(nRT) \\
&= RT\Delta n \\
&= \left(8.314\,\text{J K}^{-1}\,\text{mol}^{-1}\right)(298\,\text{K})(1) \\
&= 2.48 \times 10^3\,\text{J mol}^{-1} \\
&= 2.48\,\text{kJ mol}^{-1}
\end{aligned}$$

5.6   Certain bacteria in the soil obtain the necessary energy for growth by oxidizing nitrite to nitrate:

$$2NO_2^-(aq) + O_2(g) \rightarrow 2NO_3^-(aq)$$

Given that the standard Gibbs energies of formation of NO_2^- and NO_3^- are $-34.6\,\text{kJ mol}^{-1}$ and $-110.5\,\text{kJ mol}^{-1}$, respectively, calculate the amount of Gibbs energy released when 1 mole of NO_2^- is oxidized to 1 mole of NO_3^-.

According to the chemical equation $2NO_2^-(aq) + O_2(g) \rightarrow 2NO_3^-(aq)$,

$$\Delta_r G^\circ = 2\Delta_f \overline{G}^\circ[NO_3^-(aq)] - 2\Delta_f \overline{G}^\circ[NO_2^-(aq)] - \Delta_f \overline{G}^\circ[O_2(g)]$$

$$= 2\left(-110.5\,\text{kJ mol}^{-1}\right) - 2\left(-34.6\,\text{kJ mol}^{-1}\right) - 0\,\text{kJ mol}^{-1}$$

$$= -151.8\,\text{kJ mol}^{-1}$$

When 1 mole of NO_2^- is oxidized to 1 mole of NO_3^-,

$$\Delta_r G^\circ = \frac{-151.8\,\text{kJ mol}^{-1}}{2} = -75.9\,\text{kJ mol}^{-1}$$

5.8   This problem involves the synthesis of diamond from graphite:

$$C(\text{graphite}) \rightarrow C(\text{diamond})$$

(a) Calculate the values of $\Delta_r H^\circ$ and $\Delta_r S^\circ$ for the reaction. Will the conversion occur spontaneously at 25°C or any other temperature? **(b)** From density measurements, the molar volume of graphite is found to be 2.1 cm^3 greater than that of diamond. Can the conversion of graphite to diamond be brought about at 25°C by applying pressure on graphite? If so, estimate the pressure at which the process becomes spontaneous. [*Hint*: Starting from Equation 5.16, derive the equation $\Delta G = \left(\overline{V}_{\text{diamond}} - \overline{V}_{\text{graphite}} \right) \Delta P$ for a constant-temperature process. Next, calculate the ΔP value that would lead to the necessary decrease in Gibbs energy.]

(a)

$$\Delta_r H^\circ = \Delta_f \overline{H}^\circ[\text{C(diamond)}] - \Delta_f \overline{H}^\circ[\text{C(graphite)}]$$

$$= 1.90 \text{ kJ mol}^{-1} - 0 \text{ kJ mol}^{-1}$$

$$= 1.90 \text{ kJ mol}^{-1}$$

$$\Delta_r S^\circ = \overline{S}^\circ[\text{C(diamond)}] - \overline{S}^\circ[\text{C(graphite)}]$$

$$= 2.4 \text{ J K}^{-1} \text{ mol}^{-1} - 5.7 \text{ J K}^{-1} \text{ mol}^{-1}$$

$$= -3.3 \text{ J K}^{-1} \text{ mol}^{-1}$$

At 25°C,

$$\Delta_r G^\circ = \Delta_r H^\circ - (298.15 \text{ K}) \, \Delta_r S^\circ$$

$$= 1.90 \text{ kJ mol}^{-1} - (298.15 \text{ K}) \left(-3.3 \times 10^{-3} \text{ kJ K}^{-1} \text{ mol}^{-1} \right)$$

$$= 2.884 \text{ kJ mol}^{-1} = 2.88 \text{ kJ mol}^{-1}$$

Therefore, the conversion from graphite to diamond is not spontaneous at 25°C and when both are in their standard states. In fact, because $\Delta_r H^\circ$ is positive and $T \Delta_r S^\circ$ is negative, $\Delta_r G^\circ$ can never be negative. Thus, the conversion will not be spontaneous at any other temperature.

(b) The integration of Equation 5.16

$$\left(\frac{\partial G}{\partial P} \right)_T = V$$

gives

$$G_2 = G_1 + V \left(P_2 - P_1 \right)$$

where G_1 and G_2 are the Gibbs energies at P_1 and P_2, respectively. Using molar quantities and $\Delta P = P_2 - P_1$, the equation becomes

$$\overline{G}_2 = \overline{G}_1 + \overline{V} \Delta P$$

Apply this equation to graphite and diamond,

$$\overline{G}_2[\text{graphite}] = \overline{G}_1[\text{graphite}] + \overline{V}[\text{graphite}] \Delta P$$

$$\overline{G}_2[\text{diamond}] = \overline{G}_1[\text{diamond}] + \overline{V}[\text{diamond}] \Delta P$$

These two equations are combined to relate the values of $\Delta_r G$ for the conversion of graphite to diamond at two different pressures:

$$\Delta_r G_2 = \overline{G}_2[\text{diamond}] - \overline{G}_2[\text{graphite}]$$

$$= \left\{\overline{G}_1[\text{diamond}] + \overline{V}[\text{diamond}]\Delta P\right\} - \left\{\overline{G}_1[\text{graphite}] + \overline{V}[\text{graphite}]\Delta P\right\}$$

$$= \Delta_r G_1 + \left\{\overline{V}[\text{diamond}] - \overline{V}[\text{graphite}]\right\} \Delta P$$

If $P_1 = 1$ bar, then $\Delta_r G_1 = \Delta_r G^\circ = 2.884$ kJ mol^{-1} at 25°C. $\Delta_r G_2$ becomes

$$\Delta_r G_2 = 2.884 \text{ kJ mol}^{-1}$$

$$+ \left(-2.1 \text{ cm}^3 \text{ mol}^{-1}\right) \Delta P \left(\frac{1 \text{ L}}{1000 \text{ cm}^3}\right) \left(\frac{1 \text{ atm}}{1.013 \text{ bar}}\right) \left(\frac{101.3 \text{ J}}{1 \text{ L atm}}\right) \left(\frac{1 \text{ kJ}}{1000 \text{ J}}\right)$$

$$= 2.884 \text{ kJ mol}^{-1} - 2.1 \times 10^{-4} \Delta P \text{ kJ bar}^{-1} \text{mol}^{-1}$$

If the process is spontaneous, then

$$\Delta_r G_2 = 2.884 \text{ kJ mol}^{-1} - 2.1 \times 10^{-4} \Delta P \text{ kJ bar}^{-1} \text{mol}^{-1} < 0$$

$$2.884 \text{ kJ mol}^{-1} < 2.1 \times 10^{-4} \Delta P \text{ kJ bar}^{-1} \text{mol}^{-1}$$

$$\Delta P > \frac{2.884 \text{ kJ mol}^{-1}}{2.1 \times 10^{-4} \text{ kJ bar}^{-1} \text{mol}^{-1}} = 1.4 \times 10^4 \text{ bar}$$

Therefore, at a very high pressure, $P_2 = \Delta P + P_1 = 1.4 \times 10^4$ bar + 1 bar = 1.4×10^4 bar, the conversion from graphite to diamond is spontaneous.

5.10 From the standard molar enthalpy of combustion of benzene at 298 K, calculate the value of $\Delta_r A^\circ$ for the process. Compare the value of $\Delta_r A^\circ$ with that of $\Delta_r H^\circ$. Comment on the difference.

The chemical equation corresponding to the combustion of benzene is

$$C_6H_6(l) + \frac{15}{2}O_2(g) \to 6CO_2(g) + 3H_2O(l)$$

At constant temperature, $\Delta_r A^\circ$ can be calculated directly from $\Delta_r U^\circ$ and $\Delta_r S^\circ$. $\Delta_r U^\circ$ in turn can be calculated from $\Delta_r H^\circ$.

$$\Delta_r H^\circ = 6\Delta_f \overline{H}^\circ[\text{CO}_2(g)] + 3\Delta_f \overline{H}^\circ[\text{H}_2\text{O}(l)] - \Delta_f \overline{H}^\circ[\text{C}_6\text{H}_6(l)] - \frac{15}{2}\Delta_f \overline{H}^\circ[\text{O}_2(g)]$$

$$= 6\left(-393.5 \text{ kJ mol}^{-1}\right) + 3\left(-285.8 \text{ kJ mol}^{-1}\right) - 49.04 \text{ kJ mol}^{-1} - \frac{15}{2}\left(0 \text{ kJ mol}^{-1}\right)$$

$$= -3267.44 \text{ kJ mol}^{-1}$$

$$\Delta_r U^\circ = \Delta_r H^\circ - RT \Delta n$$

$$= -3267.44 \text{ kJ mol}^{-1} - \left(8.314 \text{ J K}^{-1} \text{mol}^{-1}\right)(298 \text{ K})(-1.5)\left(\frac{1 \text{ kJ}}{1000 \text{ J}}\right)$$

$$= -3263.72 \text{ kJ mol}^{-1} = -3263.7 \text{ kJ mol}^{-1}$$

$$\Delta_r S^\circ = 6\overline{S}^\circ[CO_2(g)] + 3\overline{S}^\circ[H_2O(l)] - \overline{S}^\circ[C_6H_6(l)] - \frac{15}{2}\overline{S}^\circ[O_2(g)]$$

$$= 6\left(213.6 \text{ J K}^{-1}\text{mol}^{-1}\right) + 3\left(69.9 \text{ J K}^{-1}\text{mol}^{-1}\right)$$

$$- 172.8 \text{ J K}^{-1}\text{mol}^{-1} - \frac{15}{2}\left(205.0 \text{ J K}^{-1}\text{mol}^{-1}\right)$$

$$= -219.0 \text{ J K}^{-1}\text{mol}^{-1}$$

$$\Delta_r A^\circ = \Delta_r U^\circ - T\Delta_r S^\circ = -3263.72 \text{ kJ mol}^{-1} - (298 \text{ K})\left(-219.0 \text{ J K}^{-1}\text{mol}^{-1}\right)\left(\frac{1 \text{ kJ}}{1000 \text{ J}}\right)$$

$$= -3198.5 \text{ kJ mol}^{-1}$$

The maximum work ($\Delta_r A^\circ$) that can be obtained from this combustion is not as great as the heat released at constant pressure ($\Delta_r H^\circ$). This is because there is an entropy decrease in the system.

5.12   Predict the signs of ΔH, ΔS, and ΔG of the system for the following processes at 1 atm: **(a)** ammonia melts at $-60°C$, **(b)** ammonia melts at $-77.7°C$, **(c)** ammonia melts at $-100°C$. (The normal melting point of ammonia is $-77.7°C$.)

Since melting is an endothermic process, ΔH is positive in all three cases. A substance in the liquid phase is more disordered than in the solid phase. Therefore, ΔS is positive in all three cases.

When the temperature is above the melting point, the melting process is spontaneous, that is, ΔG is negative. When the temperature is at the melting point, the melting process is at equilibrium, that is, ΔG is zero. When the temperature is below the melting point, the melting process is not spontaneous, that is, ΔG is positive.

In summary,

(a) ΔH: +, ΔS: +, ΔG: −

(b) ΔH: +, ΔS: +, ΔG: 0

(c) ΔH: +, ΔS: +, ΔG: +

5.14   A student looked up the $\Delta_f\overline{G}^\circ$, $\Delta_f\overline{H}^\circ$, and $\overline{S}^\circ$ values for CO_2 in Appendix B. Plugging these values into Equation 5.3, he found that $\Delta_f\overline{G}^\circ \neq \Delta_f\overline{H}^\circ - T\overline{S}^\circ$ at 298 K. What is wrong with his approach?

The equation $\Delta G^\circ = \Delta H^\circ - T\Delta S^\circ$ applies to a process. If the process is a formation reaction, such as the formation of CO_2 [C(graphite) + $O_2(g) \rightarrow CO_2(g)$], then

$$\Delta_r G^\circ = \Delta_f G^\circ = -394.4 \text{ kJ mol}^{-1}$$

$$\Delta_r H^\circ = \Delta_f H^\circ = -393.5 \text{ kJ mol}^{-1}$$

$$\Delta_r S^\circ = \overline{S}^\circ[CO_2(g)] - \overline{S}^\circ[C(graphite)] - \overline{S}^\circ[O_2(g)]$$

$$= 213.6 \text{ J K}^{-1}\text{mol}^{-1} - 5.7 \text{ J K}^{-1}\text{mol}^{-1} - 205.0 \text{ J K}^{-1}\text{mol}^{-1}$$

$$= 2.9 \text{ J K}^{-1}\text{mol}^{-1}$$

Note that $\Delta_r S^\circ$ is not the same as $\overline{S}^\circ[CO_2(g)]$ as stated in the question. At 298 K,

$$\Delta_r H^\circ - T\Delta_r S^\circ = -393.5 \text{ kJ mol}^{-1} - (298 \text{ K})\left(2.9 \times 10^{-3} \text{ kJ K}^{-1} \text{mol}^{-1}\right)$$

$$= -394.4 \text{ kJ mol}^{-1}$$

which is the same as $\Delta_r G^\circ$.

5.16 A certain reaction is known to have a $\Delta_r G^\circ$ value of -122 kJ. Will the reaction necessarily occur if the reactants are mixed together?

No, for two reasons. First, the reaction rate can be very slow. Secondly, even if the rate is rapid enough for reaction to occur, the value of $\Delta_r G$ for the reaction depends on the concentrations of reactants and products and will be different from the value $\Delta_r G^\circ$ that is appropriate when all species are present in their standard pressure or concentration.

5.18 The pressure exerted on ice by a 60.0-kg skater is about 300 atm. Calculate the depression in freezing point. The molar volumes are $\overline{V}_L = 0.0180 \text{ L mol}^{-1}$ and $\overline{V}_S = 0.0196 \text{ L mol}^{-1}$.

The depression of freezing point can be obtained from the slope of the solid-liquid curve in the phase diagram.

$$\frac{dP}{dT} = \frac{\Delta\overline{H}}{T\Delta\overline{V}} = \frac{\left(6.01 \times 10^3 \text{ J mol}^{-1}\right)\left(\frac{1\text{ L atm}}{101.3\text{ J}}\right)}{(273.15 \text{ K})\left(0.0180 \text{ L mol}^{-1} - 0.0196 \text{ L mol}^{-1}\right)}$$

$$= -136 \text{ atm K}^{-1}$$

$$dP = -136 \text{ atm K}^{-1}dT$$

$$\int_{1\text{ atm}}^{300\text{ atm}} dP = -136 \text{ atm K}^{-1}\int_{273.15\text{ K}}^{T\text{ K}} dT$$

$$299 \text{ atm} = -136 \text{ atm K}^{-1}\Delta T$$

$$\Delta T = \frac{299 \text{ atm}}{-136 \text{ atm K}^{-1}} = -2.20 \text{ K} = -2.2 \text{ K}$$

Therefore, the freezing point is depressed by 2.2 K when 300 atm is applied to ice. In other words, the freezing point at this pressure is $273.15 \text{ K} - 2.20 \text{ K} = 271.0 \text{ K}$ or $-2.2°$C.

5.20 Use the phase diagram of water (Figure 5.5) to predict the dependence of the freezing and boiling points of water on pressure.

The freezing point decreases with increasing pressure (because the S–L curve has a negative slope) whereas the boiling point increases with increasing pressure (because the L–V curve has a positive slope).

5.22 Below is a rough sketch of the phase diagram of carbon. **(a)** How many triple points are there, and what are the phases that can coexist at each triple point? **(b)** Which has a higher density, graphite or diamond? **(c)** Synthetic diamond can be made from graphite. Using the phase diagram, how would you go about making diamond?

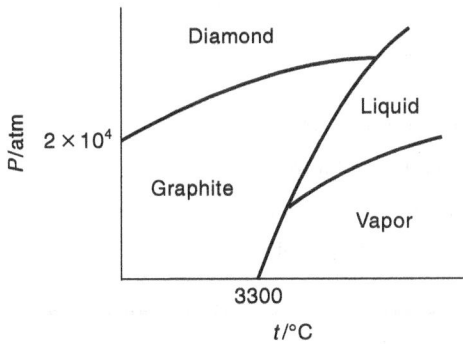

(a) There are two triple points. At one triple point, diamond, graphite and liquid carbon coexist. At the other, graphite, liquid carbon, and gaseous carbon coexist.

(b) According to the Clapeyron equation, $\frac{dP}{dT} = \frac{\Delta\overline{H}}{T\,\Delta\overline{V}}$. From the phase diagram, the graphite-diamond curve has positive slope for most of its length. From Problem 5.8, the standard molar enthalpy change for C(graphite) → C(diamond), $\Delta\overline{H}^{\circ}$, is also positive. This implies that $\Delta\overline{V}$ is positive, or that diamond has a greater molar volume than graphite. Consequently, one would conclude that diamond is less dense than graphite. In fact this conclusion is in error, and diamond is the denser phase. The problem is that by using $\Delta\overline{H}^{\circ}$, the assumption is made that ΔH is independent of pressure. Indeed, the evidence suggests that it changes sign and is negative at those pressures where graphite and diamond are in equilibrium.

(c) The phase diagram indicates that high pressures are required to convert graphite to diamond.

5.24 The plot in Figure 5.4 is no longer linear at high temperatures. Explain.

The deviation from linearity at high temperatures means that the slope is not constant over a large temperature range. This is because $\Delta_{vap}\overline{H}$, which is related to the slope, remains constant only over a limited temperature range.

5.26 The normal boiling point of ethanol is 78.3°C, and its molar enthalpy of vaporization is 39.3 kJ mol^{-1}. What is its vapor pressure at 30°C?

$$\ln\frac{P_2}{P_1} = -\frac{\Delta_{vap}\overline{H}}{R}\left(\frac{1}{T_2} - \frac{1}{T_1}\right)$$

$$\ln\frac{P_2}{1\,\text{atm}} = -\frac{39.3 \times 10^3\,\text{J mol}^{-1}}{8.314\,\text{J K}^{-1}\,\text{mol}^{-1}}\left(\frac{1}{303.15\,\text{K}} - \frac{1}{351.45\,\text{K}}\right) = -2.143$$

$$P_2 = (1\,\text{atm})\,e^{-2.143} = 0.117\,\text{atm} = 89.1\,\text{torr}$$

5.28 Give the conditions under which each of the following equations may be applied.

(a) $dA \leq 0$ (for equilibrium and spontaneity)

(b) $dG \leq 0$ (for equilibrium and spontaneity)

(c) $\ln\frac{P_2}{P_1} = \frac{\Delta\overline{H}}{R}\frac{(T_2-T_1)}{T_1T_2}$

(d) $\Delta G = nRT \ln \frac{P_2}{P_1}$.

(a) Constant volume and temperature

(b) Constant pressure and temperature

(c) $\Delta_{vap}\overline{H}$ independent of temperature, $\overline{V}_{vap} \gg \overline{V}_{liquid}$ (or $\overline{V}_{solid}$), ideal gas behavior

(d) Constant temperature, ideal gas behavior

5.30 Protein molecules are polypeptide chains made up of amino acids. In their physiologically functioning or native state, these chains fold in a unique manner such that the nonpolar groups of the amino acids are usually buried in the interior region of the proteins, where there is little or no contact with water. When a protein denatures, the chain unfolds so that these nonpolar groups are exposed to water. A useful estimate of the changes of the thermodynamic quantities as a result of denaturation is to consider the transfer of a hydrocarbon such as methane (a nonpolar substance) from an inert solvent (such as benzene or carbon tetrachloride) to the aqueous environment:

(a) CH_4(inert solvent) $\rightarrow CH_4(g)$

(b) $CH_4(g) \rightarrow CH_4(aq)$

If ΔH° and ΔG° are approximately 2.0 kJ mol^{-1} and -14.5 kJ mol^{-1}, respectively, for **(a)** and -13.5 kJ mol^{-1} and 26.5 kJ mol^{-1}, respectively, for **(b)**, then calculate the values of ΔH° and ΔG° for the transfer of 1 mole of CH_4 according to the equation

$$CH_4(\text{inert solvent}) \rightarrow CH_4(aq)$$

Comment on your results. Assume $T = 298$ K.

For the process CH_4(inert solvent) $\rightarrow CH_4(aq)$

$$\Delta H^\circ = \Delta H_a^\circ + \Delta H_b^\circ = 2.0 \text{ kJ mol}^{-1} - 13.5 \text{ kJ mol}^{-1} = -11.5 \text{ kJ mol}^{-1}$$

$$\Delta G^\circ = \Delta G_a^\circ + \Delta G_b^\circ = -14.5 \text{ kJ mol}^{-1} + 26.5 \text{ kJ mol}^{-1} = 12.0 \text{ kJ mol}^{-1}$$

Since $\Delta G^\circ > 0$, the process is not spontaneous when the reactant and product are in their standard states. Furthermore,

$$\Delta S^\circ = \frac{\Delta H^\circ - \Delta G^\circ}{T} = \frac{-11.5 \text{ kJ mol}^{-1} - 12.0 \text{ kJ mol}^{-1}}{298 \text{ K}} \left(\frac{1000 \text{ J}}{1 \text{ kJ}}\right) = -79 \text{ J K}^{-1}\text{mol}^{-1}$$

This negative value of ΔS° indicates that there is a increase in order when CH_4 is dissolved in H_2O. This is a result of order imposed on the solvent (water) molecules, due to the special arrangement of water molecules around each CH_4 molecule (see Section 17.6).

5.32 A rubber band is stretched vertically by tying a weight to one end and attaching the other end to a ring stand. When heated with a hot-air blower, the rubber band shrinks slightly in length. Account for this observation.

When the rubber band is heated, ΔH is positive. If a process occurs, ΔG must be negative.

$$\Delta G = \Delta H - T \Delta S < 0$$

$$\Delta S = \frac{\Delta H - \Delta G}{T} > 0$$

The ultimate result in the last expression follows because a positive quantity minus a negative one is necessarily positive. Thus, ΔS is positive for this process. As shown in Figure 5.9, the entropy of a rubber band increases when it contracts.

5.34 A sample of supercooled water freezes at $-10°C$. What are the signs of ΔH, ΔS, and ΔG for this process? All the changes refer to the system.

ΔG: $-$ because the process is spontaneous (constant temperature given and constant pressure assumed).

ΔH: $-$ because freezing is an exothermic process.

ΔS: $-$ because ice is more ordered than liquid water.

5.36 A chemist has synthesized a hydrocarbon compound (C_xH_y). Briefly describe what measurements are needed to determine the values of $\Delta_f \overline{H}^\circ$, $\overline{S}^\circ$, and $\Delta_f \overline{G}^\circ$ of the compound.

A determination of the enthalpy of combustion $[C_xH_y + (x + \frac{y}{4})O_2(g) \rightarrow xCO_2(g) + \frac{y}{2}H_2O(l)]$ using a calorimeter will enable a calculation of $\Delta_f \overline{H}^\circ$ from this measurement and the known $\Delta_f \overline{H}^\circ$'s for O_2, CO_2 and H_2O.

$\overline{S}^\circ$ may be found via a determination of the third-law entropy from 0 K to 298 K. This assumes no residual entropy at 0 K.

Once $\overline{S}^\circ$ is known for the compound, it is used together with the known values of $\overline{S}^\circ$ for C(graphite) and $H_2(g)$ to calculate $\Delta_r S^\circ$ for the reaction $xC(graphite) + \frac{y}{2}H_2(g) \rightarrow C_xH_y$. $\Delta_f \overline{G}^\circ$ for the compound is determined via $\Delta_f \overline{G}^\circ = \Delta_f \overline{H}^\circ - T\Delta_r S^\circ$.

5.38 A person heated water in a cup in a microwave oven for tea. After removing the cup from the oven, she added a tea bag to the hot water. To her surprise, the water started to boil violently. Explain what happened.

The water was superheated. Although thermodynamically unstable, superheated water will not boil if left undisturbed. Any mechanical disturbance, such as shaking, will cause it to boil. Adding the tea bag acts like adding boiling chips, which facilitates the boiling action.

5.40 The molar entropy of argon (Ar) is given by

$$\overline{S}^\circ = (36.4 + 20.8 \ln T/K) \text{ J K}^{-1} \text{mol}^{-1}$$

Calculate the change in Gibbs energy when 1.0 mole of Ar is heated at constant pressure from 20°C to 60°C. (*Hint*: Use the relation $\int \ln x \, dx = x \ln x - x$.)

ΔG° can be evaluated by using the relation

$$\left(\frac{\partial G}{\partial T}\right)_P = -S$$

At constant pressure,

$$dG = -SdT$$

$$\int_{G_1}^{G_2} dG = -\int_{T_1}^{T_2} SdT$$

$$\Delta G = -\int_{293.15\,\text{K}}^{333.15\,\text{K}} (1.0\ \text{mol}) \left[(36.4 + 20.8 \ln T/\text{K})\ \text{J}\,\text{K}^{-1}\text{mol}^{-1}\right] dT$$

$$= -[36.4T + 20.8\,(T \ln T - T)]_{293.15}^{333.15}\ \text{J}$$

$$= -\{[36.4\,(333.15) + 20.8\,(333.15 \ln 333.15 - 333.15)]$$

$$- [36.4\,(293.15) + 20.8\,(293.15 \ln 293.15 - 293.15)]\}\ \text{J}$$

$$= -6.24 \times 10^3\ \text{J} = -6.2\ \text{kJ}$$

CHAPTER 6

Nonelectrolyte Solutions

PROBLEMS AND SOLUTIONS

6.2 What is the molarity of a 2.12 mol kg^{-1} aqueous sulfuric acid solution? The density of this solution is 1.30 g cm^{-3}.

To find the molarity of the solution, the number of moles of solute and the volume of solution in a sample must be determined. Assume 1 kg of water is present in the solution.

$$\text{Number of moles of } H_2SO_4 = 2.12 \text{ mol}$$

$$\text{Mass of solution} = \text{mass of water} + \text{mass of } H_2SO_4$$

$$= 1000 \text{ g} + (2.12 \text{ mol } H_2SO_4) \left(\frac{98.09 \text{ g}}{1 \text{ mol } H_2SO_4} \right) = 1208.0 \text{ g}$$

$$\text{Volume of solution} = \frac{1208.0 \text{ g}}{(1.30 \text{ g cm}^{-3}) \left(\frac{1000 \text{ cm}^3}{1 \text{ L}} \right)} = 0.9292 \text{ L}$$

$$\text{Molarity of solution} = \frac{2.12 \text{ mol}}{0.9292 \text{ L}} = 2.28 \text{ } M$$

6.4 The concentrated sulfuric acid we use in the laboratory is 98.0% sulfuric acid by weight. Calculate the molality and molarity of concentrated sulfuric acid if the density of the solution is 1.83 g cm^{-3}.

Assume 100 g of solution is present. The solution contains 98.0 g H_2SO_4 and 2.0 g H_2O.

$$\text{Number of moles of } H_2SO_4 = (98.0 \text{ g}) \left(\frac{1 \text{ mol}}{98.09 \text{ g}} \right) = 0.9991 \text{ mol}$$

The molality of the solution is the ratio between the number of moles of solute and the mass of solvent:

$$\text{Molality of solution} = \frac{0.9991 \text{ mol}}{2.0 \times 10^{-3} \text{ kg}} = 5.0 \times 10^2 \text{ } m$$

The molarity of the solution is the ratio between the number of moles of solute and the volume of the solution.

$$\text{Volume of solution} = \frac{100 \text{ g}}{\left(1.83 \text{ g cm}^{-3}\right)\left(\frac{1000 \text{ cm}^3}{1 \text{ L}}\right)} = 5.464 \times 10^{-2} \text{ L}$$

$$\text{Molarity of solution} = \frac{0.9991 \text{ mol}}{5.464 \times 10^{-2} \text{ L}} = 18.3 \text{ M}$$

6.6 For dilute aqueous solutions in which the density of the solution is roughly equal to that of the pure solvent, the molarity of the solution is equal to its molality. Show that this statement is correct for a 0.010 M aqueous urea [$(NH_2)_2CO$] solution.

To convert molarity to molality, the volume of the solution has to be converted to the mass of solvent. Assume 1 L of solution is present.

$$\text{Number of moles of urea} = 0.010 \text{ mol}$$

$$\text{Mass of solvent} = \text{mass of solution} - \text{mass of solute}$$

$$= \left(1000 \text{ cm}^3\right)\left(1.00 \text{ g cm}^{-3}\right) - (0.010 \text{ mol urea})\left(\frac{60.06 \text{ g}}{1 \text{ mol urea}}\right)$$

$$= 999.4 \text{ g} = 0.9994 \text{ kg}$$

$$\text{Molality of solution} = \frac{0.010 \text{ mol}}{0.9994 \text{ kg}} = 0.010 \text{ m}$$

Therefore, for a dilute aqueous solution, such as 0.010 M urea, its molality is numerically the same as its molarity.

6.8 The strength of alcoholic beverages is usually described in terms of "proof," which is defined as twice the percentage by volume of ethanol. Calculate the number of grams of alcohol in 2 quarts of 75-proof gin. What is the molality of the gin? (The density of ethanol is 0.80 g cm^{-3} and 1 quart = 0.946 L.)

---

75 proof = 37.5% by volume of ethanol = 0.375. To find the molality of the gin, the number of moles of ethanol and the mass of water in the quantity of gin have to be calculated. In 2 quarts of gin,

$$\text{Volume of ethanol} = (0.375)\left(2 \text{ quarts}\right)\left(\frac{0.946 \text{ L}}{1 \text{ quart}}\right) = 0.7095 \text{ L} = 0.7095 \times 10^3 \text{ cm}^3$$

$$\text{Mass of ethanol} = \left(0.7095 \times 10^3 \text{ cm}^3\right)\left(0.80 \text{ g cm}^{-3}\right) = 568 \text{ g} = 5.7 \times 10^2 \text{ g}$$

$$\text{Number of moles of ethanol} = \left(568 \text{ g}\right)\left(\frac{1 \text{ mol}}{46.07 \text{ g}}\right) = 12.3 \text{ mol}$$

To evaluate the mass of water in 2 quarts of gin, two assumptions have to be made: (1) the volumes of ethanol and water are additive, and (2) the density of water is 1 g cm^{-3}. The second assumption is not bad, but the first is not particularly good. Indeed, there is a significant nonideality in water-ethanol solutions.

Volume of water in 2 quarts of gin = volume of gin − volume of ethanol

$$= (2 \text{ quarts}) \left(\frac{0.946 \text{ L}}{1 \text{ quart}} \right) - 0.7095 \text{ L} = 1.1825 \text{ L}$$

$$\text{Mass of water in 2 quarts of gin} = \left(1.1825 \times 10^3 \text{ cm}^3 \right) \left(1 \text{ g cm}^{-3} \right)$$

$$= 1.1825 \times 10^3 \text{ g} = 1.1825 \text{ kg}$$

Therefore,

$$\text{Molality of the gin} = \frac{12.3 \text{ mol}}{1.1825 \text{ kg}} = 10. \, m$$

6.10 Calculate the changes in entropy for the following processes: **(a)** mixing of 1 mole of nitrogen and 1 mole of oxygen, and **(b)** mixing of 2 moles of argon, 1 mole of helium, and 3 moles of hydrogen. Both **(a)** and **(b)** are carried out under conditions of constant temperature (298 K) and constant pressure. Assume ideal behavior.

(a)
$$\Delta_{\text{mix}} S = -nR \left(x_{\text{N}_2} \ln x_{\text{N}_2} + x_{\text{O}_2} \ln x_{\text{O}_2} \right)$$

$$= - (2 \text{ mol}) \left(8.314 \text{ J K}^{-1} \text{mol}^{-1} \right) \left(\frac{1}{2} \ln \frac{1}{2} + \frac{1}{2} \ln \frac{1}{2} \right) = 11.53 \text{ J K}^{-1}$$

(b)
$$\Delta_{\text{mix}} S = -nR \left(x_{\text{Ar}} \ln x_{\text{Ar}} + x_{\text{He}} \ln x_{\text{He}} + x_{\text{H}_2} \ln x_{\text{H}_2} \right)$$

$$= - (6 \text{ mol}) \left(8.314 \text{ J K}^{-1} \text{mol}^{-1} \right) \left(\frac{2}{6} \ln \frac{2}{6} + \frac{1}{6} \ln \frac{1}{6} + \frac{3}{6} \ln \frac{3}{6} \right) = 50.45 \text{ J K}^{-1}$$

6.12 Prove the statement that an alternative way to express Henry's law of gas solubility is to say that the volume of gas that dissolves in a fixed volume of solution is independent of pressure at a given temperature.

When Henry's law, $P_2 = K x_2$, is expressed in terms of the number of moles of gas dissolved in the solution, n_2, the result is

$$P_2 = K \frac{n_2}{n_T}$$

where n_T is the total number of moles.

The volume that would be occupied by this number of moles of gas were it in the gas phase at pressure P_2, is taken to satisfy the ideal gas law,

$$P_2 = \frac{n_2 RT}{V}$$

These two expressions for P_2 must agree,

$$K \frac{n_2}{n_T} = \frac{n_2 RT}{V}$$

$$V = \frac{n_T RT}{K}$$

Thus, V is independent of the pressure, although the amount of gas (number of moles) that occupies this volume is, of course, pressure dependent.

6.14   The Henry's law constant of oxygen in water at 25°C is 773 atm mol^{-1} kg of water. Calculate the molality of oxygen in water under a partial pressure of 0.20 atm. Assuming that the solubility of oxygen in blood at 37°C is roughly the same as that in water at 25°C, comment on the prospect for our survival without hemoglobin molecules. (The total volume of blood in the human body is about 5 L.)

$$P_2 = K'm$$

$$m = \frac{P_2}{K'} = \frac{0.20 \text{ atm}}{773 \text{ atm mol}^{-1} \text{ kg}} = 2.59 \times 10^{-4} \text{ mol kg}^{-1}$$

In a dilute aqueous solution, molality and molarity are numerically the same (see Problem 6.6). Therefore,

$$\text{Number of moles of O}_2 \text{ in blood} = \left(2.59 \times 10^{-4} M\right)(5 \text{ L}) = 1.3 \times 10^{-3} \text{ mol}$$

$$\text{Mass of O}_2 \text{ in blood} = \left(1.3 \times 10^{-3} \text{ mol}\right)\left(32.00 \text{ g mol}^{-1}\right) = 0.042 \text{ g}$$

This amount of O$_2$ is too small to sustain metabolic processes. This is the reason why we need hemoglobin molecules to transport O$_2$.

6.16   Which of the following has a higher chemical potential? If neither, answer "same." **(a)** H$_2$O(s) or H$_2$O(l) at water's normal melting point, **(b)** H$_2$O(s) at −5°C and 1 bar or H$_2$O(l) at −5°C and 1 bar, **(c)** benzene at 25°C and 1 bar or benzene in a 0.1 M toluene solution in benzene at 25°C and 1 bar.

The less stable species of a pair has a higher chemical potential. When both substances in a pair are at equilibrium, then they have the same chemical potential.

(a) Same

(b) H$_2$O(l) at −5°C and 1 bar

(c) Benzene at 25°C and 1 bar. This is because $x_{\text{benzene}} < 1$ in the toluene solution and according to the following relation:

$$\mu_{\text{benzene}}(l) = \mu^*_{\text{benzene}}(l) + RT \ln x_{\text{benzene}}$$

Therefore,

$$\mu_{\text{benzene}}(l) < \mu^*_{\text{benzene}}(l)$$

6.18   Derive the Gibbs phase rule (Equation 5.23) in terms of chemical potentials.

The total number of independent intensive variables for a system containing c components and p phases is found as in Appendix 5.2 to be $p(c-1) + 2$. At equilibrium, the chemical potential of each component is the same in all phases. Taking the phases to be α, β, γ, etc., this means that for a component present in two phases,

$$\mu_\alpha = \mu_\beta$$

which gives one equation. For a component present in three phases,

$$\mu_\alpha = \mu_\beta$$
$$\mu_\beta = \mu_\gamma$$

or two equations. Extending this process is seen to result in $(p - 1)$ independent equations relating the chemical potentials for the p phases present. Since there are c components, this gives a total of $c(p - 1)$ known variables restricting the total number of degrees of freedom, which is found via the difference

$$f = \left[p(c - 1) + 2 \right] - c(p - 1)$$
$$= c - p + 2$$

6.20　A solution is made up by dissolving 73 g of glucose ($C_6H_{12}O_6$; molar mass 180.2 g) in 966 g of water. Calculate the activity coefficient of glucose in this solution if the solution freezes at $-0.66°C$.

If the solution is ideal, the freezing point depression ΔT is $K_f m_2$. However, if it is nonideal, then the molality of the glucose solution should be replaced by the activity, a, of glucose. The activity coefficient of glucose can be calculated by comparing the activity with molality.

The freezing point depression of the solution is $0°C - (-0.66°C) = 0.66°C = 0.66$ K. The activity of the solution is

$$a = \frac{0.66 \text{ K}}{K_f} = \frac{0.66 \text{ K}}{1.86 \text{ K mol}^{-1} \text{kg}} = 0.355 \text{ mol kg}^{-1}$$

The molality of the solution is

$$m = \frac{(73 \text{ g}) \left(\frac{1 \text{ mol}}{180.16 \text{ g}} \right)}{966 \times 10^{-3} \text{ kg}} = 0.419 \text{ mol kg}^{-1}$$

The activity coefficient is

$$\gamma = \frac{a}{m} = \frac{0.355 \text{ mol kg}^{-1}}{0.419 \text{ mol kg}^{-1}} = 0.85$$

6.22　At 45°C, the vapor pressure of water is 65.76 mmHg for a glucose solution in which the mole fraction of glucose is 0.080. Calculate the activity and activity coefficient of the water in the solution. The vapor pressure of pure water at 45°C is 71.88 mmHg.

The activity is calculated by comparing the vapor pressure of the solution and that of pure water. This together with the mole fraction of water gives the activity coefficient.

$$a_{H_2O} = \frac{P_{H_2O}}{P^*_{H_2O}} = \frac{65.76 \text{ mmHg}}{71.88 \text{ mmHg}} = 0.9149$$

$$\gamma_{H_2O} = \frac{a_{H_2O}}{x_{H_2O}} = \frac{a_{H_2O}}{1 - x_{glucose}} = \frac{0.9149}{1 - 0.080} = 0.994$$

6.24   List the important assumptions in the derivation of Equation 6.39.

The derivation makes use of several assumptions. The solute is assumed to be nonvolatile. The solution is assumed to be dilute (so that $\Delta_{vap}\overline{H}$ is unchanged from the pure solvent value). The solution is assumed to be dilute-ideal (so that Equation 6.34 is valid). The solute is assumed to not be an electrolyte, and the boiling point elevation is assumed to be small. Finally, the solute is assumed to not dimerize or polymerize.

6.26   A mixture of ethanol and *n*-propanol behaves ideally at 36.4°C. **(a)** Determine graphically the mole fraction of *n*-propanol in a mixture of ethanol and *n*-propanol that boils at 36.4°C and 72 mmHg. **(b)** What is the total vapor pressure over the mixture at 36.4°C when the mole fraction of *n*-propanol is 0.60? **(c)** Calculate the composition of the vapor in **(b)**. (The equilibrium vapor pressures of ethanol and *n*-propanol at 36.4°C are 108 mmHg and 40.0 mmHg, respectively.)

(a) The vapor pressure of ethanol, $P_{ethanol}$, the vapor pressure of propanol, $P_{propanol}$ and the total vapor pressure, P_{total} can each be expressed as a function of the mole fraction of propanol, $x_{propanol}$:

$$P_{ethanol} = x_{ethanol}P^*_{ethanol} = x_{ethanol}\left(108 \text{ mmHg}\right) = \left(1 - x_{propanol}\right)\left(108 \text{ mmHg}\right)$$

$$P_{propanol} = x_{propanol}P^*_{propanol} = x_{propanol}\left(40.0 \text{ mmHg}\right)$$

$$P_{total} = P_{ethanol} + P_{propanol}$$

These relations are plotted below.

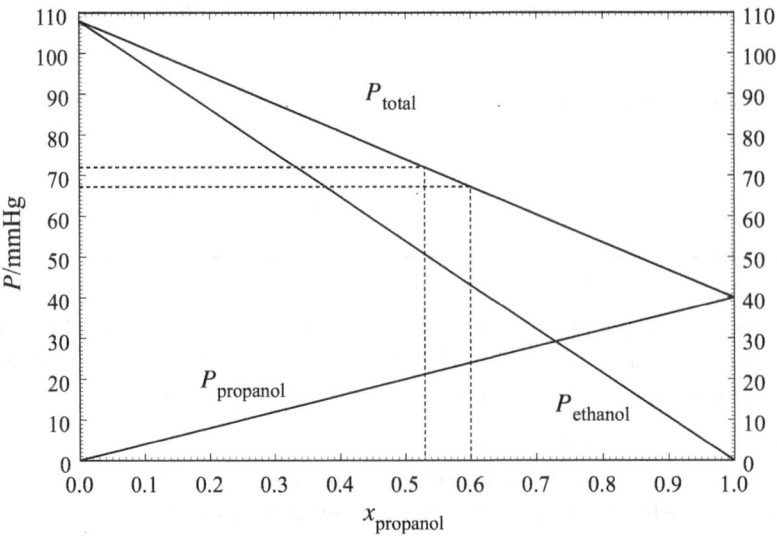

The vapor pressure of the mixture is the same as the external pressure when it boils. According to the graph, when the total vapor pressure is 72 mmHg, the mole fraction of propanol is 0.53 and the mole fraction of ethanol is $1 - 0.53 = 0.47$.

(b) According to the graph, when the mole fraction of propanol is 0.60, the total vapor pressure is 67.2 mmHg.

(c) The vapor pressure of each component can be written in terms of (1) its mole fraction in the vapor and the total pressure or (2) its mole fraction in the solution and the vapor pressure of the pure component.

$$P_{ethanol} = x^v_{ethanol} P_{total} = x_{ethanol} P^*_{ethanol}$$

$$P_{propanol} = x^v_{propanol} P_{total} = x_{propanol} P^*_{propanol}$$

Take the ratio between $P_{ethanol}$ and $P_{propanol}$,

$$\frac{P_{ethanol}}{P_{propanol}} = \frac{x^v_{ethanol} P_{total}}{x^v_{propanol} P_{total}} = \frac{x_{ethanol} P^*_{ethanol}}{x_{propanol} P^*_{propanol}}$$

$$\frac{x^v_{ethanol}}{x^v_{propanol}} = \frac{(0.40)\,(108\text{ mmHg})}{(0.60)\,(40.0\text{ mmHg})} = 1.80$$

$$x^v_{ethanol} = 1.80 x^v_{propanol} = 1.80\left(1 - x^v_{ethanol}\right)$$

$$2.80 x^v_{ethanol} = 1.80$$

$$x^v_{ethanol} = 0.64$$

$$x^v_{propanol} = 1 - 0.64 = 0.36$$

6.28 At 298 K, the vapor pressure of pure water is 23.76 mmHg and that of seawater is 22.98 mmHg. Assume that seawater contains only NaCl, estimate its concentration. (*Hint*: Sodium chloride is a strong electrolyte.)

The mole fraction of the solute is related to the vapor pressure lowering of the solvent.

$$\Delta P = x_2 P^*_1$$

$$x_2 = \frac{\Delta P}{P^*_1} = \frac{23.76\text{ mmHg} - 22.98\text{ mmHg}}{23.76\text{ mmHg}} = 3.28 \times 10^{-2}$$

Since NaCl is a strong electrolyte, it dissociates virtually completely into ions in solutions. Thus, x_2 calculated above represents the mole fraction of Na^+ and Cl^-. The mole fraction of NaCl, x_{NaCl}, is half of the calculated amount, or 1.64×10^{-2}, and the mole fraction of water is $1 - 1.64 \times 10^{-2} = 0.983586$. In other words, for every mole of molecules in the solution there are 1.64×10^{-2} mole of NaCl and 0.9836 mole of water. Assuming the density of water is 1 $kg\,L^{-1}$, the volume of water is

$$V = \frac{(0.9836\text{ mol})\left(\frac{18.02 \times 10^{-3}\text{ kg}}{1\text{ mol}}\right)}{1\text{ kg L}^{-1}} = 0.01772\text{ L}$$

The concentration of NaCl is

$$M = \frac{1.64 \times 10^{-2}\text{ mol}}{0.01772\text{ L}} = 0.93\text{ mol L}^{-1}$$

6.30 Explain why jams and honey can each be stored under atmospheric conditions for long periods of time without spoilage.

The high concentration of sugar in a jam results in a hypertonic solution. Any bacteria landing in the jam will lose its intracellular water through osmosis to the more concentrated solution (the jam). The dessicated bacteria are no longer viable.

6.32 The freezing-point-depression measurement of benzoic acid in acetone yields a molar mass of 122 g; the same measurement in benzene gives a value of 244 g. Account for this discrepancy. (*Hint*: Consider solvent–solute and solute–solute interactions.)

In benzene, a nonpolar solvent, benzoic acid exists in a dimeric form while in acetone it exists as a monomer.

6.34 For intravenous injections, great care is taken to ensure that the concentration of solutions to be injected is comparable to that of blood plasma. Why?

If the injected solution is either hypertonic or hypotonic, the resulting osmotic pressure would cause water to be transferred across cell (such as red blood cell) membranes leading to either hemolysis or crenation.

6.36 A mixture of liquids A and B exhibits ideal behavior. At 84°C, the total vapor pressure of a solution containing 1.2 moles of A and 2.3 moles of B is 331 mmHg. Upon the addition of another mole of B to the solution, the vapor pressure increases to 347 mmHg. Calculate the vapor pressures of pure A and B at 84°C.

The total vapor pressure depends on the vapor pressures of A and B in a mixture, which in turn depends on the vapor pressures of pure A and B. With the total vapor pressure of the two mixtures known, a pair of simultaneous equations can be written in terms of the vapor pressures of pure A and B.

For the solution containing 1.2 moles of A and 2.3 moles of B,

$$x_A = \frac{1.2}{1.2 + 2.3} = 0.343$$

$$x_B = 1 - 0.343 = 0.657$$

$$P_{\text{total}} = P_A + P_B = x_A P_A^* + x_B P_B^*$$

$$331\,\text{mmHg} = 0.343 P_A^* + 0.657 P_B^*$$

Solve the last equation for P_A^*:

$$P_A^* = \frac{331\,\text{mmHg} - 0.657 P_B^*}{0.343} = 965\,\text{mmHg} - 1.92 P_B^* \tag{6.36.1}$$

Now consider the solution with the additional mole of B.

$$x_A = \frac{1.2}{1.2 + 2.3 + 1.0} = 0.267$$

$$x_B = 1 - 0.267 = 0.733$$

$$P_{total} = P_A + P_B = x_A P_A^* + x_B P_B^*$$

$$347 \text{ mmHg} = 0.267 P_A^* + 0.733 P_B^* \qquad (6.36.2)$$

Substitute Eq. 6.36.1 into Eq. 6.36.2:

$$347 \text{ mmHg} = 0.267 \left(965 \text{ mmHg} - 1.92 P_B^*\right) + 0.733 P_B^*$$

$$0.220 P_B^* = 89.3 \text{ mmHg}$$

$$P_B^* = 406 \text{ mmHg} = 4.1 \times 10^2 \text{ mmHg}$$

Substitute the value of P_B^* into Eq. 6.36.1:

$$P_A^* = 965 \text{ mmHg} - 1.92 \left(406 \text{ mmHg}\right) = 1.9 \times 10^2 \text{ mmHg}$$

6.38   Liquids A (molar mass 100 g mol^{-1}) and B (molar mass 110 g mol^{-1}) form an ideal solution. At 55°C, A has a vapor pressure of 95 mmHg and B a vapor pressure of 42 mmHg. A solution is prepared by mixing equal weights of A and B. **(a)** Calculate the mole fraction of each component in the solution. **(b)** Calculate the partial pressures of A and B over the solution at 55°C. **(c)** Suppose that some of the vapor described in **(b)** is condensed to a liquid. Calculate the mole fraction of each component in this liquid and the vapor pressure of each component above this liquid at 55°C.

(a) Let m g be the mass of A and therefore the mass of B.

$$n_A = \text{number of moles of A} = \frac{m \text{ g}}{100 \text{ g mol}^{-1}}$$

$$n_B = \text{number of moles of B} = \frac{m \text{ g}}{110 \text{ g mol}^{-1}}$$

The mole fractions are

$$x_A = \frac{n_A}{n_A + n_B} = \frac{\frac{m \text{ g}}{100 \text{ g mol}^{-1}}}{\frac{m \text{ g}}{100 \text{ g mol}^{-1}} + \frac{m \text{ g}}{110 \text{ g mol}^{-1}}} = \frac{\frac{1}{100}}{\frac{1}{100} + \frac{1}{110}} = 0.5238 = 0.524$$

$$x_B = 1 - x_A = 1 - 0.5238 = 0.4762 = 0.476$$

(b) $$P_A = x_A P_A^* = (0.5238)\left(95 \text{ mmHg}\right) = 49.8 \text{ mmHg} = 50 \text{ mmHg}$$

$$P_B = x_B P_B^* = (0.4762)\left(42 \text{ mmHg}\right) = 20.0 \text{ mmHg} = 20 \text{ mmHg}$$

(c) The composition of the liquid is the same as the composition of the vapor from which it is condensed. Using Dalton's law of partial pressure,

$$x_A = x_A^v = \frac{P_A}{P_A + P_B} = \frac{49.8\ \text{mmHg}}{49.8\ \text{mmHg} + 20.0\ \text{mmHg}} = 0.713 = 0.71$$

$$x_B = x_B^v = 1 - x_A^v = 1 - 0.713 = 0.287 = 0.29$$

Note that the vapor, and therefore, the condensed liquid, contains a higher mole fraction of A, the more volatile component, than the original mixture.

The vapor pressure of each component above this liquid is

$$P_A = x_A P_A^* = (0.713)\ (95\ \text{mmHg}) = 68\ \text{mmHg}$$

$$P_B = x_B P_B^* = (0.287)\ (42\ \text{mmHg}) = 12\ \text{mmHg}$$

6.40 The following argument is frequently used to explain the fact that the vapor pressure of the solvent is lower over a solution than over the pure solvent and that lowering is proportional to the concentration. A dynamic equilibrium exists in both cases, so that the rate at which molecules of solvent evaporate from the liquid is always equal to that at which they condense. The rate of condensation is proportional to the partial pressure of the vapor, whereas that of evaporation is unimpaired in the pure solvent but is impaired by solute molecules in the surface of the solution. Hence the rate of escape is reduced in proportion to the concentration of the solute, and maintenance of equilibrium requires a corresponding lowering of the rate of condensation and therefore of the partial pressure of the vapor phase. Explain why this argument is incorrect. [*Source*: K. J. Mysels, *J. Chem. Educ.* **32**, 179 (1955).]

Two reasons suggest themselves from general principles.

1. If the explanation given were correct, then the nature of the solute molecules would have a significant effect on the vapor pressure lowering. Namely, larger molecules would be more efficient at blocking the surface and would reduce the vapor pressure to a greater extent than would smaller molecules at the same concentration. On the contrary, Raoult's law shows the same vapor pressure lowering as a function of concentration regardless of molecular identity.

2. The solute molecules would impede the incorporation of solvent molecules back into solution to the same extent as they would the escape into the vapor phase. This is a requirement of the principle of microscopic reversibility to be discussed in Chapter 15.

The following reasons rely on more specific observations.

3. Surfactant molecules are known to accumulate at the surface of a solution for dilute concentrations, but do not affect the vapor pressure lowering. (At higher concentrations, micelles are formed, which is clearly a non-ideal effect. The resulting change in the chemical potential causes significant deviations in the vapor pressure curve from ideal behavior. It should be noted, however, that this is *not* a surface effect.)

4. Certain insoluble materials, which are observed to have no effect on the equilibrium vapor pressure of the solution, are known to form tightly packed monolayers on the surface of the solution (cetyl alcohol in water is an example) that greatly inhibit the rate of evaporation. This argues in support of the second reason given above that the rate of condensation must likewise be slowed.

6.42 Two aqueous urea solutions have osmotic pressures of 2.4 atm and 4.6 atm, respectively, at a certain temperature. What is the osmotic pressure of a solution prepared by mixing equal volumes of these two solutions at the same temperature?

First calculate the number of moles of urea in a given volume, V L, of each solution. The solutions with osmotic pressures of 2.4 atm and 4.6 atm are denoted Solutions 1 and 2, respectively.

$$\Pi_1 = M_1 RT = \frac{n_1}{V} RT$$

$$n_1 = \frac{\Pi_1 V}{RT}$$

Similarly,

$$n_2 = \frac{\Pi_2 V}{RT}$$

If V L of Solution 1 is mixed with V L of Solution 2, then the total number of moles of urea is

$$n_{\text{mix}} = n_1 + n_2 = \frac{(\Pi_1 + \Pi_2)\, V}{RT}$$

and the molarity of the solution is

$$M_{\text{mix}} = \frac{\frac{(\Pi_1 + \Pi_2)V}{RT}}{2V} = \frac{\Pi_1 + \Pi_2}{2RT}$$

The osmotic pressure of the mixture is

$$\Pi_{\text{mix}} = M_{\text{mix}} RT = \frac{\Pi_1 + \Pi_2}{2RT} RT = \frac{\Pi_1 + \Pi_2}{2} = \frac{2.4\ \text{atm} + 4.6\ \text{atm}}{2} = 3.5\ \text{atm}$$

This simple result comes about because the osmotic pressure is proportional to the molarity of the solute. When equal volumes of the two solutions are mixed, the molarity will just be the mean of the molarities of the two solutions (assuming additive volume). Since the osmotic pressure is proportional to the molarity, the osmotic pressure of the solution will be the average of the osmotic pressures of the two solutions.

6.44 "Time-release" drugs have the advantage of releasing the drug to the body at a constant rate so that the drug concentration at any time is not high enough to have harmful side effects or so low as to be ineffective. A schematic diagram of a pill that works on this basis is shown below. Explain how it works.

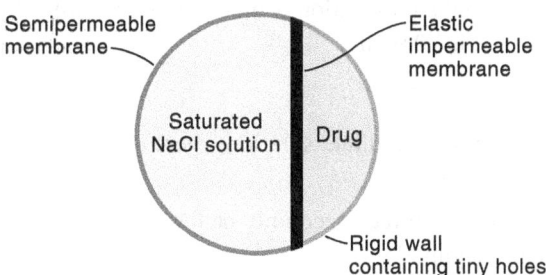

When swallowed, the pill will be in a hypotonic solution and the resulting osmotic pressure will cause water to move across the semipermeable membrane into the pill. This increases the volume of the salt solution and pushes the elastic membrane to the right, which causes the drug to exit through the small holes at a constant rate.

6.46 Acetic acid is a polar molecule that can form hydrogen bonds with water molecules. Therefore, it has a high solubility in water. Yet acetic acid is also soluble in benzene (C_6H_6), a nonpolar solvent that lacks the ability to form hydrogen bonds. A solution of 3.8 g of CH_3COOH in 80 g C_6H_6 has a freezing point of 3.5°C. Calculate the molar mass of the solute, and suggest what its structure might be. (*Hint*: Acetic acid molecules can form hydrogen bonds among themselves.)

The freezing point of C_6H_6 is 5.5°C (Problem 6.43). Thus, the acetic acid depresses the freezing point of benzene by

$$\Delta T = 5.5°C - 3.5°C = 2.0°C = 2.0 \text{ K}$$

The molality of acetic acid is

$$m_2 = \frac{\Delta T}{K_f} = \frac{2.0 \text{ K}}{5.12 \text{ K mol}^{-1} \text{ kg}} = 0.391 \text{ mol kg}^{-1}$$

The number of moles of acetic acid is

$$n_2 = \left(0.391 \text{ mol kg}^{-1}\right)\left(80 \times 10^{-3} \text{ kg}\right) = 0.0313 \text{ mol}$$

which gives a molar mass of

$$\mathcal{M} = \frac{3.8 \text{ g}}{0.0313 \text{ mol}} = 1.2 \times 10^2 \text{ g mol}^{-1}$$

The calculated molar mass is twice that of the molar mass of acetic acid. This suggests that acetic acid, when placed in benzene, exists as a dimer held together by hydrogen bonds.

6.48 Comment on whether each of the following statements is true or false, and briefly explain your answer: **(a)** If one component of a solution obeys Raoult's law, then the other component must also obey the same law. **(b)** Intermolecular forces are small in ideal solutions. **(c)** When 15.0 mL of an aqueous 3.0 *M* ethanol solution is mixed with 55.0 mL of an aqueous 3.0 *M* ethanol solution, the total volume is 70.0 mL.

(a) False, consider the case of a dilute solution where the solvent obeys Raoult's law and the solute obeys Henry's law.

(b) False, the intermolecular forces need only be the same between like and unlike molecules regardless of their magnitude.

(c) False, this is not an ideal solution and the final volume will in general be either smaller or larger than the sum of the two initial volumes. (For this specific case, the final volume is less than 70.0 mL.)

6.50 Nonideal solutions are the result of unequal intermolecular forces between components. Based on this knowledge, comment on whether a racemic mixture of a liquid compound would behave as an ideal solution.

The solution will not behave as an ideal solution. The two enantiomers interact differently with each other than they do with themselves. That is, although the interactions between the members of a $(+)(+)$ pair are the same as those in a $(-)(-)$ pair, these will be different from the interactions in a $(+)(-)$ pair.

6.52 Explain the following phenomena. **(a)** A cucumber placed in concentrated brine (saltwater) shrivels into a pickle. **(b)** A carrot placed in fresh water swells in volume.

(a) The cucumber is in a hypertonic solution, it loses water through osmosis to the concentrated brine.

(b) The carrot is in a hypotonic solution. There is a higher concentration of salts, etc. in the cells in the carrot, and water enters into the cells.

6.54 **(a)** Which of the following expressions is incorrect as a representation of the partial molar volume of component A in a two-component solution? Why? How would you correct it?

$$\left(\frac{\partial V_m}{\partial n_A}\right)_{T,P,n_B} \qquad \left(\frac{\partial V_m}{\partial x_A}\right)_{T,P,x_B}$$

(b) Given that the molar volume of this mixture (V_m) of A and B is given by

$$V_m = \left[0.34 + 3.6 x_A x_B + 0.4 x_B \left(1 - x_A\right)\right] \text{ L mol}^{-1}$$

derive an expression for the partial molar volume for A and calculate its value at $x_A = 0.20$.

(a) The second expression is incorrect, since it is impossible to vary x_A while holding x_B constant. The correct expression is $\left(\dfrac{\partial V_m}{\partial x_A}\right)_{T,P}$.

(b) Write V_m in terms of x_A.

$$V_m = \left[0.34 + 3.6 x_A \left(1 - x_A\right) + 0.4 \left(1 - x_A\right) \left(1 - x_A\right)\right] \text{ L mol}^{-1}$$

$$= \left[0.34 + 3.6 x_A - 3.6 x_A^2 + 0.4 - 0.8 x_A + 0.4 x_A^2\right] \text{ L mol}^{-1}$$

$$= \left(0.38 + 2.8 x_A - 3.2 x_A^2\right) \text{ L mol}^{-1}$$

Differentiate V_m with respect to x_A.

$$\left(\frac{\partial V_m}{\partial x_A}\right)_{P,T} = \left(2.8 - 6.4x_A\right)\ \text{L mol}^{-1}$$

At $x_A = 0.20$,

$$\left(\frac{\partial V_m}{\partial x_A}\right)_{P,T} = [2.8 - 6.4\,(0.20)]\ \text{L mol}^{-1} = 1.5\ \text{L mol}^{-1}$$

6.56   The osmotic pressure of poly(methyl methacrylate) in toluene has been measured at a series of concentrations at 298 K. Determine graphically the molar mass of the polymer.

Π/atm	8.40×10^{-4}	1.72×10^{-3}	2.52×10^{-3}	3.23×10^{-3}	7.75×10^{-3}
c/g L^{-1}	8.10	12.31	15.00	18.17	28.05

In the limit of a dilute solution, where all virial coefficients except the second may be ignored, the osmotic pressure Π is given by

$$\frac{\Pi}{c} = \frac{RT}{\mathcal{M}}\left(1 + Bc\right)$$

Thus, a graph of Π/c vs. c will extrapolate to a y intercept of $\frac{RT}{\mathcal{M}}$, where $\mathcal{M}$ is the molar mass of the solute.

The data for the graph is given in the table below.

(Π/c)/atm L g^{-1}	1.037×10^{-4}	1.397×10^{-4}	1.680×10^{-4}	1.778×10^{-4}	2.763×10^{-4}
c/g L^{-1}	8.10	12.31	15.00	18.17	28.05

As seen in the graph, the extrapolation to zero concentration gives an intercept of 3.4×10^{-5} atm L g^{-1}.

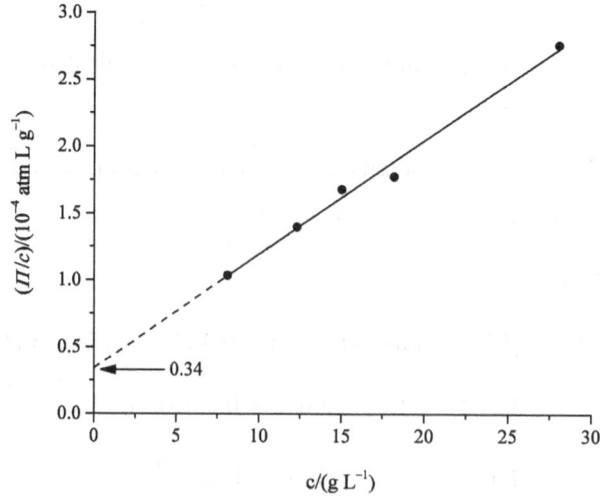

Thus,

$$3.4 \times 10^{-5} \text{ atm L g}^{-1} = \frac{RT}{\mathcal{M}}$$

$$= \frac{\left(0.08206 \text{ L atm K}^{-1} \text{ mol}^{-1}\right)(298 \text{ K})}{\mathcal{M}}$$

$$\mathcal{M} = 7.2 \times 10^5 \text{ g mol}^{-1}$$

6.58 Suppose 2.6 moles of He at 0.80 atm and 25°C are mixed with 4.1 moles of Ne at 2.7 atm and 25°C. Calculate the Gibbs energy change for the process. Assume ideal behavior.

The expression $\Delta_{\text{mix}}G = nRT\left(x_1 \ln x_1 + x_2 \ln x_2\right)$ is valid only for mixing two gases originally at the same pressure and is not directly applicable here. G is a state function, however, so consider a path in which the two gases are first individually brought to the final pressure of the mixture and then mixed. The overall ΔG is the sum of the ΔG's for the steps. The process is assumed to take place with the gases originally in two containers connected by a valve which is opened. The final volume of the mixture is then the sum of the original volumes. Since $V_{\text{He}} = \frac{n_{\text{He}}RT}{P_{\text{He}}}$, with an similar expression for Ne,

$$P_{\text{f}} = \frac{n_{\text{total}}RT}{V_{\text{f}}}$$

$$= \frac{\left(n_{\text{He}} + n_{\text{Ne}}\right)RT}{\frac{n_{\text{He}}RT}{P_{\text{He}}} + \frac{n_{\text{Ne}}RT}{P_{\text{Ne}}}}$$

$$= \frac{n_{\text{He}} + n_{\text{Ne}}}{\frac{n_{\text{He}}}{P_{\text{He}}} + \frac{n_{\text{Ne}}}{P_{\text{Ne}}}}$$

$$= \frac{2.6 \text{ mol} + 4.1 \text{ mol}}{\frac{2.6 \text{ mol}}{0.80 \text{ atm}} + \frac{4.1 \text{ mol}}{2.7 \text{ atm}}}$$

$$= 1.41 \text{ atm}$$

For the isothermal compression (expansion) of an ideal gas, $\Delta G = nRT \ln \frac{P_{\text{f}}}{P_{\text{i}}}$, so that

$$\Delta G_{\text{He}} = (2.6 \text{ mol})\left(8.314 \text{ J K}^{-1} \text{ mol}^{-1}\right)(298.15 \text{ K}) \ln \frac{1.41 \text{ atm}}{0.80 \text{ atm}} = 3.65 \times 10^3 \text{ J}$$

$$\Delta G_{\text{Ne}} = (4.1 \text{ mol})\left(8.314 \text{ J K}^{-1} \text{ mol}^{-1}\right)(298.15 \text{ K}) \ln \frac{1.41 \text{ atm}}{2.7 \text{ atm}} = -6.60 \times 10^3 \text{ J}$$

$$\Delta_{\text{mix}}G = n_{\text{total}}RT\left(x_{\text{He}} \ln x_{\text{He}} + x_{\text{Ne}} \ln x_{\text{Ne}}\right)$$

$$= (2.6 \text{ mol} + 4.1 \text{ mol})\left(8.314 \text{ J K}^{-1} \text{ mol}^{-1}\right)(298.15 \text{ K})$$

$$\times \left[\frac{2.6 \text{ mol}}{2.6 \text{ mol} + 4.1 \text{ mol}} \ln \left(\frac{2.6 \text{ mol}}{2.6 \text{ mol} + 4.1 \text{ mol}}\right)\right.$$

$$\left. + \frac{4.1 \text{ mol}}{2.6 \text{ mol} + 4.1 \text{ mol}} \ln \left(\frac{4.1 \text{ mol}}{2.6 \text{ mol} + 4.1 \text{ mol}}\right)\right]$$

$$= -1.11 \times 10^4 \text{ J}$$

$$\Delta G_{\text{total}} = \Delta G_{\text{He}} + \Delta G_{\text{Ne}} + \Delta_{\text{mix}} G$$

$$= 3.65 \times 10^3 \text{ J} - 6.60 \times 10^3 \text{ J} - 1.11 \times 10^4 \text{ J}$$

$$= -1.4 \times 10^4 \text{ J} = -14 \text{ kJ}$$

6.60 As an after dinner party trick, the host brings out a glass of water with an ice cube floating on top and a thread. Then he asks the guests to remove the ice cube with the thread, but they are not allowed to tie a loop around the ice cube. Describe how the guests might accomplish this task.

The guests should place a portion of the thread on top of the ice cube floating in the glass and then put some salt (presumably still in the shaker on the table) over the thread and the ice cube. The salt will depress the freezing point sufficiently that some of the ice will melt, wetting the thread and allowing it to "sink" into the melted water on the surface of the ice cube. As more water melts, the concentration of the salt water decreases, and the water refreezes around the thread. Now, just pick up the loose end of the thread and remove the ice cube, frozen to the other end.

6.62 **(a)** Derive the equation relating the molality (m) of a solution to its molarity (M)

$$m = \frac{M}{d - \frac{M\mathcal{M}}{1000}}$$

where d is the density of the solution (g/mL) and $\mathcal{M}$ is the molar mass of the solute (g/mol). (*Hint*: Start by expressing the solvent in kilograms in terms of the difference between the mass of the solution and the mass of the solute.) **(b)** Show that, for dilute aqueous solutions, m is approximately equal to M.

(a) Let V be the volume of the solution in L. Note that the numerical values of d are the same for the two sets of units: $g \, mL^{-1}$ and $kg \, L^{-1}$.

$$m = \frac{\text{number of moles of solute}}{\text{mass of solvent in kg}}$$

$$= \frac{\text{number of moles of solute}}{\text{mass of solution in kg} - \text{mass of solute in kg}}$$

$$= \frac{\text{number of moles of solute}}{d(\text{in kg L}^{-1}) \, V(\text{in L}) - (\text{number of moles of solute}) \left[\mathcal{M}(\text{in g mol}^{-1})\right] \left(\frac{1 \text{ kg}}{1000 \text{ g}}\right)}$$

$$= \frac{\frac{\text{number of moles of solute}}{V(\text{in L})}}{d(\text{in kg L}^{-1}) - \left(\frac{\text{number of moles of solute}}{V(\text{in L})}\right) \left[\mathcal{M}(\text{in g mol}^{-1})\right] \left(\frac{1 \text{ kg}}{1000 \text{ g}}\right)}$$

$$= \frac{M}{d - \frac{M\mathcal{M}}{1000}}$$

(b) Since the density of water is $1.0 \, g \, mL^{-1}$ over a large range of temperatures, 1 kg of water has a volume of 1 liter. For a dilute aqueous solution, the volumes of the solvent and the solution are practically the same. Thus,

$$\text{molality} = \frac{\text{number of moles of solute}}{\text{mass of solvent in kg}} \approx \frac{\text{number of moles of solute}}{\text{volume of solution in L}} = \text{Molarity}$$

This conclusion can also be reached using the expression derived in part **(a)**. In a dilute solution, the density of the solution is approximately $1 \text{ g mL}^{-1} = 1 \text{ kg L}^{-1}$, and because the concentration (M) is small, then $\frac{M\mathcal{M}}{1000} << d$. Thus,

$$m = \frac{M}{d - \frac{M\mathcal{M}}{1000}} \approx \frac{M}{d} \approx \frac{M}{1 \text{ kg L}^{-1}}$$

The density allows for the unit conversion from molarity (moles per liter) to molality (moles per kilogram), but the two quantities have the same numerical value.

6.64   The mole fractions of dry air are approximately 21% O_2 and 79% N_2. Calculate the masses of these two gases dissolved in 1000 g of water at 25°C and 1 atm.

First calculate the mole fractions of the gases in water using the Henry's law constants from Table 6.1:

$$x_{O_2} = \frac{x_{O_2}^v P}{K_{O_2}} = \frac{P_{O_2}}{K} = \frac{(0.21)(760 \text{ torr})}{3.27 \times 10^7 \text{ torr}} = 4.88 \times 10^{-6}$$

$$x_{N_2} = \frac{x_{N_2}^v P}{K_{N_2}} = \frac{P_{N_2}}{K} = \frac{(0.79)(760 \text{ torr})}{6.80 \times 10^7 \text{ torr}} = 8.83 \times 10^{-6}$$

Since these mole fractions are very small, the amounts of O_2 and N_2 in water are negligible compared to the amount of water. The number of moles of water is $1000g/18.02 \text{ g mol}^{-1} = 55.49$ mol. Thus,

$$n_{O_2} = x_{O_2}\left(n_{H_2O} + n_{O_2} + n_{N_2}\right) \approx x_{O_2} n_{H_2O}$$

$$= \left(4.88 \times 10^{-6}\right)(55.49 \text{ mol}) = 2.71 \times 10^{-4} \text{ mol}$$

$$m_{O_2} = \left(2.71 \times 10^{-4} \text{ mol}\right)\left(32.00 \text{ g mol}^{-1}\right) = 8.7 \times 10^{-3} \text{ g}$$

$$n_{N_2} = x_{N_2}\left(n_{H_2O} + n_{O_2} + n_{N_2}\right) \approx x_{N_2} n_{H_2O}$$

$$= \left(8.83 \times 10^{-6}\right)(55.49 \text{ mol}) = 4.90 \times 10^{-4} \text{ mol}$$

$$m_{N_2} = \left(4.90 \times 10^{-4} \text{ mol}\right)\left(28.02 \text{ g mol}^{-1}\right) = 1.4 \times 10^{-2} \text{ g}$$

CHAPTER 7

Electrolyte Solutions

PROBLEMS AND SOLUTIONS

7.2 Using the cell described in Problem 7.1, a student determined the resistance of a 0.086 M KCl solution to be 20.4 Ω. Calculate the molar conductance of this solution.

Rearranging the relation between resistance and molar conductance

$$\frac{1}{R} = \Lambda c \frac{A}{l}$$

obtained in Problem 7.1, the molar conductance is

$$\Lambda = \frac{1}{Rc}\frac{l}{A}$$

$$= \frac{1}{(20.4\ \Omega)\ (0.086\ \text{mol L}^{-1})}\left(0.263\ \text{cm}^{-1}\right)\left(\frac{1000\ \text{cm}^3}{1\ \text{L}}\right)$$

$$= 1.5 \times 10^2\ \Omega^{-1}\,\text{mol}^{-1}\,\text{cm}^2$$

7.4 Given that the measurement of Λ_0 for weak electrolytes is generally difficult, how would you deduce the value of Λ_0 for CH_3COOH from the data listed in Table 7.1? (*Hint:* Consider CH_3COONa, HCl, and NaCl.)

At infinite dilution, any electrolyte is completely dissociated. Therefore, Λ_0 for CH_3COOH can be obtained from Λ_0 for CH_3COO^- and Λ_0 for H^+, which in turn can be derived by combining Λ_0's for CH_3COONa and HCl and subtracting Λ_0 for NaCl.

$$\Lambda_0(CH_3COOH) = \Lambda_0(CH_3COONa) + \Lambda_0(HCl) - \Lambda_0(NaCl)$$

$$= (91.00 + 426.16 - 126.45)\ \Omega^{-1}\,\text{mol}^{-1}\,\text{cm}^2$$

$$= 390.71\ \Omega^{-1}\,\text{mol}^{-1}\,\text{cm}^2$$

7.6 A conductance cell consists of two electrodes, each with an area of 4.2×10^{-4} m^2, separated by 0.020 m. The resistance of the cell when filled with a 6.3×10^{-4} M KNO$_3$ solution is 26.7 Ω. What is the molar conductance of the solution?

Rearranging the relation between resistance and molar conductance

$$\frac{1}{R} = \Lambda c \frac{A}{l}$$

obtained in Problem 7.1, the molar conductance of the solution is

$$\Lambda = \frac{1}{Rc}\frac{l}{A} = \left[\frac{1}{(26.7\ \Omega)\ (6.3 \times 10^{-4}\ \text{mol L}^{-1})}\right]\left(\frac{0.020\ \text{m}}{4.2 \times 10^{-4}\ \text{m}^2}\right)\left(\frac{1\ \text{m}^3}{1000\ \text{L}}\right)$$

$$= 2.8\ \Omega^{-1}\ \text{mol}^{-1}\ \text{m}^2$$

7.8 Calculate the solubility of $BaSO_4$ (in $g\ L^{-1}$) in **(a)** water and **(b)** a $6.5 \times 10^{-5}\ M\ MgSO_4$ solution. The solubility product of $BaSO_4$ is 1.1×10^{-10}. Assume ideal behavior.

The equation for the dissolution of $BaSO_4$ is

$$BaSO_4(s) \rightleftharpoons Ba^{2+}(aq) + SO_4^{2-}(aq)$$

(a) If the molar solubility of $BaSO_4$ is x, then there are x M of Ba^{2+} and x M SO_4^{2-} in the solution.

$$K_{sp} = 1.1 \times 10^{-10} = [Ba^{2+}]\,[SO_4^{2-}] = x \cdot x$$

$$x = 1.05 \times 10^{-5}$$

Therefore,

The solubility of $BaSO_4 = \left(1.05 \times 10^{-5}\ \text{mol L}^{-1}\right)\left(\frac{233.4\ \text{g}}{1\ \text{mol}}\right) = 2.5 \times 10^{-3}\ \text{g L}^{-1}$

(b) The $MgSO_4$ solution contains $6.5 \times 10^{-5}\ M$ of SO_4^{2-}. If the molar solubility of $BaSO_4$ is x, then there are x M of Ba^{2+} and $(x + 6.5 \times 10^{-5})\ M\ SO_4^{2-}$ in the solution.

$$K_{sp} = 1.1 \times 10^{-10} = [Ba^{2+}]\,[SO_4^{2-}] = x\left(x + 6.5 \times 10^{-5}\right)$$

$$x^2 + 6.5 \times 10^{-5}x - 1.1 \times 10^{-10} = 0$$

Using the quadratic equation,

$$x = 1.65 \times 10^{-6} \quad \text{or} \quad x = -6.67 \times 10^{-5}\ \text{(nonphysical)}$$

Therefore,

The solubility of $BaSO_4 = \left(1.65 \times 10^{-6}\ \text{mol L}^{-1}\right)\left(\frac{233.4\ \text{g}}{1\ \text{mol}}\right) = 3.9 \times 10^{-4}\ \text{g L}^{-1}$

7.10 Referring to Problem 7.9, calculate the value of ΔG° for the process

$$AgCl(s) \rightarrow Ag^+(aq) + Cl^-(aq)$$

to yield a saturated solution at 298 K. (*Hint*: Use the well-known equation $\Delta G^\circ = -RT \ln K$.)

The value of K for this reaction is the thermodynamic solubility product, K^o_{sp}, for AgCl from Problem 7.9.

$$\Delta G^o = -RT \ln K = -\left(8.314\,\text{J}\,\text{K}^{-1}\,\text{mol}^{-1}\right)(298\,\text{K})\ln 1.6 \times 10^{-10} = 5.6 \times 10^4\,\text{J}\,\text{mol}^{-1}$$

7.12 Oxalic acid, $(COOH)_2$, is a poisonous compound present in many plants and vegetables, including spinach. Calcium oxalate is only slightly soluble in water ($K_{sp} = 3.0 \times 10^{-9}$ at 25°C) and its ingestion can result in kidney stones. Calculate **(a)** the apparent and thermodynamic solubility of calcium oxalate in water, and **(b)** the concentrations of calcium and oxalate ions in a 0.010 M Ca(NO$_3$)$_2$ solution. Assume ideal behavior in **(b)**.

(a) The reaction corresponding to the dissolution of calcium oxalate, CaOx, in water is

$$CaOx(s) \rightleftharpoons Ca^{2+}(aq) + Ox^{2-}(aq)$$

The apparent solubility is readily calculated from the given (apparent) solubility product and the stoichiometry of the dissociation. Let the solubility of CaOx be S M, then $[Ca^{2+}] = [Ox^{2-}] = S$.

$$K_{sp} = [Ca^{2+}][Ox^{2-}] = S^2 = 3.0 \times 10^{-9}$$

$$S = 5.48 \times 10^{-5}\,M = 5.5 \times 10^{-5}\,M$$

The thermodynamic solubility is determined from the thermodynamic solubility product, $K^o_{sp} = \gamma^2_{\pm} K_{sp}$, which is calculated from the apparent concentrations and the mean ionic activity obtained from the Debye–Hückel limiting law.

Since this is a very dilute solution, the molarities of the ionic species are numerically equal to the molalities required in determining the ionic strength.

$$I = \frac{1}{2}\sum_i m_i z_i^2$$

$$= \frac{1}{2}\left[\left(5.48 \times 10^{-5}\,m\right)(2)^2 + \left(5.48 \times 10^{-5}\,m\right)(-2)^2\right]$$

$$= 2.19 \times 10^{-4}\,m$$

This is then used to determine the mean ionic activity

$$\log \gamma_{\pm} = -0.509 \left|z_+ z_-\right| \sqrt{I}$$

$$= -0.509 \left|(2)(-2)\right| \sqrt{\left(2.19 \times 10^{-4}\right)}$$

$$= -3.01 \times 10^{-2}$$

$$\gamma_{\pm} = 0.933$$

Thus, the thermodynamic solubility constant and thermodynamic solubility, S^o, are

$$K^o_{sp} = \gamma^2_{\pm} K_{sp} = (0.933)^2 \left(3.0 \times 10^{-9} \right)$$

$$= 2.61 \times 10^{-9} = 2.6 \times 10^{-9} = \left(S^o \right)^2$$

$$S^o = \sqrt{2.61 \times 10^{-9}} = 5.1 \times 10^{-5}\ M$$

(b) The solubility in the $Ca(NO_3)_2$ solution is decreased due to the presence of Ca^{2+} ions. If the concentration of dissolved calcium oxalate is taken as S, then in the solution $[Ca^{2+}] = (0.010 + S)$ and $[Ox^{2-}] = S$.

$$K_{sp} = [Ca^{2+}][Ox^{2-}] = 3.0 \times 10^{-9}$$

$$(0.010 + S)\ S = 3.0 \times 10^{-9}$$

$$S = 3.0 \times 10^{-7}\ M = [Ox^{2-}]$$

$$[Ca^{2+}] = 0.010\ M + 3.0 \times 10^{-7}\ M = 0.010\ M$$

7.14 Calculate the ionic strength and the mean activity coefficient for the following solutions at 298 K: **(a)** 0.10 m NaCl, **(b)** 0.010 m MgCl$_2$, and **(c)** 0.10 m K$_4$Fe(CN)$_6$.

The ionic strength can be obtained from the equation

$$I = \frac{1}{2} \sum_i m_i z_i^2$$

and subsequently the mean activity from the Debye–Hückel limiting law

$$\log \gamma_{\pm} = -0.509\ |z_+ z_-|\ \sqrt{I}$$

(a) 0.10 m NaCl: $z_+ = 1$, $z_- = -1$, $m_+ = 0.10\ m$, $m_- = 0.10\ m$

$$I = \frac{1}{2} \left[(0.10\ m)\ (1)^2 + (0.10\ m)\ (-1)^2 \right] = 0.10\ m$$

$$\log \gamma_{\pm} = -0.509\ |(1)\ (-1)|\ \sqrt{0.10} = -0.161$$

$$\gamma_{\pm} = 0.69$$

(b) 0.010 m MgCl$_2$: $z_+ = 2$, $z_- = -1$, $m_+ = 0.010\ m$, $m_- = 0.020\ m$

$$I = \frac{1}{2} \left[(0.010\ m)\ (2)^2 + (0.020\ m)\ (-1)^2 \right] = 0.030\ m$$

$$\log \gamma_{\pm} = -0.509\ |(2)\ (-1)|\ \sqrt{0.030} = -0.176$$

$$\gamma_{\pm} = 0.67$$

(c) 0.10 m K$_4$Fe(CN)$_6$: $z_+ = 1$, $z_- = -4$, $m_+ = 0.40\ m$, $m_- = 0.10\ m$

$$I = \frac{1}{2}\left[(0.40\,m)\,(1)^2 + (0.10\,m)\,(-4)^2\right] = 1.0\,m$$

$$\log \gamma_\pm = -0.509\,|(1)\,(-4)|\,\sqrt{1.0} = -2.04$$

$$\gamma_\pm = 9.1 \times 10^{-3}$$

7.16 A $0.20\,m$ $Mg(NO_3)_2$ solution has a mean ionic activity coefficient of 0.13 at 25°C. Calculate the mean molality, the mean ionic activity, and the activity of the compound.

For the $Mg(NO_3)_2$ solution, $\nu_+ = 1$, $\nu_- = 2$, $\nu = 3$, $m_+ = 0.20\,m$, $m_- = 0.40\,m$. The mean molality is

$$m_\pm = \left[m_+^{\nu_+} m_-^{\nu_-}\right]^{1/\nu} = \left[(0.20\,m)\,(0.40\,m)^2\right]^{1/3} = 0.317\,m = 0.32\,m$$

The mean ionic activity is

$$a_\pm = \gamma_\pm m_\pm = (0.13)\,(0.317) = 0.0412 = 0.041$$

The activity is

$$a = a_\pm^\nu = (0.0412)^3 = 7.0 \times 10^{-5}$$

7.18 The size of the ionic atmosphere, called the Debye radius, is $1/\kappa$, where κ is given by

$$\kappa = \left(\frac{e^2 N_A}{\varepsilon_0 \varepsilon k_B T}\right)^{1/2} \sqrt{I}$$

where e is the electronic charge, N_A Avogadro's constant, ε_0 the permittivity of vacuum ($8.854 \times 10^{-12}\,C^2\,N^{-1}\,m^{-2}$), ε the dielectric constant of the solvent, k_B the Boltzmann constant, T the absolute temperature, and I the ionic strength (see the physical chemistry texts listed in Chapter 1). Calculate the Debye radius in a $0.010\,m$ aqueous Na_2SO_4 solution at 25°C.

The ionic strength of a $0.010\,m$ Na_2SO_4 solution is

$$I = \frac{1}{2}\sum_i m_i z_i^2$$

$$= \frac{1}{2}\left[(0.020\,m)\,(1)^2 + (0.010\,m)\,(-2)^2\right]$$

$$= 0.030\,m$$

Since this is a dilute solution, the ionic strength in $mol\,L^{-1}$ can be taken as numerically equal to that in $mol\,kg^{-1}$, or $0.030\,M$.

For water, $\varepsilon = 78.54$. Thus,

$$\kappa = \left(\frac{e^2 N_A}{\varepsilon_0 \varepsilon k_B T}\right)^{1/2} \sqrt{I}$$

$$= \left[\frac{\left(1.602 \times 10^{-19}\,\text{C}\right)^2 \left(6.022 \times 10^{23}\,\text{mol}^{-1}\right)}{\left(8.854 \times 10^{-12}\,\text{C}^2\,\text{N}^{-1}\,\text{m}^{-2}\right)(78.54)\left(1.381 \times 10^{-23}\,\text{J K}^{-1}\right)(298.15\,\text{K})}\right]^{1/2}$$

$$\times \sqrt{\left(0.030\,\text{mol L}^{-1}\right)\left(\frac{1000\,\text{L}}{1\,\text{m}^3}\right)}$$

$$= 4.02 \times 10^8\,\text{m}^{-1}$$

$$\frac{1}{\kappa} = 2.49 \times 10^{-9}\,\text{m} = 25\,\text{Å}$$

7.20 The freezing-point depression of a 0.010 m acetic acid solution is 0.0193 K. Calculate the degree of dissociation for acetic acid at this concentration.

The degree of dissociation is related to the van't Hoff factor, i, which can be obtained from the freezing point depression.

$$\Delta T = K_f \left(i m_2\right)$$

$$i = \frac{\Delta T}{K_f m_2} = \frac{0.0193\,\text{K}}{\left(1.86\,\text{K mol}^{-1}\,\text{kg}\right)\left(0.010\,\text{mol kg}^{-1}\right)} = 1.04$$

The degree of dissociation of the acetic acid is

$$\alpha = \frac{i-1}{\nu-1} = \frac{1.04-1}{2-1} = 0.04 = 4\%$$

7.22 The osmotic pressure of blood plasma is about 7.5 atm at 37°C. Estimate the total concentration of dissolved species and the freezing point of blood plasma.

The total concentration of dissolved species is

$$c = \frac{\Pi}{RT} = \frac{7.5\,\text{atm}}{\left(0.08206\,\text{L atm K}^{-1}\,\text{mol}^{-1}\right)(310.15\,\text{K})} = 0.295\,M = 0.30\,M$$

Assuming that molality can be approximated by molarity, the freezing point depression caused by the dissolved species is

$$\Delta T = K_f m_2 = \left(1.86\,\text{K mol}^{-1}\,\text{kg}\right)\left(0.295\,\text{mol kg}^{-1}\right) = 0.55\,\text{K}$$

Therefore, blood plasma freezes at −0.55°C or 272.60 K.

7.24 Referring to Figure 7.13, calculate the osmotic pressure for the following cases at 298 K: **(a)** The left compartment contains 200 g of hemoglobin in 1 liter of solution; the right compartment contains pure water. **(b)** The left compartment contains the same hemoglobin solution as in part **(a)**, and the right compartment initially contains 6.0 g of NaCl in 1 liter of solution. Assume that the pH of the solution is such that the hemoglobin molecules are in the Na^+Hb^- form. (The molar mass of hemoglobin is 65,000 g mol^{-1}.)

(a) To maintain electrical neutrality, all the Na^+ ions remain in the left compartment with the Hb^- anionic form of the protein. The total concentration is twice the hemoglobin concentration.

$$c = 2 \left(\frac{\frac{200 \text{ g}}{65000 \text{ g mol}^{-1}}}{1 \text{ L}} \right) = 2 \left(3.077 \times 10^{-3} M \right) = 6.154 \times 10^{-3} M$$

From the osmotic pressure equation,

$$\Pi = cRT$$

$$= \left(6.154 \times 10^{-3} M \right) \left(0.08206 \text{ L atm K}^{-1} \text{mol}^{-1} \right) (298 \text{ K})$$

$$= 0.150 \text{ atm}$$

(b) Na^+ and Cl^- will diffuse through the membrane from right to left, maintaining electrical neutrality, until the chemical potentials of NaCl on both sides of the membrane are equal. According to Equation 7.35, the concentration of the NaCl, x, that diffuses from right to left depends on the concentration of NaCl initially in the right compartment, b, and the concentration of the nondiffusible ion, c.

$$b = \frac{\frac{6.0 \text{ g}}{58.44 \text{ g mol}^{-1}}}{1 \text{ L}} = 0.103 M$$

$$c = 3.077 \times 10^{-3} M \text{ (from above)}$$

$$x = \frac{b^2}{c + 2b} = \frac{(0.103 \, M)^2}{3.077 \times 10^{-3} M + 2 \, (0.103 \, M)} = 0.0507 \, M$$

The osmotic pressure is determined by the difference between the number of particles in the left compartment and that in the right compartment.

$$c = \left([Hb^-] + [Na^+] + [Cl^-] \right)_L - \left([Na^+] + [Cl^-] \right)_R$$

$$= \{ \left(3.077 \times 10^{-3} M + 3.077 \times 10^{-3} M + 0.0507 M + 0.0507 M \right)$$

$$- [(0.103 \, M - 0.0507 \, M) + (0.103 \, M - 0.0507 \, M)] \}$$

$$= 2.95 \times 10^{-3} M$$

Using the osmotic pressure equation,

$$\Pi = cRT$$

$$= \left(2.95 \times 10^{-3} M \right) \left(0.08206 \text{ L atm K}^{-1} \text{mol}^{-1} \right) (298 \text{ K})$$

$$= 0.072 \text{ atm}$$

7.26   From the data in Table 7.2, determine Λ_0 for H_2O. Given that the specific conductance (κ) for water is $5.7 \times 10^{-8} \, \Omega^{-1} \text{cm}^{-1}$, calculate the ion product (K_w) of water at 298 K.

For water,

$$\Lambda_0 = \lambda_0^{H^+} + \lambda_0^{OH^-}$$

$$= 349.81\ \Omega^{-1}\,mol^{-1}\,cm^2 + 198.3\ \Omega^{-1}\,mol^{-1}\,cm^2$$

$$= 548.11\ \Omega^{-1}\,mol^{-1}\,cm^2 = 548.1\ \Omega^{-1}\,mol^{-1}\,cm^2$$

The concentration of H^+ is the same as that of OH^-, which is

$$c = \frac{\kappa}{\Lambda_0} = \frac{5.7 \times 10^{-8}\ \Omega^{-1}\,cm^{-1}}{548.11\ \Omega^{-1}\,mol^{-1}\,cm^2}$$

$$= 1.04 \times 10^{-10}\ mol\,cm^{-3} = 1.04 \times 10^{-7}\ mol\,L^{-1}$$

Therefore,

$$K_w = \left(1.04 \times 10^{-7}\right)\left(1.04 \times 10^{-7}\right) = 1.1 \times 10^{-14}$$

7.28 **(a)** The root cells of plants contain a solution that is hypertonic in relation to water in the soil. Thus water can move into the roots by osmosis. Explain why salts (NaCl and $CaCl_2$) spread on roads to melt ice can be harmful to nearby trees. **(b)** Just before urine leaves the human body, the collecting ducts in the kidney (which contain the urine) pass through a fluid whose salt concentration is considerably greater than is found in the blood and tissues. Explain how this action helps to conserve water in the body.

(a) When the road salts get into the soil water, the concentration there becomes greater than the concentration in the plant root cells, making the solution in the cells hypotonic. Thus the osmotic pressure difference will be reversed, and water will flow out of the plant roots into the soil. This action is harmful and potentially fatal to the plant. Even if the effect is not as severe, there will be a reduction in the osmotic pressure, which will limit the height to which the water can rise in the plant.

(b) The high-salt fluid is hypertonic relative to urine. Thus, some of the water in the urine flows into the fluid by osmosis. This action concentrates the waste products in the urine and helps to conserve water in the body.

7.30 **(a)** Using the Debye–Hückel limiting law, calculate the value of $\gamma_\pm$ for a $2.0 \times 10^{-3}\ m$ Na_3PO_4 solution at 25°C. **(b)** Calculate the values of γ_+ and γ_- for the Na_3PO_4 solution, and show that they give the same value for $\gamma_\pm$ as that obtained in **(a)**.

For the Na_3PO_4 solution, $z_+ = 1$, $z_- = -3$, $m_+ = 6.0 \times 10^{-3}\ m$, and $m_- = 2.0 \times 10^{-3}\ m$.

The ionic strength of the solution is

$$I = \frac{1}{2}\sum_i m_i z_i^2 = \frac{1}{2}\left[\left(6.0 \times 10^{-3}\ m\right)(1)^2 + \left(2.0 \times 10^{-3}\ m\right)(-3)^2\right] = 1.2 \times 10^{-2}\ m$$

(a)
$$\log \gamma_\pm = -0.509\,|z_+ z_-|\,\sqrt{I} = -0.509\,|(1)\,(-3)|\,\sqrt{1.2 \times 10^{-2}} = -0.167$$

$$\gamma_\pm = 0.681 = 0.68$$

(b) The activity coefficients of Na^+ and PO_4^{3-} can be evaluated using

$$\log \gamma_i = -0.509 z_i^2 \sqrt{I}$$

For Na^+,

$$\log \gamma_+ = -0.509 \,(1)^2 \sqrt{1.2 \times 10^{-2}} = -5.58 \times 10^{-2}$$

$$\gamma_+ = 0.879 = 0.88$$

For PO_4^{3-},

$$\log \gamma_- = -0.509 \,(-3)^2 \sqrt{1.2 \times 10^{-2}} = -0.502$$

$$\gamma_- = 0.315 = 0.32$$

The mean ionic coefficient is

$$\gamma_\pm = \left(\gamma_+^{\nu_+} \gamma_-^{\nu_-}\right)^{1/\nu} = \left[(0.879)^3 \,(0.315)\right]^{1/4} = 0.680$$

which is the same as that obtained in **(a)**.

CHAPTER 8

Chemical Equilibrium

PROBLEMS AND SOLUTIONS

8.2 At 1024°C, the pressure of oxygen gas from the decomposition of copper(II) oxide (CuO) is 0.49 bar:

$$4CuO(s) \rightleftharpoons 2Cu_2O(s) + O_2(g)$$

(a) What is the value of K_P for the reaction? **(b)** Calculate the fraction of CuO that will decompose if 0.16 mole of it is placed in a 2.0-L flask at 1024°C. **(c)** What would the fraction be if a 1.0-mole sample of CuO were used? **(d)** What is the smallest amount of CuO (in moles) that would establish the equilibrium?

(a) At 1024°C,

$$K_P = \frac{P_{O_2}}{P^o} = \frac{0.49 \text{ bar}}{1 \text{ bar}} = 0.49$$

(b) First calculate the number of moles of O_2 formed by the reaction, from which the number of moles of CuO decomposed is determined. Assume O_2 behaves ideally.

$$\text{Number of moles of } O_2 \text{ formed} = \frac{PV}{RT} = \frac{(0.49 \text{ bar}) \left(\frac{1 \text{ atm}}{1.013 \text{ bar}}\right) (2.0 \text{ L})}{(0.08206 \text{ L atm K}^{-1} \text{ mol}^{-1}) (1297.15 \text{ K})}$$

$$= 9.09 \times 10^{-3} \text{ mol}$$

$$\text{Number of moles of CuO decomposed} = \left(9.09 \times 10^{-3} \text{ mol O}_2\right) \left(\frac{4 \text{ mol CuO}}{1 \text{ mol O}_2}\right)$$

$$= 0.0364 \text{ mol}$$

Therefore,

$$\text{Fraction of CuO decomposed} = \frac{0.0364 \text{ mol}}{0.16 \text{ mol}} = 0.23$$

(c) If 1.0 mole sample of CuO were used, the pressure of O_2 would still be the same (0.49 bar), and it would be due to the same quantity of O_2. (A pure solid does not affect the equilibrium position, as long as it is in excess at equilibrium.) The number of moles of CuO lost would still be 0.0364 mol. Therefore,

$$\text{Fraction of CuO decomposed} = \frac{0.0364 \text{ mol}}{1.0 \text{ mol}} = 0.036$$

(d) If the number of moles of CuO were less than 0.036 mole, the equilibrium could not be established because the pressure of O_2 would be less than 0.49 bar. Therefore, the smallest number of moles of CuO needed to establish equilibrium would be slightly greater than 0.036 mole.

8.4 About 75% of the hydrogen produced for industrial use is produced by the *steam-reforming* process. This process is carried out in two stages called primary and secondary reforming. In the primary stage, a mixture of steam and methane at about 30 atm is heated over a nickel catalyst at 800°C to give hydrogen and carbon monoxide:

$$CH_4(g) + H_2O(g) \rightleftharpoons CO(g) + 3H_2(g) \quad \Delta_r H^\circ = 206 \text{ kJ mol}^{-1}$$

The secondary stage is carried out at about 1000°C, in the presence of air, to convert the remaining methane to hydrogen:

$$CH_4(g) + \frac{1}{2}O_2(g) \rightleftharpoons CO(g) + 2H_2(g) \quad \Delta_r H^\circ = 35.7 \text{ kJ mol}^{-1}$$

(a) What conditions of temperature and pressure would favor the formation of products in both the primary and secondary stage? **(b)** The equilibrium constant, K_c, for the primary stage is 18 at 800°C. **(i)** Calculate the value of K_P for the reaction. **(ii)** If the partial pressures of methane and steam were both 15 atm at the start, what would the pressures of all the gases be at equilibrium?

(a) Since both reactions are endothermic ($\Delta_r H^\circ$ is positive for each), according to LeChatelier's principle the products would be favored at high temperatures. Indeed, the steam-reforming process is carried out at very high temperatures (between 800°C and 1000°C). It is interesting to note that in a plant that uses natural gas (methane) for both hydrogen generation and heating, about one-third of the gas is burned to maintain the high temperatures.

In each reaction there are more moles of products than reactants; therefore, products are favored at low pressures. In reality, the reactions are carried out at high pressures. The reason is that when the hydrogen gas produced is used captively (usually in the synthesis of ammonia), high pressure leads to higher yields of ammonia.

(b) (i) According to Problem 8.1,

$$K_P = K_c (RT)^{\Delta n} \left(P^\circ\right)^{-\Delta n}$$

In this equation, if P° is in bar, R has to be in L bar K^{-1} mol^{-1}:

$$R = \left(0.08206 \text{ L atm K}^{-1} \text{ mol}^{-1}\right)\left(\frac{1.013 \text{ bar}}{1 \text{ atm}}\right) = 0.083127 \text{ L bar K}^{-1} \text{ mol}^{-1}$$

and $\Delta n = (1 + 3) - (1 + 1) = 2$. Therefore,

$$K_P = (18) [(0.083127) (1073.15)]^2 (1)^{-2} = 1.43 \times 10^5 = 1.4 \times 10^5$$

(ii) The pressures need to be converted to bars in order to use the K_P expression where P° = 1 bar. The partial pressures of the reactants are

$$(15 \text{ atm}) \left(\frac{1.013 \text{ bar}}{1 \text{ atm}} \right) = 15.2 \text{ bar}$$

Let $x = P_{CO}$ at equilibrium. The initial and equilibrium pressures of all species are shown in the following.

	CH_4	+	H_2O	$\rightleftharpoons$	CO	+	$3H_2$	
Initial	15.2		15.2					bar
At equilibrium	$15.2 - x$		$15.2 - x$		x		$3x$	bar

$$K_P = \frac{\left(P_{CO}/P^\circ \right) \left(P_{H_2}/P^\circ \right)^3}{\left(P_{CH_4}/P^\circ \right) \left(P_{H_2O}/P^\circ \right)}$$

$$1.43 \times 10^5 = \frac{x \, (3x)^3}{(15.2 - x)^2} = \frac{27x^4}{(15.2 - x)^2}$$

Take the square root of K_P and the last term in the above expression.

$$378 = \frac{5.20x^2}{15.2 - x}$$

$$5.20x^2 + 378x - 5746 = 0$$

Using the quadration equation,

$$x = 12.9 \quad \text{or} \quad -85.6 \text{ (nonphysical)}$$

Therefore, at equilibrium,

$$P_{CH_4} = (15.2 - x) \text{ bar} = 2 \text{ bar} = 2 \text{ atm}$$

$$P_{H_2O} = (15.2 - x) \text{ bar} = 2 \text{ bar} = 2 \text{ atm}$$

$$P_{CO} = x \text{ bar} = 13 \text{ bar} = 13 \text{ atm}$$

$$P_{H_2} = 3x \text{ bar} = 39 \text{ bar} = 38 \text{ atm}$$

8.6 The vapor pressure of mercury is 0.002 mmHg at 26°C. **(a)** Calculate the vaue of K_c and K_P for the process $Hg(l) \rightleftharpoons Hg(g)$. **(b)** A chemist breaks a thermometer and spills mercury onto the floor of a laboratory measuring 6.1 m long, 5.3 m wide, and 3.1 m high. Calculate the mass of mercury (in grams) vaporized at equilibrium and the concentration of mercury vapor in mg m^{-3}. Does this concentration exceed the safety limit of 0.05 mg m^{-3}? (Ignore the volume of furniture and other objects in the laboratory.)

(a)

$$K_P = \frac{P_{Hg}}{P^\circ} = \frac{(0.002 \text{ mmHg}) \left(\frac{1 \text{ atm}}{760 \text{ mmHg}} \right) \left(\frac{1.013 \text{ bar}}{1 \text{ atm}} \right)}{1 \text{ bar}} = 2.67 \times 10^{-6}$$

From Problem 8.1,

$$K_c = \frac{K_P}{(RT)^{\Delta n} (P^\circ)^{-\Delta n}}$$

To use this expression, R has to be expressed in L bar K^{-1} mol^{-1}, which has been done in Problem 8.4. Δn is 1.

$$K_c = \frac{2.67 \times 10^{-6}}{[(0.083127)(299.15)]^1 (1)^{-1}} = 1.07 \times 10^{-7}$$

(b) The volume occupied by Hg vapor is the same as the volume of the room.

$$V = (6.1 \text{ m})(5.3 \text{ m})(3.1 \text{ m}) = \left(100 \text{ m}^3\right)\left(\frac{1000 \text{ L}}{1 \text{ m}^3}\right) = 1.00 \times 10^5 \text{ L}$$

At equilibrium, the equilibrium vapor pressure is 0.002 mmHg. The number of moles of Hg vapor is calculated by using the ideal gas law.

$$n = \frac{PV}{RT} = \frac{(0.002 \text{ mmHg})\left(\frac{1 \text{ atm}}{760 \text{ mmHg}}\right)(1.00 \times 10^5 \text{ L})}{(0.08206 \text{ L atm K}^{-1} \text{ mol}^{-1})(299.15 \text{ K})} = 1.07 \times 10^{-2} \text{ mol}$$

Therefore, the mass of Hg vapor is

$$m = \left(1.07 \times 10^{-2} \text{ mol}\right)\left(200.6 \text{ g mol}^{-1}\right) = 2.15 \text{ g}$$

and the concentration of Hg is

$$\text{Concentration} = \frac{2.15 \text{ g}}{100 \text{ m}^3} = 2.2 \times 10^{-2} \text{ g m}^{-3} = 22 \text{ mg m}^{-3}$$

This concentration greatly exceeds the safety limit.

8.8 Consider the thermal decomposition of $CaCO_3$:
$$CaCO_3(s) \rightleftharpoons CaO(s) + CO_2(g)$$

The equilibrium vapor pressures of CO_2 are 22.6 mmHg at 700°C and 1829 mmHg at 950°C. Calculate the standard enthalpy of the reaction.

The van't Hoff equation,

$$\ln \frac{K_2}{K_1} = \frac{\Delta_r H^\circ}{R}\left(\frac{1}{T_1} - \frac{1}{T_2}\right),$$

is used to solve this problem. Since

$$K_P = \frac{P_{CO_2}}{P^\circ},$$

K_P is proportional to P_{CO_2}. Thus,

$$\ln \frac{K_2}{K_1} = \ln \frac{P_{CO_2,2}}{P_{CO_2,1}} = \frac{\Delta_r H^\circ}{R}\left(\frac{1}{T_1} - \frac{1}{T_2}\right)$$

$$\ln \frac{1829 \text{ mmHg}}{22.6 \text{ mmHg}} = \frac{\Delta_r H^\circ}{8.314 \text{ J K}^{-1} \text{ mol}^{-1}}\left(\frac{1}{973.15 \text{ K}} - \frac{1}{1223.15 \text{ K}}\right)$$

$$\Delta_r H^\circ = 1.74 \times 10^5 \text{ J mol}^{-1}$$

8.10 The vapor pressure of dry ice (solid CO_2) is 672.2 torr at $-80°C$ and 1486 torr at $-70°C$. Calculate the molar heat of sublimation of CO_2.

The van't Hoff equation

$$\ln \frac{K_2}{K_1} = \frac{\Delta_r H°}{R} \left(\frac{1}{T_1} - \frac{1}{T_2} \right)$$

is used to solve this problem. The process is

$$CO_2(s) \rightleftharpoons CO_2(g),$$

and the equilibrium constant is

$$K_P = \frac{P_{CO_2}}{P°}.$$

Since K_P is proportional to P_{CO_2}, the van't Hoff equation becomes

$$\ln \frac{K_2}{K_1} = \ln \frac{P_{CO_2,2}}{P_{CO_2,1}} = \frac{\Delta_r H°}{R} \left(\frac{1}{T_1} - \frac{1}{T_2} \right)$$

$$\ln \frac{1486 \text{ torr}}{672.2 \text{ torr}} = \frac{\Delta_r H°}{8.314 \text{ J K}^{-1} \text{mol}^{-1}} \left(\frac{1}{193.15 \text{ K}} - \frac{1}{203.15 \text{ K}} \right)$$

$$\Delta_r H° = 2.59 \times 10^4 \text{ J mol}^{-1}$$

8.12 Calculate the values of $\Delta_r G°$ for each of the following equilibrium constants: 1.0×10^{-4}, 1.0×10^{-2}, 1.0, 1.0×10^2, 1.0×10^4 at 298 K.

$\Delta_r G°$ can be evaluated using

$$\Delta_r G° = -RT \ln K = - \left(8.314 \text{ J K}^{-1} \text{mol}^{-1} \right) (298 \text{ K}) \ln K$$

K	$\Delta_r G°/\text{J mol}^{-1}$
1.0×10^{-4}	2.3×10^4
1.0×10^{-2}	1.1×10^4
1.0	0
1.0×10^2	-1.1×10^4
1.0×10^4	-2.3×10^4

8.14 The dissociation of N_2O_4 into NO_2 is 16.7% complete at 298 K and 1 atm:

$$N_2O_4(g) \rightleftharpoons 2NO_2(g)$$

Calculate the equilibrium constant and the standard Gibbs energy change for the reaction. [*Hint*: Let α be the degree of dissociation and show that $K_P = 4\alpha^2 P/(1 - \alpha^2)$, where P is the total pressure.]

The equilibrium constant is related to the degree of dissociation, α. In this case $\alpha = 16.7\% = 0.167$. The exact relation is derived as follows.

The dissociation reaction is

$$N_2O_4(g) \rightleftharpoons 2NO_2(g)$$

Let the initial number of moles of N_2O_4 be n. At equilibrium, $n\alpha$ moles of N_2O_4 dissociates, giving $2n\alpha$ moles of NO_2. The number of moles of N_2O_4 remaining is $n - n\alpha = n(1 - \alpha)$. The total number of moles of gases are $n(1 - \alpha) + 2n\alpha = n(1 + \alpha)$.

The equilibrium constant can be calculated once the partial pressures of the gases are known. The partial pressures are calculated from the mole fractions of the gases.

$$x_{N_2O_4} = \frac{n_{N_2O_4}}{n_{total}} = \frac{n(1 - \alpha)}{n(1 + \alpha)} = \frac{1 - \alpha}{1 + \alpha}$$

$$x_{NO_2} = \frac{n_{NO_2}}{n_{total}} = \frac{2n\alpha}{n(1 + \alpha)} = \frac{2\alpha}{1 + \alpha}$$

The partial pressures are

$$P_{N_2O_4} = x_{N_2O_4}P = \frac{1 - \alpha}{1 + \alpha}P$$

$$P_{NO_2} = x_{NO_2}P = \frac{2\alpha}{1 + \alpha}P$$

The equilibrium constant can now be calculated using the partial pressures in bars. The total pressure is 1 atm, which is equivalent to 1.013 bar.

$$K_P = \frac{P_{NO_2}^2}{P_{N_2O_4}} = \frac{\left(\frac{2\alpha}{1+\alpha}P\right)^2}{\frac{1-\alpha}{1+\alpha}P} = \frac{4\alpha^2 P}{(1+\alpha)(1-\alpha)} = \frac{4\alpha^2}{1-\alpha^2}P = \frac{4(0.167)^2}{1-0.167^2}(1.013)$$

$$= 0.1162 = 0.116$$

The standard Gibbs energy change is

$$\Delta_r G^\circ = -RT \ln K = -\left(8.314\,\text{J K}^{-1}\,\text{mol}^{-1}\right)(298\,\text{K})\ln 0.1162 = 5.33 \times 10^3\,\text{J mol}^{-1}$$

8.16   Consider the decomposition of magnesium carbonate:

$$MgCO_3(s) \rightleftharpoons MgO(s) + CO_2(g)$$

(a) Write an equilibrium constant expression (K_P) for the reaction. **(b)** The rate of decomposition is slow until the partial pressure of carbon dioxide is equal to 1 bar. Calculate the temperature at which the decomposition becomes spontaneous. Assume that $\Delta_r H^\circ$ and $\Delta_r S^\circ$ are temperature independent. Use the data in Appendix B for your calculation.

(a) $K_P = \dfrac{P_{CO_2}}{P^\circ}$

(b) When the reactants and products are in their standard states, the decomposition becomes spontaneous when $\Delta_r G^\circ < 0$. The temperature at which this occurs can be calculated using the relation

$$\Delta_r G^\circ = \Delta_r H^\circ - T\Delta_r S^\circ < 0$$

$\Delta_r H^\circ$ and $\Delta_r S^\circ$ are

$$\Delta_r H^\circ = \Delta_f \overline{H}^\circ[MgO(s)] + \Delta_f \overline{H}^\circ[CO_2(g)] - \Delta_f \overline{H}^\circ[MgCO_3(s)]$$

$$= -601.8 \text{ kJ mol}^{-1} + \left(-393.5 \text{ kJ mol}^{-1}\right) - \left(-1095.8 \text{ kJ mol}^{-1}\right)$$

$$= 100.5 \text{ kJ mol}^{-1}$$

$$\Delta_r S^\circ = \overline{S}^\circ[MgO(s)] + \overline{S}^\circ[CO_2(g)] - \overline{S}^\circ[MgCO_3(s)]$$

$$= 26.78 \text{ J K}^{-1} \text{mol}^{-1} + 213.6 \text{ J K}^{-1} \text{mol}^{-1} - 65.7 \text{ J K}^{-1} \text{mol}^{-1}$$

$$= 174.68 \text{ J K}^{-1} \text{mol}^{-1}$$

Therefore, for the reaction to become spontaneous,

$$\Delta_r H^\circ - T\Delta_r S^\circ = 100.5 \times 10^3 \text{ J mol}^{-1} - T\left(174.68 \text{ J K}^{-1}\text{mol}^{-1}\right) < 0$$

$$T > \frac{100.5 \times 10^3 \text{ J mol}^{-1}}{174.68 \text{ J K}^{-1}\text{mol}^{-1}}$$

$$T > 575.3 \text{ K}$$

8.18   Consider the reaction

$$2NO_2(g) \rightleftharpoons N_2O_4(g) \quad \Delta_r H^\circ = -58.04 \text{ kJ mol}^{-1}$$

Predict what happens to the system at equilibrium if **(a)** the temperature is raised, **(b)** the pressure on the system is increased, **(c)** an inert gas is added to the system at constant pressure, **(d)** an inert gas is added to the system at constant volume, and **(e)** a catalyst is added to the system.

(a) The equilibrium will shift from right to left. The equilibrium constant K_P will decrease.

(b) The equilibrium will shift from left to right. K_P remains unchanged.

(c) The volume of the system expands, and the gases are "diluted." K_P remains unchanged, but the equilibrium shifts from right to left.

(d) The pressure of the system will increase, but the partial pressures of NO_2 and N_2O_4 remain constant. K_P remains unchanged, and the position of equilibrium will not shift.

(e) The catalyst has no effect on either K_P or the position of equilibrium.

8.20   At a certain temperature the equilibrium pressures of NO_2 and N_2O_4 are 1.6 bar and 0.58 bar, respectively. If the volume of the container is doubled at constant temperature, what would be the partial pressures of the gases when equilibrium is re-established?

The equilibrium process is

$$N_2O_4(g) \rightleftharpoons 2NO_2(g)$$

The equilibrium constant can be calculated using the equilibrium pressures.

$$K_P = \frac{P^2_{NO_2}}{P_{N_2O_4}} = \frac{1.6^2}{0.58} = 4.41$$

When the volume of the container is doubled at constant temperature, the partial pressures of the gases are halved. That is,

$$P_{NO_2} = 0.80 \text{ bar}$$

$$P_{N_2O_4} = 0.29 \text{ bar}$$

According to LeChatelier's principle, when the volume is increased, the reaction will shift to produce more molecules. In the case at hand, more NO_2 will be produced.

Assume the equilibrium pressure of N_2O_4 to be $(0.29 - x)$ bar. The equilibrium pressure of NO_2 is obtained using the stoichiometric relationship between the two compounds.

$$N_2O_4 \quad \rightleftharpoons \quad 2NO_2$$

At equilibrium $0.29 - x$ $0.80 + 2x$ bar

Solve for x using the equilibrium expression and discarding the non-physical root.

$$K_P = 4.41 = \frac{(0.80 + 2x)^2}{0.29 - x}$$

$$4.41 (0.29 - x) = (0.80 + 2x)^2$$

$$4x^2 + 7.61x - 0.639 = 0$$

$$x = 8.06 \times 10^{-2}$$

Therefore, when equilibrium is reestablished,

$$P_{N_2O_4} = 0.29 - x = 0.21 \text{ bar}$$

$$P_{NO_2} = 0.80 + 2x = 0.96 \text{ bar}$$

8.22 Photosynthesis can be represented by

$$6CO_2(g) + 6H_2O(l) \rightleftharpoons C_6H_{12}O_6(s) + 6O_2(g) \quad \Delta_r H^\circ = 2801 \text{ kJ mol}^{-1}$$

Explain how the equilibrium would be affected by the following changes: **(a)** the partial pressure of CO_2 is increased, **(b)** O_2 is removed from the mixture, **(c)** $C_6H_{12}O_6$ (sucrose) is removed from the mixture, **(d)** more water is added, **(e)** a catalyst is added, **(f)** the temperature is decreased, and **(g)** more sunlight shines on the plants.

(a) The equilibrium would shift from left to right.

(b) The equilibrium would shift from left to right.

(c) The equilibrium would be unaffected, since sucrose is a solid. (As long as excess sucrose remains.)

(d) The equilibrium would be unaffected, since water is a liquid. (As long as excess water remains.)

(e) The catalyst has no effect on the position of equilibrium.

(f) The equilibrium would shift from right to left.

(g) Assuming constant temperature, the position of equilibrium is unaffected by the amount of sunlight falling on the plant, although the rate of attaining equilibrium is increased.

8.24 Industrially, sodium metal is obtained by electrolyzing molten sodium chloride. The reaction at the cathode is $Na^+ + e^- \rightarrow Na$. We might expect that the potassium metal could also be prepared by electrolyzing molten potassium chloride. Potassium metal is soluble in molten potassium chloride, however, and is therefore hard to recover. Furthermore, potassium vaporizes readily at the operating temperature, creating hazardous conditions. Instead, potassium is prepared by the distillation of molten potassium chloride in the presence of sodium vapor at 892° C:

$$Na(g) + KCl(l) \rightleftharpoons NaCl(l) + K(g)$$

Considering that potassium is a stronger reducing agent than sodium, explain why this approach works. (The boiling points of sodium and potassium are 892°C and 770°C, respectively.)

Potassium is more volatile than sodium and is removed from the system more rapidly, causing the equilibrium to shift from left to right.

8.26 Derive Equation 8.23 from 8.21.

Equation 8.21 is

$$Y = \frac{[PL]}{[P] + [PL]}$$

Rearrange

$$K_d = \frac{[P][L]}{[PL]}$$

to give

$$[PL] = \frac{[P][L]}{K_d}$$

and substitute into Equation 8.21:

$$Y = \frac{\frac{[P][L]}{K_d}}{[P] + \frac{[P][L]}{K_d}}$$

$$= \frac{[P][L]}{[P]K_d + [P][L]}$$

Inverting this last equation:

$$\frac{1}{Y} = \frac{[P]K_d}{[P][L]} + \frac{[P][L]}{[P][L]}$$

$$= \frac{K_d}{[L]} + 1$$

which is Equation 8.23

8.28 An equilibrium dialysis experiment showed that the concentrations of the free ligand, bound ligand, and protein are $1.2 \times 10^{-5}\ M$, $5.4 \times 10^{-6}\ M$, and $4.9 \times 10^{-6}\ M$, respectively. Calculate the dissociation constant for the reaction PL $\rightleftharpoons$ P + L. Assume there is one binding site per protein molecule.

Since there is one binding site per protein molecule, $[PL] = [L]_{bound}$.

$$K_d = \frac{[P][L]}{[PL]} = \frac{\left(4.9 \times 10^{-6}\right)\left(1.2 \times 10^{-5}\right)}{5.4 \times 10^{-6}} = 1.1 \times 10^{-5}$$

8.30 Based on the material covered so far in the text, describe as many ways as you can for calculating the $\Delta_r G^\circ$ value of a process.

There are at least three methods.

(a) From $\Delta_r H^\circ$, $\Delta_r S^\circ$ and T via $\Delta_r G^\circ = \Delta_r H^\circ - T\Delta_r S^\circ$

(b) From $\Delta_f \overline{G}^\circ$ values for the reactants and products

(c) Using the value of the equilibrium constant for the process to give $\Delta_r G^\circ = -RT \ln K$

8.32 In physical chemistry, the standard state for a solution is $1\ M$. In biological systems, however, we define the standard state of H^+ as $1 \times 10^{-7}\ M$ because the physiological pH is about 7. Consequently, the change in the standard Gibbs energy according to these two conventions will be different involving uptake or release of H^+ ions, depending on which convention is used. We will therefore replace $\Delta_r G^\circ$ with $\Delta_r G^{\circ\prime}$, where the prime denotes that it is the standard Gibbs energy change for a biological process. **(a)** Consider the reaction

$$A + B \rightarrow C + xH^+$$

where x is a stoichiometric coefficient. Derive a relation between $\Delta_r G^\circ$ and $\Delta_r G^{\circ\prime}$, keeping in mind that $\Delta_r G$ is the same for a process regardless of which convention is used. Repeat the derivation for the reverse process:

$$C + xH^+ \rightarrow A + B$$

(b) NAD$^+$ and NADH are the oxidized and reduced forms of nicotinamide adenine dinucleotide, two key compounds in the metabolic pathways. For the oxidation of NADH,

$$NADH + H^+ \rightarrow NAD^+ + H_2$$

$\Delta_r G^\circ$ is -21.8 kJ mol^{-1} at 298 K. Calculate $\Delta_r G^{\circ\prime}$. Also calculate $\Delta_r G$ using both the chemical and biological conventions when $[NADH] = 1.5 \times 10^{-2}\ M$, $[H^+] = 3.0 \times 10^{-5}\ M$, $[NAD^+] = 4.6 \times 10^{-3}\ M$, and $P_{H_2} = 0.010$ bar.

(a) For the reaction A + B $\rightarrow$ C + xH$^+$, using $1\ M$ as standard states for solutions,

$$\Delta_r G = \Delta_r G^\circ + RT \ln \frac{([C]/1\,M)\,([H^+]/1\,M)^x}{([A]/1\,M)\,([B]/1\,M)}$$

Using the biochemical standard states,

$$\Delta_r G = \Delta_r G^{\circ\prime} + RT \ln \frac{([C]/1\,M)\,([H^+]/1 \times 10^{-7}\,M)^x}{([A]/1\,M)\,([B]/1\,M)}$$

Since the value of $\Delta_r G$ does not depend on the standard states chosen, the two expressions above can be equated.

$$\Delta_r G^\circ + RT \ln \frac{([C]/1\,M)\,([H^+]/1\,M)^x}{([A]/1\,M)\,([B]/1\,M)} = \Delta_r G^{\circ\prime} + RT \ln \frac{([C]/1\,M)\,([H^+]/1 \times 10^{-7}\,M)^x}{([A]/1\,M)\,([B]/1\,M)}$$

$$\Delta_r G^\circ = \Delta_r G^{\circ\prime} + RT \ln \left(\frac{1\,M}{1 \times 10^{-7}\,M}\right)^x$$

$$\Delta_r G^\circ = \Delta_r G^{\circ\prime} + x\,RT \ln \frac{1}{1 \times 10^{-7}}$$

For the reaction $C + xH^+ \rightarrow A + B$, using the respective standard states,

$$\Delta_r G = \Delta_r G^\circ + RT \ln \frac{([A]/1\,M)\,([B]/1\,M)}{([C]/1\,M)\,([H^+]/1\,M)^x}$$

$$\Delta_r G = \Delta_r G^{\circ\prime} + RT \ln \frac{([A]/1\,M)\,([B]/1\,M)}{([C]/1\,M)\,([H^+]/1 \times 10^{-7}\,M)^x}$$

Equating these two expressions,

$$\Delta_r G^\circ + RT \ln \frac{([A]/1\,M)\,([B]/1\,M)}{([C]/1\,M)\,([H^+]/1\,M)^x} = \Delta_r G^{\circ\prime} + RT \ln \frac{([A]/1\,M)\,([B]/1\,M)}{([C]/1\,M)\,([H^+]/1 \times 10^{-7}\,M)^x}$$

$$\Delta_r G^\circ = \Delta_r G^{\circ\prime} + RT \ln \left(\frac{1 \times 10^{-7}\,M}{1\,M}\right)^x$$

$$\Delta_r G^\circ = \Delta_r G^{\circ\prime} + x\,RT \ln \left(1 \times 10^{-7}\right)$$

$$\Delta_r G^\circ = \Delta_r G^{\circ\prime} - x\,RT \ln \frac{1}{1 \times 10^{-7}}$$

(b) The expression derived above for the reaction $C + xH^+ \rightarrow A + B$ is appropriate for the oxidation of NADH. In this case, $x = 1$. Using the biological convention,

$$\Delta_r G^\circ = \Delta_r G^{\circ\prime} + RT \ln \left(1 \times 10^{-7}\right)$$

$$\Delta_r G^{\circ\prime} = \Delta_r G^\circ - RT \ln \left(1 \times 10^{-7}\right)$$

$$= -21.8 \times 10^3 \,J\,mol^{-1} - \left(8.314\,J\,K^{-1}\,mol^{-1}\right)(298\,K) \ln \left(1 \times 10^{-7}\right)$$

$$= 18.13 \times 10^3 \,J\,mol^{-1} = 18.1\,kJ\,mol^{-1}$$

Now calculate $\Delta_r G^\circ$. Using the chemical conventions,

$$\Delta_r G = \Delta_r G^\circ + RT \ln \frac{\left([\text{NAD}^+]/1\,M\right)\left([P_{H_2}/1\,\text{bar}\right)}{\left([\text{NADH}]/1\,M\right)\left([\text{H}^+]/1\,M\right)}$$

$$= -21.8 \times 10^3 \,\text{J mol}^{-1} + \left(8.314\,\text{J K}^{-1}\,\text{mol}^{-1}\right)(298\,\text{K})\ln\frac{\left(4.6 \times 10^{-3}\right)(0.010)}{\left(1.5 \times 10^{-2}\right)\left(3.0 \times 10^{-5}\right)}$$

$$= -10.3 \times 10^3 \,\text{J mol}^{-1} = -10.3\,\text{kJ mol}^{-1}$$

Using the biological convention,

$$\Delta_r G = \Delta_r G^{\circ\prime} + RT \ln \frac{\left([\text{NAD}^+]/1\,M\right)\left([P_{H_2}/1\,\text{bar}\right)}{\left([\text{NADH}]/1\,M\right)\left([\text{H}^+]/1 \times 10^{-7}\,M\right)}$$

$$= 18.13 \times 10^3 \,\text{J mol}^{-1} + \left(8.314\,\text{J K}^{-1}\,\text{mol}^{-1}\right)(298\,\text{K})\ln\frac{\left(4.6 \times 10^{-3}\right)(0.010)}{\left(1.5 \times 10^{-2}\right)\left(\frac{3.0 \times 10^{-5}}{1 \times 10^{-7}}\right)}$$

$$= -10.3 \times 10^3 \,\text{J mol}^{-1} = -10.3\,\text{kJ mol}^{-1}$$

As expected (and indeed required), the result for $\Delta_r G$ is the same regardless of the choice of standard state.

8.34 The K_{sp} value of AgCl is 1.6×10^{-10} at 25°C. What is its value at 60°C?

The reaction is

$$\text{AgCl}(s) \rightleftharpoons \text{Ag}^+(aq) + \text{Cl}^-(aq)$$

First calculate $\Delta_r H^\circ$ for the reaction.

$$\Delta_r H^\circ = \Delta_f \overline{H}^\circ[\text{Ag}^+(aq)] + \Delta_f \overline{H}^\circ[\text{Cl}^-(aq)] - \Delta_f \overline{H}^\circ[\text{AgCl}(s)]$$

$$= 105.9\,\text{kJ mol}^{-1} + \left(-167.2\,\text{kJ mol}^{-1}\right) - \left(-127.0\,\text{kJ mol}^{-1}\right)$$

$$= 65.7\,\text{kJ mol}^{-1}$$

The van't Hoff equation is now used to calculate K_{sp} at 60° C.

$$\ln \frac{K_2}{K_1} = \frac{\Delta_r H^\circ}{R}\left(\frac{1}{T_1} - \frac{1}{T_2}\right)$$

$$\ln \frac{K_2}{1.6 \times 10^{-10}} = \frac{65.7 \times 10^3 \,\text{J mol}^{-1}}{8.314\,\text{J K}^{-1}\,\text{mol}^{-1}}\left(\frac{1}{298.15\,\text{K}} - \frac{1}{333.15\,\text{K}}\right)$$

$$K_2 = 2.6 \times 10^{-9}$$

The increase in the value of K at higher temperature indicates that the solubility of AgCl increases with temperature.

8.36 Consider the equilibrium system $3A \rightleftharpoons B$. Sketch the change in the concentrations of A and B with time for the following situations: **(a)** initially only A is present; **(b)** initially only B is present; **(c)** initially both A and B are present (with A in higher concentration). In each case, assume that the concentration of B is higher than that of A at equilibrium.

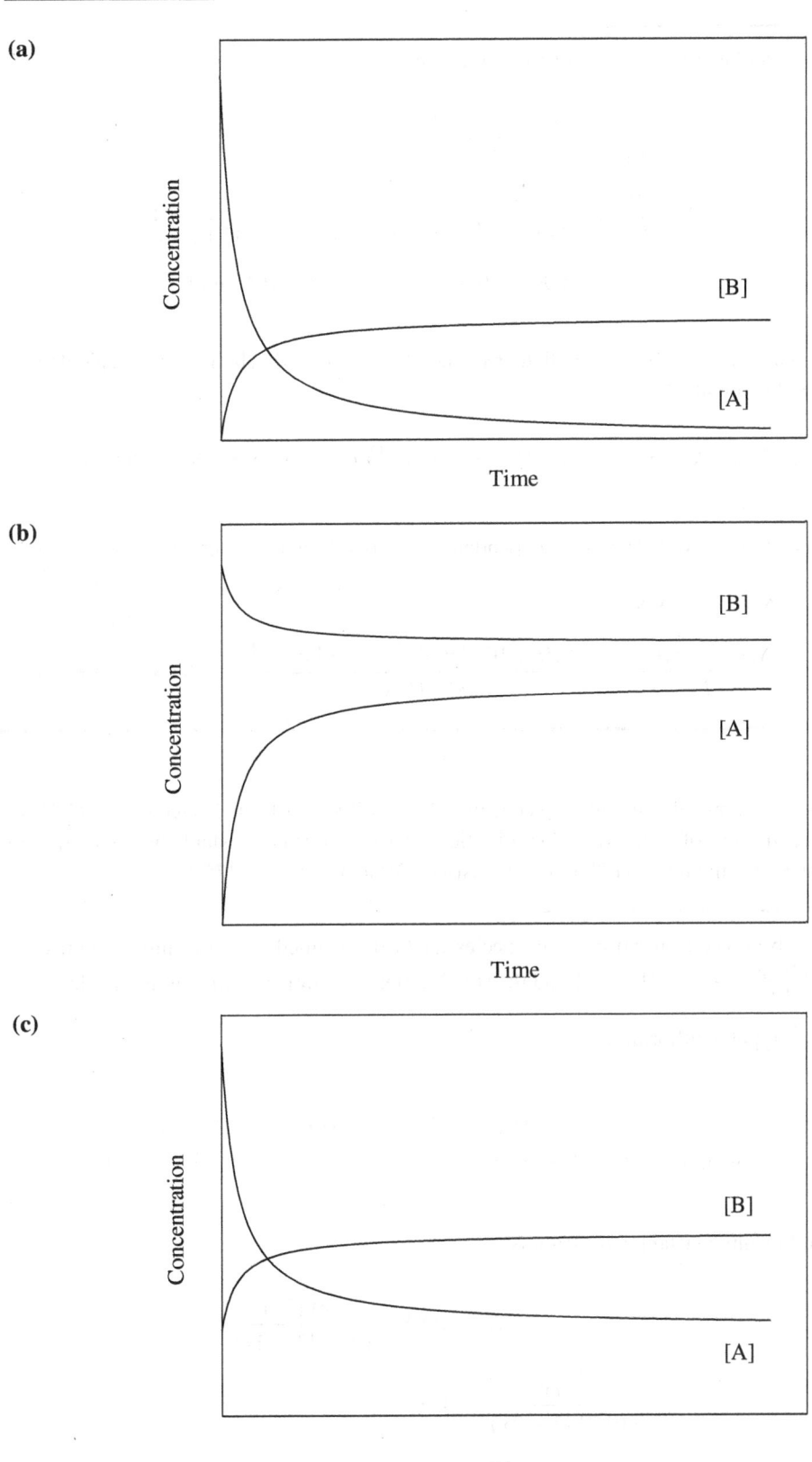

8.38 A polypeptide can exist in either the helical or random coil forms. The equilibrium constant for the equilibrium reaction of the helix to random coil transition is 0.86 at 40°C and 0.35 at 60°C. Calculate the values of $\Delta_r H°$ and $\Delta_r S°$ for the reaction.

$\Delta_r H°$ is calculated from the van't Hoff equation.

$$\ln \frac{K_2}{K_1} = \frac{\Delta_r H°}{R} \left(\frac{1}{T_1} - \frac{1}{T_2} \right)$$

$$\ln \frac{0.35}{0.86} = \frac{\Delta_r H°}{8.314 \text{ J K}^{-1} \text{mol}^{-1}} \left(\frac{1}{313.15 \text{ K}} - \frac{1}{333.15 \text{ K}} \right)$$

$$\Delta_r H° = -3.90 \times 10^4 \text{ J mol}^{-1} = -3.9 \times 10^4 \text{ J mol}^{-1}$$

To calculate $\Delta_r S°$, $\Delta_r G°$ at a particular temperature is needed. The following calculations are carried out using 40° C.

$$\Delta_r G° = -RT \ln K = - \left(8.314 \text{ J K}^{-1} \text{mol}^{-1} \right) (313.15 \text{ K}) \ln 0.86 = 393 \text{ J mol}^{-1}$$

Assuming $\Delta_r H°$ and $\Delta_r S°$ to be independent of temperature, the latter can be determined.

$$\Delta_r G° = \Delta_r H° - T \Delta_r S°$$

$$\Delta_r S° = \frac{\Delta_r H° - \Delta_r G°}{T} = \frac{-3.90 \times 10^4 \text{ J mol}^{-1} - 393 \text{ J mol}^{-1}}{313.15 \text{ K}} = -1.3 \times 10^2 \text{ J K}^{-1} \text{mol}^{-1}$$

8.40 A 14.6-g sample of ammonia is placed in a closed 4.00-L flask and heated to 375°C. Calculate the concentrations of all the gases in molarities when equilibrium is established. The equilibrium constant K_c for the reaction $2NH_3(g) \rightleftharpoons N_2(g) + 3H_2(g)$ is 0.83 at 375°C.

The equilibrium concentrations of all species can be determined using the initial quantity of NH_3 $\left(\frac{14.6 \text{ g}/17.03 \text{ g mol}^{-1}}{4.00 \text{ L}} = 0.2143 \ M \right)$ and the stoichiometic relationship among the species.

Let $x = \left[N_2 \right]$ at equilibrium.

	$2NH_3(g)$	$\rightleftharpoons$	$N_2(g)$	$+$	$3H_2(g)$	
At equilibrium	$0.2143 - 2x$		x		$3x$	M

The equilibrium constant expression is

$$K_c = 0.83 = \frac{(x)(3x)^3}{(0.2143 - 2x)^2}$$

$$\frac{27x^4}{(0.2143 - 2x)^2} = 0.83$$

First take the square root of both sides of the equation, then solve for x and discard the non-physical roots.

$$\frac{5.20x^2}{0.2143 - 2x} = 0.911$$

$$5.20x^2 + 1.822x - 0.1952 = 0$$

$$x = 0.0860$$

Therefore, the equilibrium concentrations are

$$[NH_3] = 0.2143 - 2x = 0.042\ M$$

$$[N_2] = x = 0.086\ M$$

$$[H_2] = 3x = 0.26\ M$$

8.42 The equilibrium constant (K_P) for the reaction

$$C(s) + CO_2(g) \rightleftharpoons 2CO(g)$$

is 1.9 at 727°C. What total pressure must be applied to the reacting system to obtain 0.012 mole of CO_2 and 0.025 mole of CO?

The mole fractions of CO_2 and CO are

$$x_{CO_2} = \frac{0.012\ \text{mol}}{0.012\ \text{mol} + 0.025\ \text{mol}} = 0.324$$

$$x_{CO} = 1 - x_{CO_2} = 0.676$$

The equilibrium constant expression is

$$K_P = \frac{(P_{CO}/P^\circ)^2}{P_{CO_2}/P^\circ}$$

$$= \frac{(x_{CO}P_{total}/P^\circ)^2}{x_{CO_2}P_{total}/P^\circ}$$

$$1.9 = \frac{(0.676P_{total})^2}{0.324P_{total}}$$

$$P_{total} = 1.35\ \text{bar} = 1.3\ \text{atm}$$

8.44 Consider the following reaction at a certain temperature

$$A_2 + B_2 \rightleftharpoons 2AB$$

The mixing of 1 mole of A_2 with 3 moles of B_2 gives rise to x mole of AB at equilibrium. The addition of 2 more moles of A_2 produces another x mole of AB. What is the equilibrium constant for the reaction?

Assume the volume of the container is V. When 1 mole of A_2 and 3 moles of B_2 react, the equilibrium concentrations of all species are summarized in the following table:

	A_2	+	B_2	$\rightleftharpoons$	2AB	
At equilibrium	$(1 - \frac{x}{2})/V$		$(3 - \frac{x}{2})/V$		x/V	M

The equilibrium constant expression is

$$K_c = \frac{(x/V)^2}{\left[\left(1 - \frac{x}{2}\right)/V\right]\left[\left(3 - \frac{x}{2}\right)/V\right]} = \frac{x^2}{\left(1 - \frac{x}{2}\right)\left(3 - \frac{x}{2}\right)}$$

With the addition of 2 more moles of A_2, the equilibrium amount of AB is $2x$ moles, and the concentrations of all species are:

$$\begin{array}{ccccc} & A_2 & + & B_2 & \rightleftharpoons & 2AB \\ \text{At equilibrium} & (3-x)/V & & (3-x)/V & & 2x/V \quad M \end{array}$$

and the corresponding equilibrium expression is

$$K_c = \frac{(2x/V)^2}{\left[(3-x)/V\right]\left[(3-x)/V\right]} = \frac{4x^2}{(3-x)^2}$$

Since K_c is a constant (when there is no change in temperature), the two expressions can be equated.

$$\frac{x^2}{\left(1 - \frac{x}{2}\right)\left(3 - \frac{x}{2}\right)} = \frac{4x^2}{(3-x)^2}$$

$$(3-x)^2 = 4\left(1 - \frac{x}{2}\right)\left(3 - \frac{x}{2}\right)$$

$$9 - 6x + x^2 = 12 - 8x + x^2$$

$$2x = 3$$

$$x = 1.5$$

To calculate K_c, substitute x into either one of the two equilibrium expressions. The following shows the result when x is substituted in the second expression.

$$K_c = \frac{4(1.5)^2}{(3 - 1.5)^2} = 4.0$$

8.46 Iodine is sparingly soluble in water but much more so in carbon tetrachloride. The equilibrium constant, also called the partition coefficient, for the distribution of I_2 between these two phases

$$I_2(aq) \rightleftharpoons I_2(CCl_4)$$

is 83 at 20°C. **(a)** A student adds 0.030 L of CCl_4 to 0.200 L of an aqueous solution containing 0.032 g I_2. The mixture is shaken and the two phases are then allowed to separate. Calculate the fraction of I_2 remaining in the aqueous phase. **(b)** The student now repeats the extraction of I_2 with another 0.030 L of CCl_4. Calculate the fraction of I_2 from the original solution that remains in the aqueous phase. **(c)** Compare the result in **(b)** with a single extraction using 0.060 L of CCl_4. Comment on the difference.

(a) The total number of moles of I_2 is 0.032 g/253.8 g mol^{-1} = 1.26×10^{-4} mol. Let x be the number of moles of I_2 that remains in the aqueous phase. Then, the number of moles of I_2 dissolved in CCl_4 is $\left(1.26 \times 10^{-4} - x\right)$ mol.

$$K = \frac{[I_2(CCl_4)]}{[I_2(aq)]}$$

$$83 = \frac{(1.26 \times 10^{-4} - x) / 0.030}{x / 0.200}$$

$$83 = \frac{1.26 \times 10^{-4} - x}{x} (6.667)$$

$$12.45x = 1.26 \times 10^{-4} - x$$

$$x = 9.37 \times 10^{-6}$$

Therefore, the fraction of I_2 remaining in the aqueous phase is

$$\frac{9.37 \times 10^{-6} \text{ mol}}{1.26 \times 10^{-4} \text{ mol}} = 0.074$$

(b) The first extraction leaves 7.4% of I_2 in water. The next extraction with the same amount (0.030 L) of CCl_4 will once again leave 7.4% of I_2 in the aqueous phase resulting from the first extraction. Thus, after two extractions, the fraction of I_2 from the original solution that remains in the aqueous phase is (0.074) (0.074) = 0.0055 = 0.55%.

(c) With a single 0.060-L extraction,

$$K = \frac{[I_2(CCl_4)]}{[I_2(aq)]}$$

$$83 = \frac{(1.26 \times 10^{-4} - x) / 0.060}{x / 0.200}$$

$$83 = \frac{1.26 \times 10^{-4} - x}{x} (3.333)$$

$$24.90x = 1.26 \times 10^{-4} - x$$

$$x = 4.86 \times 10^{-6}$$

The fraction of I_2 remaining in the aqueous phase is

$$\frac{4.86 \times 10^{-6}}{1.26 \times 10^{-4}} = 0.039$$

This value is greater than the fraction after two extractions (0.0055). Thus, the extraction with 0.060 L of CCl_4 is not as effective as two separate extractions of 0.030 L each.

8.48 Use the appropriate equation in this chapter to estimate the vapor pressure of water at 60°C.

Consider the reaction $H_2O(l) \rightleftharpoons H_2O(g)$. The equilibrium constant is the numerical value of the vapor pressure of water in bars:

$$K_P = P_{H_2O(g)} / P^\circ$$

Recognizing that the vapor pressure of water at 100°C, its normal boiling point, is 1 atm (1.013 bar), the van't Hoff equation can be utilized to estimate the vapor pressure of water at 60°C. First find $\Delta_r H^\circ$:

$$\Delta_r H^\circ = \Delta_f \overline{H}^\circ \left[H_2O(g) \right] - \Delta_f \overline{H}^\circ \left[H_2O(l) \right]$$

$$= -241.8 \text{ kJ mol}^{-1} - \left(-285.8 \text{ kJ mol}^{-1} \right)$$

$$= 44.0 \text{ kJ mol}^{-1}$$

It follows then from Equation 8.18,

$$\ln \frac{K_2}{K_1} = \ln \frac{P_{H_2O(g),2}/P^\circ}{P_{H_2O(g),1}/P^\circ} = \frac{\Delta_r H^\circ}{R} \left(\frac{1}{T_1} - \frac{1}{T_2} \right)$$

$$\ln \frac{P_{H_2O(g),2}}{1.013} = \frac{44.0 \times 10^3 \text{ J mol}^{-1}}{8.314 \text{ J K}^{-1} \text{ mol}^{-1}} \left(\frac{1}{373.15 \text{ K}} - \frac{1}{333.15 \text{ K}} \right) = -1.703$$

$$P_{H_2O(g),2} = 0.1845 \text{ bar} \left(\frac{1 \text{ atm}}{1.013 \text{ bar}} \right) \left(\frac{760 \text{ torr}}{1 \text{ atm}} \right) = 138 \text{ torr}$$

CHAPTER 9

Electrochemistry

PROBLEMS AND SOLUTIONS

9.2 Calculate the emf of the Daniell cell at 298 K when the concentrations of $CuSO_4$ and $ZnSO_4$ are 0.50 M and 0.10 M, respectively. What would the emf be if activities were used instead of concentrations? (The $\gamma_\pm$ values for $CuSO_4$ and $ZnSO_4$ at their respective concentrations are 0.068 and 0.15, respectively.)

The half reactions for the Daniell cell are

$$\text{Anode:} \quad Zn \rightarrow Zn^{2+} + 2e^-$$
$$\text{Cathode:} \quad Cu^{2+} + 2e^- \rightarrow Cu$$

Thus, for the cell,

$$E^\circ = E^\circ_{cathode} - E^\circ_{anode} = 0.342 \text{ V} - (-0.762 \text{ V}) = 1.104 \text{ V}$$

The emf at the specified Cu^{2+} and Zn^{2+} concentrations is found using Equation 9.8 and setting activities equal to concentrations.

$$E = E^\circ - \frac{0.0257 \text{ V}}{\nu} \ln \frac{[Zn^{2+}]}{[Cu^{2+}]} = 1.104 \text{ V} - \frac{0.0257 \text{ V}}{2} \ln \frac{0.10}{0.50} = 1.125 \text{ V}$$

Using activities,

$$E = E^\circ - \frac{0.0257 \text{ V}}{\nu} \ln \frac{a_{Zn^{2+}}}{a_{Cu^{2+}}}$$

$$= E^\circ - \frac{0.0257 \text{ V}}{\nu} \ln \frac{\gamma_{\pm,ZnSO_4} [Zn^{2+}]}{\gamma_{\pm,CuSO_4} [Cu^{2+}]}$$

$$= 1.104 \text{ V} - \frac{0.0257 \text{ V}}{2} \ln \frac{(0.15)(0.10)}{(0.068)(0.50)}$$

$$= 1.115 \text{ V}$$

A difference of 0.010 V in 1.100 V is easily measured.

9.4   Consider a Daniell cell operating under non-standard-state conditions. Suppose that the cell's reaction is multiplied by 2. What effect does this have on each of the following quantities in the Nernst equation? **(a)** E, **(b)** E°, **(c)** Q, **(d)** $\ln Q$, and **(e)** ν

(a) None. E is an intensive property, **(b)** none. E° is an intensive property, **(c)** squared, **(d)** doubled, **(e)** doubled.

9.6   From the standard reduction potentials listed in Table 9.1 for $Cu^{2+}|Cu$ and $Pt|Cu^{2+}$, Cu^+, calculate the standard reduction potential for $Cu^+|Cu$.

The reduction half reactions and the standard reduction potentials for $Cu^{2+}|Cu$ and $Pt|Cu^{2+}$, Cu^+ are

$$Cu^{2+} + 2e^- \rightarrow Cu \qquad E_1^\circ = 0.342\ V$$
$$Cu^{2+} + e^- \rightarrow Cu^+ \qquad E_2^\circ = 0.153\ V$$

The reduction half reaction for $Cu^+|Cu$ is

$$Cu^+ + e^- \rightarrow Cu.$$

Let its reduction potential be denoted as E_3°. This half reaction can be derived from the previous two half reactions. The Gibbs energy changes are related by

$$\Delta_r G_3^\circ = \Delta_r G_1^\circ - \Delta_r G_2^\circ$$

Since $\Delta_r G^\circ = -\nu F E^\circ$, the above relation becomes

$$\left(-\nu_3 F E_3^\circ\right) = \left(-\nu_1 F E_1^\circ\right) - \left(-\nu_2 F E_2^\circ\right)$$
$$\nu_3 E_3^\circ = \nu_1 E_1^\circ - \nu_2 E_2^\circ$$
$$E_3^\circ = \frac{\nu_1 E_1^\circ - \nu_2 E_2^\circ}{\nu_3}$$
$$= \frac{2\,(0.342\ V) - 1\,(0.153\ V)}{1}$$
$$= 0.531\ V$$

9.8   Calculate the values of E°, $\Delta_r G^\circ$, and K for the following reactions at 25° C:

(a) $Zn + Sn^{4+} \rightleftharpoons Zn^{2+} + Sn^{2+}$

(b) $Cl_2 + 2I^- \rightleftharpoons 2Cl^- + I_2$

(c) $5Fe^{2+} + MnO_4^- + 8H^+ \rightleftharpoons Mn^{2+} + 4H_2O + 5Fe^{3+}$

(a)   The half reactions are

$$\begin{aligned} \text{Anode:} \qquad & Zn \rightarrow Zn^{2+} + 2e^- \\ \text{Cathode:} \qquad & Sn^{4+} + 2e^- \rightarrow Sn^{2+} \end{aligned}$$

$$E^\circ = 0.151 \text{ V} - (-0.762 \text{ V}) = 0.913 \text{ V}$$

$$\Delta_r G^\circ = -\nu F E^\circ = -2\left(96500 \text{ C mol}^{-1}\right)(0.913 \text{ V})$$

$$= -1.762 \times 10^5 \text{ J mol}^{-1} = -1.76 \times 10^5 \text{ J mol}^{-1}$$

$$K = \exp\left(-\frac{\Delta_r G^\circ}{RT}\right) = \exp\left[-\frac{\left(-1.762 \times 10^5 \text{ J mol}^{-1}\right)}{\left(8.314 \text{ J K}^{-1} \text{ mol}^{-1}\right)(298.15 \text{ K})}\right] = 7.42 \times 10^{30}$$

(b) The half reactions are

$$\begin{aligned} \text{Anode:} \quad & 2I^- \rightarrow I_2 + 2e^- \\ \text{Cathode:} \quad & Cl_2 + 2e^- \rightarrow 2Cl^- \end{aligned}$$

$$E^\circ = 1.36 \text{ V} - 0.536 \text{ V} = 0.824 \text{ V}$$

$$\Delta_r G^\circ = -\nu F E^\circ = -2\left(96500 \text{ C mol}^{-1}\right)(0.824 \text{ V})$$

$$= -1.59 \times 10^5 \text{ J mol}^{-1} = -1.6 \times 10^5 \text{ J mol}^{-1}$$

$$K = \exp\left(-\frac{\Delta_r G^\circ}{RT}\right) = \exp\left[-\frac{\left(-1.59 \times 10^5 \text{ J mol}^{-1}\right)}{\left(8.314 \text{ J K}^{-1} \text{ mol}^{-1}\right)(298.15 \text{ K})}\right] = 7.2 \times 10^{27}$$

(c) The half reactions are

$$\begin{aligned} \text{Anode:} \quad & Fe^{2+} \rightarrow Fe^{3+} + e^- \\ \text{Cathode:} \quad & MnO_4^- + 8H^+ + 5e^- \rightarrow Mn^{2+} + 4H_2O \end{aligned}$$

$$E^\circ = 1.507 \text{ V} - 0.771 \text{ V} = 0.736 \text{ V}$$

$$\Delta_r G^\circ = -\nu F E^\circ = -5\left(96500 \text{ C mol}^{-1}\right)(0.736 \text{ V})$$

$$= -3.551 \times 10^5 \text{ J mol}^{-1} = -3.55 \times 10^5 \text{ J mol}^{-1}$$

$$K = \exp\left(-\frac{\Delta_r G^\circ}{RT}\right) = \exp\left[-\frac{\left(-3.551 \times 10^5 \text{ J mol}^{-1}\right)}{\left(8.314 \text{ J K}^{-1} \text{ mol}^{-1}\right)(298.15 \text{ K})}\right] = 1.64 \times 10^{62}$$

9.10   Consider a concentration cell consisting of two hydrogen electrodes. At 25°C, the cell emf is found to be 0.0267 V. If the pressure of hydrogen gas at the anode is 4.0 bar, what is the pressure of hydrogen gas at the cathode?

Let the pressure of hydrogen gas at the cathode be x bars. The half reactions are

$$\begin{aligned} \text{Anode:} \quad & H_2(4.0 \text{ bar}) \rightarrow 2H^+ + 2e^- \\ \text{Cathode:} \quad & 2H^+ + 2e^- \rightarrow H_2(x \text{ bar}) \end{aligned}$$

The overall reaction is

$$H_2(4.0 \text{ bar}) \rightarrow H_2(x \text{ bar})$$

Since this is a concentration cell, $E^\circ = 0$ V. The emf for the cell is

$$E = E^\circ - \frac{0.0257 \text{ V}}{\nu} \ln \frac{x \text{ bar}/P^\circ}{4.0 \text{ bar}/P^\circ}$$

$$0.0267 \text{ V} = 0 \text{ V} - \frac{0.0257 \text{ V}}{2} \ln \frac{x}{4.0}$$

$$x = 0.50$$

At the cathode, $P_{H_2} = 0.50$ bar.

9.12 From the standard reduction potentials listed in Table 9.1 for $Sn^{2+} \mid Sn$ and $Pb^{2+} \mid Pb$, calculate the ratio of $[Sn^{2+}]$ to $[Pb^{2+}]$ at equilibrium at 25°C and the $\Delta_r G^\circ$ value for the reaction.

The reduction potentials for the half cells are

$$Sn^{2+} + 2e^- \rightarrow Sn \quad E^\circ = -0.138 \text{ V}$$
$$Pb^{2+} + 2e^- \rightarrow Pb \quad E^\circ = -0.126 \text{ V}$$

The standard emf for the reaction $Sn + Pb^{2+} \rightarrow Sn^{2+} + Pb$ is

$$E^\circ = -0.126 \text{ V} - (-0.138 \text{ V}) = 0.012 \text{ V}$$

The ratio of $[Sn^{2+}]$ to $[Pb^{2+}]$ at equilibrium is directly related to the equilibrium constant, which can be calculated from E°.

$$K = \frac{[Sn^{2+}]}{[Pb^{2+}]} = \exp\left(\frac{\nu F E^\circ}{RT}\right) = \exp\left[\frac{2\,(96500 \text{ C mol}^{-1})\,(0.012 \text{ V})}{(8.314 \text{ J K}^{-1}\text{ mol}^{-1})\,(298.15 \text{ K})}\right] = 2.55$$

The standard Gibbs energy is

$$\Delta_r G^\circ = -\nu F E^\circ = -2\left(96500 \text{ C mol}^{-1}\right)(0.012 \text{ V}) = -2.3 \times 10^3 \text{ J mol}^{-1}$$

9.14 Calculate the emf of the following concentration cell at 298 K:

$$Mg(s)|Mg^{2+}(0.24 \ M) \parallel Mg^{2+}(0.53 \ M)|Mg(s)$$

$$\text{Anode:} \quad Mg \rightarrow Mg^{2+}(0.24 \ M) + 2e^-$$
$$\text{Cathode:} \quad Mg^{2+}(0.53 \ M) + 2e^- \rightarrow Mg$$

The overall reaction is

$$Mg^{2+}(0.53 \ M) \rightarrow Mg^{2+}(0.24 \ M)$$

The emf of the cell depends on the concentrations of Mg^{2+} at both the anode and the cathode.

$$E = E^\circ - \frac{0.0257 \text{ V}}{\nu} \ln \frac{0.24}{0.53} = 0 \text{ V} - \frac{0.0257 \text{ V}}{2} \ln \frac{0.24}{0.53} = 0.010 \text{ V}$$

9.16 Describe an experiment that would show that the nerve cell membrane is much more permeable to K^+ than Na^+.

In separate experiments with a nerve cell, add the same concentrations of Na^+ and K^+ and monitor the changes in the membrane potential. The potential will be more susceptible to changes in the K^+ concentration.

9.18 Referring to Figure 9.8b, carry out the following operations: **(a)** Calculate the membrane potential due to K^+ ions at 25°C. **(b)** Given that biological membranes typically have a capacitance of approximately 1 $\mu F\,cm^{-2}$, calculate the charge in coulombs on a unit area (1 cm^2) of the membrane. (See Appendix 7.1 for units of capacitance.) **(c)** Convert the charge in **(b)** to number of K^+ ions. **(d)** Compare the result in **(c)** with the number of K^+ ions in 1 cm^3 of the solution in the left compartment. What can you conclude about the relative number of K^+ ions needed to establish the membrane potential?

(a)
$$E_{K^+} = \frac{0.0257\ V}{\nu}\ \ln\frac{[K^+]_r}{[K^+]_l} = (0.0257\ V)\ \ln\frac{0.010}{0.10} = -5.918 \times 10^{-2}\ V = -59.2\ mV$$

(b)
$$Q = CV = \left(1 \times 10^{-6}\ F\,cm^{-2}\right)\left(5.918 \times 10^{-2}\ V\right)\left(\frac{1\,C\,V^{-1}}{1\,F}\right)$$
$$= 5.918 \times 10^{-8}\ C\,cm^{-2} = 5.92 \times 10^{-8}\ C\,cm^{-2}$$

(c)
$$\text{Number of } K^+ \text{ ions} = \frac{(5.918 \times 10^{-8}\ C\,cm^{-2})(6.022 \times 10^{23}\ mol^{-1})}{96500\ C\,mol^{-1}}$$
$$= 3.69 \times 10^{11}\ K^+ \text{ ions } cm^{-2}$$

(d) The number of K^+ ions in 1 cm^3 of the solution in the left compartment is

$$(0.1\,M)\left(1\,cm^3\right)\left(\frac{1\,L}{1000\,cm^3}\right)\left(6.022 \times 10^{23}\ \text{ions } mol^{-1}\right) = 6.022 \times 10^{19}\ \text{ions}$$

Very few of the K^+ ions in the solution are involved in establishing the membrane potential.

9.20 A well-known organic redox system is the quinone–hydroquinone couple. In an aqueous solution at a pH below 8, we have

Quinone (Q)   Hydroquinone (HQ)

$+\ 2H^+\ +\ 2e^- \longrightarrow$ $E° = 0.699\ V$

This system can be prepared by dissolving quinhydrone, QH (a complex consisting of equimolar amounts of Q and HQ), in water. A quinhydrone electrode can be constructed by immersing a piece of platinum wire in a quinhydrone solution. **(a)** Derive an expression for the electrode potential of this couple in terms of E° and the hydrogen-ion concentration. **(b)** When the quinone–hydroquinone couple is joined to a saturated calomel electrode, the emf of the cell is found to be 0.18 V. In this arrangement, the saturated calomel electrode acts as the anode. Calculate the pH of the quinhydrone solution. Assume the temperature is 25°C.

(a)

$$E_{QH} = E^\circ - \frac{0.0257\text{ V}}{\nu}\ln\frac{[HQ]}{[Q][H^+]^2}$$

$$= 0.699\text{ V} - \frac{0.0257\text{ V}}{2}\ln\frac{1}{[H^+]^2}$$

$$= 0.699\text{ V} - \frac{0.0257\text{ V}}{2}(2)(2.303)\left(-\log[H^+]\right)$$

$$= 0.699\text{ V} - (0.0592\text{ pH})\text{ V}$$

(b) The standard reduction potential for calomel electrode is

$$Hg_2Cl_2 + 2e^- \rightarrow 2Hg + 2Cl^- \qquad E^\circ = 0.268\text{ V}$$

The quinhydrone electrode serves as the cathode, and the calomel electrode serves as the anode. The measured emf for the cell can be expressed using the emf of these half-cells.

$$E_{QH} - 0.268\text{ V} = 0.18\text{ V}$$

$$0.699\text{ V} - (0.0592\text{ pH})\text{ V} - 0.268\text{ V} = 0.18\text{ V}$$

$$pH = 4.2$$

9.22 Aluminum has a more negative standard reduction potential than iron. Yet aluminum does not form rust or corrode as easily as iron. Explain.

In fact aluminum does rust to form Al_2O_3. Al_2O_3, however, forms a thin, tough layer over the metallic aluminum underneath, isolating it from the environment and protecting it from further corrosion.

9.24 For years it was not clear whether mercury(I) ions existed in solution as Hg^+ or as Hg_2^{2+}. To distinguish between these two possibilities, we could set up the following system:

$$Hg(l) \mid \text{soln A} \parallel \text{soln B} \mid Hg(l)$$

where solution A contained 0.263 g mercury(I) nitrate per liter and solution B contained 2.63 g mercury(I) nitrate per liter. If the measured emf of such a cell is 0.0289 V at 18°C, what can you deduce about the nature of the mercury ions?

The cell described in the question is a concentration cell (that is, $E^\circ = 0$ V) and the emf depends on the nature of the mercury ions. Note that

$$\frac{[\text{mercury ions}]_A}{[\text{mercury ions}]_B} = 0.100$$

Assuming mercury(I) ions exist in solution as Hg^+, the half reactions are

$$\text{Anode:} \quad Hg \rightarrow Hg^+(\text{Soln A}) + e^-$$
$$\text{Cathode:} \quad Hg^+(\text{Soln B}) + e^- \rightarrow Hg$$

The overall reaction is

$$Hg^+(\text{Soln B}) \rightarrow Hg^+(\text{Soln A})$$

and the cell emf is

$$E = E^\circ - \frac{RT}{\nu F} \ln \frac{[Hg^+]_A}{[Hg^+]_B} = 0 \text{ V} - \frac{\left(8.314 \text{ J K}^{-1}\text{ mol}^{-1}\right)(291.15 \text{ K})}{96500 \text{ C mol}^{-1}} \ln 0.100 = 0.0578 \text{ V}$$

This emf does not match the measured emf.

Assuming mercury(I) ions exist in solution as Hg_2^{2+}, the half reactions are

$$\text{Anode:} \quad Hg \rightarrow Hg_2^{2+}(\text{Soln A}) + 2e^-$$
$$\text{Cathode:} \quad Hg_2^{2+}(\text{Soln B}) + 2e^- \rightarrow Hg$$

The overall reaction is

$$Hg_2^{2+}(\text{Soln B}) \rightarrow Hg_2^{2+}(\text{Soln A})$$

and the cell emf is

$$E = E^\circ - \frac{RT}{\nu F} \ln \frac{[Hg^+]_A}{[Hg^+]_B} = 0 \text{ V} - \frac{\left(8.314 \text{ J K}^{-1}\text{ mol}^{-1}\right)(291.15 \text{ K})}{2\left(96500 \text{ C mol}^{-1}\right)} \ln 0.100 = 0.0289 \text{ V}$$

This emf matches the measured emf. Thus, mercury(I) ions exist as Hg_2^{2+}.

9.26 Given that

$$2Hg^{2+}(aq) + 2e^- \rightarrow Hg_2^{2+}(aq) \quad E^\circ = 0.920 \text{ V}$$

$$Hg_2^{2+}(aq) + 2e^- \rightarrow 2Hg(l) \quad E^\circ = 0.797 \text{ V}$$

Calculate the values of $\Delta_r G^\circ$ and K for the following process at 25°C:

$$Hg_2^{2+}(aq) \rightarrow Hg^{2+}(aq) + Hg(l)$$

(The above reaction is an example of a *disproportionation reaction*, in which an element in one oxidation state is both oxidized and reduced.)

The sum of the two half-reactions

$$Hg_2^{2+}(aq) \rightarrow 2Hg^{2+}(aq) + 2e^-$$

$$Hg_2^{2+}(aq) + 2e^- \rightarrow 2Hg(l)$$

gives

$$2Hg_2^{2+}(aq) \rightarrow 2Hg^{2+}(aq) + 2Hg(l)$$

Therefore, the emf for the reaction is

$$E^\circ = 0.797 \text{ V} - 0.920 \text{ V} = -0.123 \text{ V}$$

Since emf is an intensive property, this is also the emf for

$$Hg_2^{2+}(aq) \rightarrow Hg^{2+}(aq) + Hg(l)$$

In this particular reaction, only 1 mole of electrons is transferred. The standard Gibbs energy and equilibrium constant are

$$\Delta_r G^\circ = -\nu F E^\circ = -\left(96500 \text{ C mol}^{-1}\right)(-0.123 \text{ V}) = 1.187 \times 10^4 \text{ J mol}^{-1}$$

$$= 1.19 \times 10^4 \text{ J mol}^{-1}$$

$$K = \exp\left(-\frac{\Delta_r G^\circ}{RT}\right) = \exp\left[-\frac{1.187 \times 10^4 \text{ J mol}^{-1}}{\left(8.314 \text{ J K}^{-1} \text{mol}^{-1}\right)(298.15 \text{ K})}\right] = 8.32 \times 10^{-3}$$

9.28   An electrochemical cell is constructed as follows. One half-cell consists of a platinum wire immersed in a solution containing 1.0 M Sn^{2+} and 1.0 M Sn^{4+}, and the other half-cell has a thallium rod immersed in a solution of 1.0 M Tl$^+$. **(a)** Write the half-cell reactions and the overall reaction. **(b)** What is the equilibrium constant at 25°C? **(c)** What is the cell voltage if the Tl$^+$ concentration is increased tenfold?

(a) The reduction potentials of the relevant half-reactions are

$$Sn^{4+} + 2e^- \rightarrow Sn^{2+} \qquad E^\circ = 0.151 \text{ V}$$
$$Tl^+ + e^- \rightarrow Tl \qquad E^\circ = -0.336 \text{ V}$$

To obtain a positive standard emf, the overall reaction must be

$$Sn^{4+} + 2Tl \rightarrow Sn^{2+} + 2Tl^+ \qquad E^\circ = 0.151 \text{ V} - (-0.336 \text{ V}) = 0.487 \text{ V}$$

(b)
$$K = \exp\left(\frac{\nu F E^\circ}{RT}\right) = \exp\left[\frac{2\left(96500 \text{ C mol}^{-1}\right)(0.487 \text{ V})}{\left(8.314 \text{ J K}^{-1} \text{mol}^{-1}\right)(298.15 \text{ K})}\right] = 2.93 \times 10^{16}$$

(c) [Tl$^+$] = 10.0 M

$$E = E^\circ - \frac{0.0257 \text{ V}}{\nu} \ln \frac{\left[Sn^{2+}\right]\left[Tl^+\right]^2}{\left[Sn^{4+}\right]}$$

$$= 0.487 \text{ V} - \frac{0.0257 \text{ V}}{2} \ln \frac{(1.0)(10.0)^2}{1.0} = 0.428 \text{ V}$$

9.30   Consider the Daniell cell shown in Figure 9.1. In the diagram, the anode appears to be negative and the cathode positive (electrons are flowing from the anode to the cathode). Yet the anions in solution are moving toward the anode, which must therefore seem positive to the anions. Because the anode cannot simultaneously be negative and positive, give an explanation for this apparently contradictory situation.

At the Zn electrode (anode)/solution interface, Zn atoms are being oxidized to form Zn^{2+} ions and electrons. The electrons enter the bulk metal of the electrode, making it negative and driving electrons towards the cathode. The cations in the solution immediately surrounding the electrode give this region of the solution a net positive charge, and it is to this that the anions are attracted.

9.32 Use the data in Appendix B and the convention that $\Delta_f \overline{G}^{\circ}[H^+(aq)] = 0$ to determine the standard reduction potentials for sodium and fluorine. (Like sodium, fluorine also reacts violently with water.)

Note that as on page 360 of the text, we will be using a more accurate value for the Faraday constant in working this problem. To determine the standard reduction potential for sodium, consider the following reaction

$$Na^+(aq) + \frac{1}{2} H_2(g) \rightarrow Na(s) + H^+(aq)$$

The standard Gibbs energy for this reaction is

$$\Delta_r G^{\circ} = \Delta_f \overline{G}^{\circ}[Na(s)] + \Delta_f \overline{G}^{\circ}[H^+(aq)] - \Delta_f \overline{G}^{\circ}[Na^+(aq)] - \frac{1}{2}\Delta_f \overline{G}^{\circ}[H_2(g)]$$

$$= 0 \text{ kJ mol}^{-1} + 0 \text{ kJ mol}^{-1} - \left(-261.9 \text{ kJ mol}^{-1}\right) - \frac{1}{2}\left(0 \text{ kJ mol}^{-1}\right)$$

$$= 261.9 \text{ kJ mol}^{-1}$$

The standard emf for the reaction is

$$E^{\circ} = -\frac{\Delta_r G^{\circ}}{\nu F} = -\frac{261.9 \times 10^3 \text{ J mol}^{-1}}{96485.3 \text{ C mol}^{-1}} = -2.714 \text{ V}$$

Since $E^{\circ} = E^{\circ}_{cathode} - E^{\circ}_{anode}$, where the half-reactions are

Anode: $H_2 \rightarrow 2H^+ + e^-$
Cathode: $Na^+ + e^- \rightarrow Na$

the reduction potential of Na is -2.714 V.

To determine the standard reduction potential for fluorine, consider the following reaction

$$\frac{1}{2} F_2(g) + \frac{1}{2} H_2(g) \rightarrow F^-(aq) + H^+ (aq)$$

The standard Gibbs energy for this reaction is

$$\Delta_r G^{\circ} = \Delta_f \overline{G}^{\circ}[F^-(aq)] + \Delta_f \overline{G}^{\circ}[H^+(aq)] - \frac{1}{2}\Delta_f \overline{G}^{\circ}[F_2(g)] - \frac{1}{2}\Delta_f \overline{G}^{\circ}[H_2(g)]$$

$$= -276.5 \text{ kJ mol}^{-1} + 0 \text{ kJ mol}^{-1} - \frac{1}{2}\left(0 \text{ kJ mol}^{-1}\right) - \frac{1}{2}\left(0 \text{ kJ mol}^{-1}\right)$$

$$= -276.5 \text{ kJ mol}^{-1}$$

The standard emf for the reaction is

$$E^\circ = -\frac{\Delta_r G^\circ}{\nu F} = -\frac{\left(-276.5 \times 10^3 \, \text{J mol}^{-1}\right)}{96485.3 \, \text{C mol}^{-1}} = 2.866 \, \text{V}$$

Since $E^\circ = E^\circ_{\text{cathode}} - E^\circ_{\text{anode}}$, where the half-reactions are

$$\text{Anode:} \quad H_2 \rightarrow 2H^+ + e^-$$

$$\text{Cathode:} \quad F_2 + 2e^- \rightarrow 2F^-$$

the reduction potential of F_2 is 2.866 V.

9.34 Consider the following cell:

$$\text{Pt} \mid H_2(1 \text{ bar}) \mid HCl(m) \mid AgCl(s) \mid Ag$$

At 25°C, the emf values at various molalities are given by

m/mol kg^{-1}	0.124	0.0539	0.0256	0.0134	0.00914	0.00562	0.00322
E/V	0.342	0.382	0.418	0.450	0.469	0.493	0.521

(a) Determine the value of E° graphically. Compare your value of E° with that listed in Table 9.1. **(b)** Calculate the mean activity coefficient ($\gamma_\pm$) for HCl at 0.124 m.

(a) Referring to the discussion in Section 9.6, prepare a graph of $y = E + 0.0514 \, \text{V} \ln m$ versus m.

$m/\left(\text{mol kg}^{-1}\right)$	0.124	0.0539	0.0256	0.0134	0.00914	0.00562	0.00322
y/V	0.2347	0.2319	0.2296	0.2283	0.2277	0.2267	0.2260

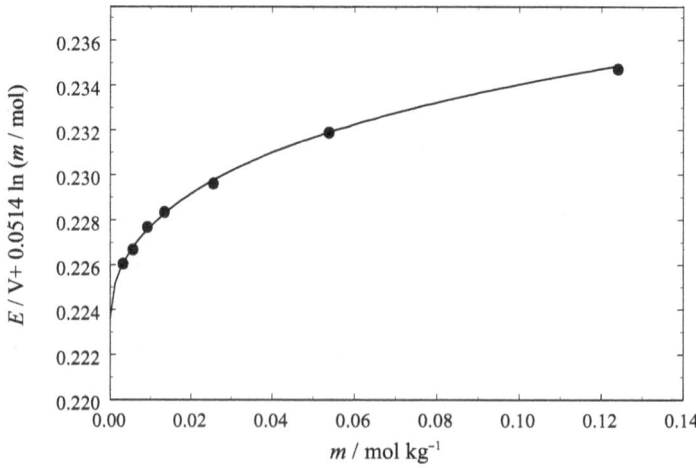

Extrapolating to zero molality gives $E^\circ = 0.223$ V, although there is considerable uncertainty in determining E° by this method. A reasonable estimate would be $E^\circ = 0.223 \pm 0.002$ V. This does compare well, however, with the value from Table 9.1, $E^\circ = 0.222$ V.

(b) At $m = 0.124$ mol kg^{-1}, $E = 0.342$ V. From Section 9.6,

$$E + (0.0514 \text{ V}) \ln m = E^\circ - (0.0514 \text{ V}) \ln \gamma_\pm$$

$$0.342 \text{ V} + (0.0514 \text{ V}) \ln 0.124 = 0.222 \text{ V} - (0.0514 \text{ V}) \ln \gamma_\pm$$

$$\ln \gamma_\pm = -0.247$$

$$\gamma_\pm = 0.78$$

where the more precisely determined value from Table 9.1 is used for E°. [Using the value determined in part (a) results in $\gamma_\pm = 0.80$.] These values compare well with the data shown in Figure 7.8.

9.36   Consider the following reaction:

$$Mg(s) + 2AgNO_3(aq) \rightarrow Mg(NO_3)_2(aq) + 2Ag(s)$$

Describe how you would measure the $\Delta_r G^\circ$, $\Delta_r H^\circ$, and $\Delta_r S^\circ$ of the reaction (a) thermochemically and (b) electrochemically. Compare the two methods.

(a) Thermochemical determination of these parameters would require measurement of equilibrium concentrations for the reaction in combination with calorimetry. One could prepare a solution containing both dissolved $AgNO_3$ and $Mg(NO_3)_2$ in contact with solid magnesium and silver metals. This solution would be allowed to reach equilibrium and the concentrations of both $AgNO_3$ and $Mg(NO_3)_2$ determined. This might be performed spectroscopically or perhaps gravimetrically through precipitation of appropriate salts. From the equilibrium concentrations, the equilibrium constant could be determined, and from this $\Delta_r G^\circ$.

$$\Delta_r G^\circ = -RT \ln K = -RT \ln \frac{\left[Mg(NO_3)_2 \right]}{\left[Ag(NO_3) \right]^2}$$

Starting with pure silver nitrate solution and magnesium metal, the same reaction could be carried out in a constant pressure calorimeter. (These are most convenient for aqueous phase reactions.) From the temperature in the calorimeter (see Section 3.3), $\Delta_r H^\circ$ is determined. Once both $\Delta_r G^\circ$ and $\Delta_r H^\circ$ are in hand, $\Delta_r S^\circ$ is found using

$$\Delta_r S^\circ = \frac{\Delta_r H^\circ - \Delta_r G^\circ}{T}$$

(b) The electrochemical determination of these quantities would use the measurement of an appropriate electrochemical cell potential and its temperature dependence. The cell would consist of one compartment containing a silver rod in contact with a silver nitrate solution of known concentration and a second compartment containing a magnesium rod in contact with a magnesium nitrate solution of known concentration. Once the cell potential is measured, the Nernst equation is used to determine E°.

$$E = E^\circ - \frac{0.0257 \text{ V}}{\nu} \ln \frac{\left[Mg(NO_3)_2 \right]}{\left[Ag(NO_3) \right]^2}$$

The standard Gibbs energy of reaction is determined using $\Delta_r G^\circ = -\nu F E^\circ$. The temperature coefficient, $\left(\partial E^\circ / \partial T \right)_P$ of the cell gives $\Delta_r S^\circ$.

$$\Delta_r S^\circ = \nu F \left(\frac{\partial E^\circ}{\partial T} \right)_P$$

Having found both $\Delta_r G^\circ$ and $\Delta_r S^\circ$, $\Delta_r H^\circ$ is determined using $\Delta_r H^\circ = \Delta_r G^\circ + T \Delta_r S^\circ$.

The electochemical determination avoids the requirement of ensuring that chemical equilibrium is attained and the need to measure potentially small (even vanishingly so) equilibrium concentrations.

9.38 Calculate the E° value for the propane–oxygen fuel cell at 298 K. The $\Delta_f \overline{G}^\circ$ value for propane is -23.49 kJ mol^{-1}.

Note that as on page 360 of the text, we will be using a more accurate value for the Faraday constant in working this problem. The propane–oxygen fuel cell utilizes the following half-reactions:

Anode: $C_3H_8(g) + 6H_2O(l) \rightarrow 3CO_2(g) + 20H^+(aq) + 20e^-$
Cathode: $5O_2(g) + 20H^+(aq) + 20e^- \rightarrow 10H_2O(l)$

The overall equation is

$$C_3H_8(g) + 5O_2(g) \rightarrow 3CO_2(g) + 4H_2O(l)$$

and the corresponding standard Gibbs energy of reaction is

$$\Delta_r G^\circ = 3\Delta_f \overline{G}^\circ [CO_2(g)] + 4\Delta_f \overline{G}^\circ [H_2O(l)] - \Delta_f \overline{G}^\circ [C_3H_8(g)] - 5\Delta_f \overline{G}^\circ [O_2(g)]$$

$$= 3\left(-394.4 \text{ kJ mol}^{-1}\right) + 4\left(-237.2 \text{ kJ mol}^{-1}\right) - \left(-23.49 \text{ kJ mol}^{-1}\right) + 5\left(0 \text{ kJ mol}^{-1}\right)$$

$$= -2108.51 \text{ kJ mol}^{-1}$$

Because $\Delta_r G^\circ = -\nu F E^\circ$,

$$E^\circ = -\frac{\Delta_r G^\circ}{\nu F}$$

$$= -\frac{\left(-2108.51 \times 10^3 \text{ J mol}^{-1}\right)}{20 \left(96485.3 \text{ C mol}^{-1}\right)}$$

$$= 1.0927 \text{ V}$$

9.40 Consider a concentration cell that consists of Co/Co^{2+} compartments of concentrations 0.10 M and 2.0 M, respectively. **(a)** Calculate the E_{cell} at 25°C. **(b)** What are the concentrations in the compartments when E_{cell} drops to 0.020 V? Assume volumes remain constant at 1.00 L in each compartment.

(a) The half reactions are

Anode: Co $\rightarrow$ Co^{2+}(0.10 M) + 2e$^-$
Cathode: Co^{2+}(2.0 M) + 2e$^-$ $\rightarrow$ Co

The overall reaction is

$$Co^{2+}(2.0\ M) \rightarrow Co^{2+}(0.10\ M)$$

The emf of the cell depends on the concentrations of Co^{2+} at both the anode and the cathode.

$$E = E° - \frac{0.0257\ V}{\nu}\ \ln\frac{0.10}{2.0}$$

$$= 0\ V - \frac{0.0257\ V}{2}\ \ln\frac{0.10}{2.0}$$

$$= 0.0385\ V = 0.039\ V$$

(b) Let x be the molarity of Co^{2+} that is produced in the anode when $E_{cell} = 0.020\ V$. It is also the concentration that is consumed in the cathode.

$$E = E° - \frac{0.0257\ V}{\nu}\ \ln\frac{0.10 + x}{2.0 - x}$$

$$0.020\ V = 0 - \frac{0.0257\ V}{2}\ \ln\frac{0.10 + x}{2.0 - x}$$

$$\frac{0.10 + x}{2.0 - x} = 0.211$$

$$0.10 + x = 0.422 - 0.211x$$

$$1.211x = 0.322$$

$$x = 0.266$$

The concentration of Co^{2+} in the anode compartment is $0.10 + x = 0.37\ M$ and the concentration in the cathode compartment is $2.0 - x = 1.73\ M$.

CHAPTER 10

Quantum Mechanics

PROBLEMS AND SOLUTIONS

10.2 The threshold frequency for dislodging an electron from a zinc metal surface is 8.54×10^{14} Hz. Calculate the minimum amount of energy required to remove an electron from the metal.

At the threshold frequency, the speed of electron, u, is 0. That is,

$$h\nu = \Phi + \frac{1}{2}m_e u^2 = \Phi$$

Therefore, the minimum amount of energy required to remove an electron from the metal is

$$\Phi = \left(6.626 \times 10^{-34}\,\text{J s}\right)\left(8.54 \times 10^{14}\,\text{Hz}\right) = 5.66 \times 10^{-19}\,\text{J}$$

Multiplying by Avogadro's constant gives the equivalent $\Phi = 341\,\text{kJ mol}^{-1}$.

10.4 Calculate the frequency and wavelength associated with the transition from the $n = 5$ to the $n = 3$ level in atomic hydrogen.

The wavenumber of the emitted radiation is

$$\tilde{\nu} = \left(109737\,\text{cm}^{-1}\right)\left|\left(\frac{1}{n_i^2} - \frac{1}{n_f^2}\right)\right| = \left(109737\,\text{cm}^{-1}\right)\left|\left(\frac{1}{5^2} - \frac{1}{3^2}\right)\right| = 7803.520\,\text{cm}^{-1}$$

A wavenumber is necessarily a positive quantity, as are wavelengths and frequencies, too. The negative sign that would arise in this and similar calculations indicates that an emission process is being considered. Therefore, we take the absolute value. The wavelength and the frequency of this radiation are

$$\lambda = \frac{1}{\tilde{\nu}} = \frac{1}{7803.520\,\text{cm}^{-1}} = 1.28147 \times 10^{-4}\,\text{cm} = 1.28147 \times 10^3\,\text{nm}$$

$$\nu = c\tilde{\nu} = \left(2.998 \times 10^{10}\,\text{cm s}^{-1}\right)\left(7803.520\,\text{cm}^{-1}\right) = 2.340 \times 10^{14}\,\text{Hz}$$

10.6 A photoelectric experiment was performed by separately shining a laser at 450 nm (blue light) and a laser at 560 nm (yellow light) on a clean metal surface and measuring the number and kinetic energy of the ejected electrons. Which light would generate more electrons? Which light would eject electrons with greater kinetic energy? Assume that the same number of photons is delivered to the metal surface by each laser and that the frequencies of the laser lights exceed the threshold frequency.

Since each laser is delivering the same number of photons to the metal surface, the two experiments will each generate the same number of electrons. Each blue light photon has greater energy than each yellow light photon. Consequently, those electrons ejected by the blue light will have greater kinetic energy.

10.8 In a photoelectric experiment, a student uses a light source whose frequency is greater than that needed to eject electrons from a certain metal. However, after continuously shining the light on the same area of the metal for a long period of time, the student notices that the maximum kinetic energy of ejected electrons begins to decrease, even though the frequency of the light is held constant. How would you account for this behavior?

In a photoelectric experiment where the ejected electrons are not replaced (perhaps by making the photocathode part of an electric circuit), the metal surface will become positively charged due to the loss of the negatively charged electrons. Eventually, this positive charge is sufficient to cause a noticeable attraction between the surface and the ejected electrons which lowers their kinetic energy.

10.10 Suppose that the uncertainty in determining the position of an electron circling an atom in an orbit is 0.4 Å. What is the uncertainty in its velocity?

The uncertainty in momentum is

$$\Delta p \geq \frac{h}{4\pi \Delta x} = \frac{6.626 \times 10^{-34} \text{ J s}}{4\pi \left(0.4 \times 10^{-10} \text{ m}\right)} = 1.3 \times 10^{-24} \text{ kg m s}^{-1}$$

Therefore, the uncertainty in the velocity of the electron is

$$\Delta v = \frac{\Delta p}{m} \geq \frac{1.3 \times 10^{-24} \text{ kg m s}^{-1}}{9.109 \times 10^{-31} \text{ kg}} = 1 \times 10^{6} \text{ m s}^{-1}$$

The uncertainty principle, when applied to a microscopic object such as an electron, results in a significant uncertainty in velocity.

10.12 The diffraction phenomenon can be observed whenever the wavelength is comparable in magnitude to the size of the slit opening. To be "diffracted," how fast must a person weighing 84 kg move through a door 1 m in width?

The person would need a wavelength comparable to 1 m to be diffracted. The momentum of the person would be

$$p = \frac{h}{\lambda} = \frac{6.626 \times 10^{-34} \text{ J s}}{1 \text{ m}} = 6.626 \times 10^{-34} \text{ kg m s}^{-1}$$

The velocity of the person is therefore

$$u = \frac{p}{m} = \frac{6.626 \times 10^{-34} \text{ kg m s}^{-1}}{84 \text{ kg}} = 7.9 \times 10^{-36} \text{ m s}^{-1}$$

At this rate, it would take 1.3×10^{35} s or 4.0×10^{27} years to move 1 m!

10.14 Spectral lines of the Lyman and Balmer series do not overlap. Verify this statement by calculating the longest wavelength associated with the Lyman series and the shortest wavelength associated with the Balmer series (in nm).

The longest wavelength associated with the Lyman series corresponds to $n_i = 2$ and $n_f = 1$. First calculate the wavenumber of the emitted radiation.

$$\tilde{\nu} = \left(109737 \text{ cm}^{-1}\right)\left|\left(\frac{1}{n_i^2} - \frac{1}{n_f^2}\right)\right| = \left(109737 \text{ cm}^{-1}\right)\left|\left(\frac{1}{2^2} - \frac{1}{1^2}\right)\right| = 82302.75 \text{ cm}^{-1}$$

Therefore, the wavelength of the radiation is

$$\lambda = \frac{1}{\tilde{\nu}} = \frac{1}{82302.75 \text{ cm}^{-1}} = 1.21503 \times 10^{-5} \text{ cm} = 121.503 \text{ nm}$$

The shortest wavelength associated with the Balmer series corresponds to $n_i = \infty$ and $n_f = 2$. First calculate the wavenumber of the emitted radiation.

$$\tilde{\nu} = \left(109737 \text{ cm}^{-1}\right)\left|\left(\frac{1}{n_i^2} - \frac{1}{n_f^2}\right)\right| = \left(109737 \text{ cm}^{-1}\right)\left|\left(\frac{1}{\infty^2} - \frac{1}{2^2}\right)\right| = 27434.25 \text{ cm}^{-1}$$

Therefore, the wavelength of the radiation is

$$\lambda = \frac{1}{\tilde{\nu}} = \frac{1}{27434.25 \text{ cm}^{-1}} = 3.64508 \times 10^{-5} \text{ cm} = 364.508 \text{ nm}$$

Therefore, spectral lines of the two series do not overlap.

10.16 An electron in an excited state in a hydrogen atom can return to the ground state in two different ways: **(a)** via a direct transition in which a photon of wavelength λ_1 is emitted and **(b)** via an intermediate excited state reached by the emission of a photon of wavelength λ_2. This intermediate excited state then decays to the ground state by emitting another photon of wavelength λ_3. Derive an equation that relates λ_1 to λ_2 and λ_3.

The energy of the photon of wavelength λ_1, E_1, equals the sum of the energy of the photon of wavelength λ_2, E_2, and the energy of the photon of wavelength λ_3, E_3. That is,

$$E_1 = E_2 + E_3$$
$$\frac{hc}{\lambda_1} = \frac{hc}{\lambda_2} + \frac{hc}{\lambda_3}$$
$$\frac{1}{\lambda_1} = \frac{1}{\lambda_2} + \frac{1}{\lambda_3}$$

10.18 A 368-g sample of water absorbs infrared radiation at 1.06×10^4 nm from a carbon dioxide laser. Suppose all the absorbed radiation is converted to heat. Calculate the number of photons at this wavelength required to raise the temperature of the water by 5.00°C.

The energy required to raise the temperature of water by 5.00°C is

$$ms\,\Delta T = (368 \text{ g}) \left(4.184 \text{ J g}^{-1}\,{}^{\circ}\text{C}^{-1}\right) (5.00{}^{\circ}\text{C}) = 7.699 \times 10^3 \text{ J}$$

The energy of one photon is

$$h\nu = \frac{hc}{\lambda} = \frac{(6.626 \times 10^{-34} \text{ J s})\,(2.998 \times 10^8 \text{ m s}^{-1})}{1.06 \times 10^{-5} \text{ m}} = 1.874 \times 10^{-20} \text{ J}$$

The number of photons required to raise the temperature of water by 5.00°C is

$$\frac{7.699 \times 10^3 \text{ J}}{1.874 \times 10^{-20} \text{ J}} = 4.11 \times 10^{23}$$

10.20 Scientists have found interstellar hydrogen atoms with quantum number n in the hundreds. Calculate the wavelength of light emitted when a hydrogen atom undergoes a transition from $n = 236$ to $n = 235$. In what region of the electromagnetic spectrum does this wavelength fall?

The wavenumber of the emitted radiation is

$$\tilde{\nu} = \left(109737 \text{ cm}^{-1}\right) \left| \left(\frac{1}{n_i^2} - \frac{1}{n_f^2} \right) \right| = \left(109737 \text{ cm}^{-1}\right) \left| \left(\frac{1}{236^2} - \frac{1}{235^2} \right) \right|$$

$$= 1.680406 \times 10^{-2} \text{ cm}^{-1}$$

Therefore, the wavelength of this radiation is

$$\lambda = \frac{1}{\tilde{\nu}} = \frac{1}{1.680406 \times 10^{-2} \text{ cm}^{-1}} = 59.5094 \text{ cm}$$

This wavelength is in the radio frequency region.

10.22 In the mid-nineteenth century, physicists studying the solar emission spectrum (a continuum) noticed a set of dark lines that did not match any of the emission lines (bright lines) on Earth. They concluded that the lines came from a yet unknown element. Later this element was identified as helium. **(a)** What is the origin of the dark lines? How were these lines correlated with the emission lines of helium? **(b)** Why was helium so difficult to detect in Earth's atmosphere? **(c)** Where is the most likely place to detect helium on Earth?

(a) The lines are absorption lines. For atomic systems, the lines occur at the same wavelengths as the emission lines (of helium in this case).

(b) Because of the light mass of helium, its thermal velocity is actually greater than the escape velocity of Earth's gravitational field. (The same is true for H_2, but unlike hydrogen, helium does not form compounds, whose greater mass traps the element on Earth.)

(c) α decay produces helium, so uranium mines are a likely source of helium. Helium is also found in natural gas deposits, where it has been trapped by layers of rock.

10.24   An alternative to the Copenhagen interpretation of the wave function is the so-called "many worlds" interpretation. Both interpretations are consistent with observation. Do some research on each of these interpretations and decide for yourself which one you prefer.

Some of the key features of the Copenhagen interpretation are (1) that although its square may be interpreted as a probability density, the wave function itself has no intrinsic meaning or existence even though it provides all the information we can know about the system it describes, (2) that quantum theory does not describe an objective reality, but rather describes the state of the observer's knowledge of a system, and (3) that the wave function of a system provides the probabilities for obtaining specific results of a measurement, but evolves independently of the observer and is not connected to the observer until the moment of measurement at which point it instantaneously, and by an unspecified mechanism, "collapses" to the eigenfunction associated with the obtained result. For much of the history of quantum mechanics, this has been the most widely utilized interpretation. The primary objections to the interpretation lie in the probabilistic nature of its predictions (the theory is not *deterministic*), the awkwardness of the wave function collapse, and the avoidance of the question of the true nature of the objective reality of the system.

The many worlds interpretation instead postulates a real existence for the wave function, but requires that the observer also be described by a wave function. The wave functions of the system and the observer evolve independently only until the moment of measurement, at which point they are coupled ("entangled") and the state of one can only be specified with respect to the state of the other. The measurement causes a branching of the universe into separate, non-communicating realities corresponding to each possible result. In the many world interpretation, (1) the wave function is real and is equivalent to the object, (2) although the quantum weight of each universe is proportional to a probability, the many worlds interpretation is deterministic in that each universe is "created" in a completely predictive manner, (3) there is no special role for the observer or for the process of measurement, and there is no need to invoke the collapse of the wave function. The primary objection to the many worlds interpretation is the necessity of invoking an ever increasing number of universes, although the strict many worlds interpretation would insist there is only one universe that contains all the possible outcomes simultaneously. Others have questioned whether the many worlds interpretation is "testable," or subjection to verification or contradiction, although recent advances in quantum computing have suggested potential possibilities for such an experiment.

The debate continues despite the unqualified success of quantum mechanics in describing the results of experiments. The decision about which interpretation one prefers should not compromise one's understanding and mastery of the use of quantum mechanics in chemistry.

10.26   Given the function $f(x)$, which of the following operators are Hermitian?

(a) $\dfrac{d^2}{dx^2}$, (b) $\dfrac{d}{dx}$, (c) the identity operator, (d) multiply by a real constant, (e) multiply by x.

An operator acting on a function in one-dimensional Cartesian space, $\hat{A}$, is Hermitian if $\int_{-\infty}^{\infty} f^*\hat{A}g \, dx = \int_{-\infty}^{\infty} g\hat{A}^* f^* \, dx$. The functions f and g depend on x and are well-behaved; thus, their limits as $x \to \pm\infty$ are 0 and the limits of their first derivatives as $x \to \pm\infty$ are also 0.

(a) For $\hat{A} = \dfrac{d^2}{dx^2}$,

$$\int_{-\infty}^{\infty} f^* \hat{A} g \, dx = \int_{-\infty}^{\infty} f^* \frac{d^2 g}{dx^2} \, dx$$

$$= f^* \frac{dg}{dx} \bigg|_{-\infty}^{\infty} - \int_{-\infty}^{\infty} \frac{df^*}{dx} \frac{dg}{dx} \, dx$$

$$= 0 - \int_{-\infty}^{\infty} \frac{df^*}{dx} \frac{dg}{dx} \, dx$$

$$= - \left(\frac{df^*}{dx} g \bigg|_{-\infty}^{\infty} - \int_{-\infty}^{\infty} \frac{d^2 f^*}{dx} g \, dx \right)$$

$$= - \left(0 - \int_{-\infty}^{\infty} g \hat{A}^* f^* \, dx \right)$$

$$= \int_{-\infty}^{\infty} g \hat{A}^* f^* \, dx$$

The operator d^2/dx^2 is Hermitian. Notice that $\hat{A}^* = \hat{A}$, and that because the operator contained a second derivative, integration by parts was used twice.

(b) For $\hat{A} = \dfrac{d}{dx}$,

$$\int_{-\infty}^{\infty} f^* \hat{A} g \, dx = \int_{-\infty}^{\infty} f^* \frac{dg}{dx} \, dx$$

$$= f^* g \big|_{-\infty}^{\infty} - \int_{-\infty}^{\infty} \frac{df^*}{dx} g \, dx$$

$$= 0 - \int_{-\infty}^{\infty} g \hat{A} f^* \, dx$$

$$= - \int_{-\infty}^{\infty} g \hat{A}^* f^* \, dx$$

$$\neq \int_{-\infty}^{\infty} g \hat{A}^* f^* \, dx$$

The operator d/dx is not Hermitian. Once again, $\hat{A}^* = \hat{A}$, but because integration by parts is only used once, a minus sign is introduced.

(c) For $\hat{A} = $ identity,

$$\int_{-\infty}^{\infty} f^* \hat{A} g \, dx = \int_{-\infty}^{\infty} f^* g \, dx$$

$$= \int_{-\infty}^{\infty} g f^* \, dx$$

$$= \int_{-\infty}^{\infty} g \hat{A}^* f^* \, dx$$

For identity operator, $\hat{A}^* = \hat{A}$, and identity operator is Hermitian.

(d) For $\hat{A} = a$, where a is a real constant,

$$\int_{-\infty}^{\infty} f^* \hat{A} g \, dx = \int_{-\infty}^{\infty} f^* a g \, dx$$

$$= \int_{-\infty}^{\infty} g a f^* \, dx$$

$$= \int_{-\infty}^{\infty} g a^* f^* \, dx$$

$$= \int_{-\infty}^{\infty} g \hat{A}^* f^* \, dx$$

Since a is real, $a = a^*$, and the operator $\hat{A} = a$ is Hermitian.

(e) For $\hat{A} = x$,

$$\int_{-\infty}^{\infty} f^* \hat{A} g \, dx = \int_{-\infty}^{\infty} f^* x g \, dx$$

$$= \int_{-\infty}^{\infty} g x f^* \, dx$$

$$= \int_{-\infty}^{\infty} g x^* f^* \, dx$$

$$= \int_{-\infty}^{\infty} g \hat{A}^* f^* \, dx$$

Since x is real, $x = x^*$, and the operator x is Hermitian.

10.28 Which of the following pairs of functions are orthogonal over the interval indicated?

(a) $f(x) = \sin(x)$ and $g(x) = \cos x$ $[-\infty, \infty]$, **(b)** $f(x) = \sin(x)$ and $g(x) = \cos x$ $[0, 2\pi]$, **(c)** $f(x) = e^{i\pi x}$ and $g(x) = e^{i2\pi x}$ $[-1, 1]$, **(d)** $f(x) = e^{i\pi x}$ and $g(x) = e^{-i\pi x}$ $[-1, 1]$, **(e)** $f(x) = ix$ and $g(x) = -ix$ $[-1, 1]$.

Two functions, $f(x)$ and $g(x)$ are orthogonal if $\int_{\text{all space}} f^* g \, dx = 0$. All space is the interval under consideration.

(a) $\int_{-\infty}^{\infty} f^* g \, dx = \int_{-\infty}^{\infty} \sin(x) \cos(x) \, dx$.

Since $\sin(x)$ is an odd function and $\cos(x)$ is an even function, the integrand is an odd function integrated over an interval symmetic about $x = 0$. Thus, the integral is 0. The functions are orthogonal.

(b) $\int_{0}^{2\pi} f^* g \, dx = \int_{0}^{2\pi} \sin(x) \cos(x) \, dx = 0$

The functions are orthogonal.

(c) $\int_{-1}^{1} f^* g \, dx = \int_{-1}^{1} e^{-i\pi x} e^{i2\pi x} dx = \int_{-1}^{1} e^{i\pi x} dx = 0$

The functions are orthogonal.

(d) $\int_{-1}^{1} f^* g \, dx = \int_{-1}^{1} e^{-i\pi x} e^{-i\pi x} dx = \int_{-1}^{1} e^{-i2\pi x} dx = 0$

The functions are orthogonal.

(e) $\int_{-1}^{1} f^* g \, dx = \int_{-1}^{1} (-ix)(-ix) \, dx = \int_{-1}^{1} \left(-x^2\right) dx = -\frac{2}{3} \neq 0$

The functions are not orthogonal.

10.30   Which of the following functions are normalized over the interval indicated?

(a) $f(x) = \dfrac{1}{\sqrt{a}}$ [0, a], **(b)** $f(x) = x$ [0, 1], **(c)** $f(x) = \sqrt{\dfrac{2}{a}} \sin \dfrac{5\pi x}{a}$ [0, a],

(d) $f(x) = \dfrac{1}{\sqrt{\pi}} e^{-ax^2}$ [−∞, ∞], **(e)** $f(x) = \left(\dfrac{a}{\pi}\right)^{1/4} e^{-ax^2/2}$ [−∞, ∞]

The function $f(x)$ is normalized if $\int_{\text{all space}} f^* f \, dx = 1$. All space is the interval under consideration. Assuming that the constant a is a real number, all the functions under consideration in this question are real. (This is self-evident from the specification of the interval in parts **(a)** and **(c)**). Thus, $f^* f = f^2$.

(a) $\int_0^a f^* f \, dx = \int_0^a \dfrac{1}{a} \, dx = \dfrac{1}{a} \int_0^a dx = 1$

The function is normalized.

(b) $\int_0^1 f^* f \, dx = \int_0^1 x^2 \, dx = \dfrac{1}{3}$

The function is not normalized.

(c) $\int_0^a f^* f \, dx = \int_0^a \left(\dfrac{2}{a}\right) \sin^2 \dfrac{5\pi x}{a} \, dx = \dfrac{2}{a} \int_0^a \sin^2 \dfrac{5\pi x}{a} \, dx = 1$

The function is normalized.

(d) $\int_{-\infty}^{\infty} f^* f \, dx = \int_{-\infty}^{\infty} \left(\dfrac{1}{\pi}\right) e^{-2ax^2} \, dx = \dfrac{1}{\pi} \int_{-\infty}^{\infty} e^{-2ax^2} \, dx = \dfrac{1}{\sqrt{2a\pi}}$

The integral is evaluated assuming that $a > 0$. In general, the function is not normalized, although it is for the specific case $a = \dfrac{1}{2\pi}$.

(e) $\int_{-\infty}^{\infty} f^* f \, dx = \int_{-\infty}^{\infty} \left(\dfrac{a}{\pi}\right)^{1/2} e^{-ax^2} \, dx = \left(\dfrac{a}{\pi}\right)^{1/2} \int_{-\infty}^{\infty} e^{-ax^2} \, dx = 1$

The integral is evaluated assuming that $a > 0$. The function is normalized.

10.32   According to Equation 10.50, the energy is inversely proportional to the square of the length of the box. How would you account for this dependence in terms of the Heisenberg uncertainty principle?

As the length of the box is decreased, the particle is located with greater certainty. Therefore, Δx decreases. According to the uncertainty principle,

$$\Delta x \, \Delta p \geq \frac{h}{4\pi}$$

requires an increase in Δp and hence p itself. Consequently, the kinetic energy of the particle, $p^2/2m$ must also increase.

10.34 Derive Equation 10.50 using de Broglie's relation. (*Hint*: First you must express the wavelength of the particle in the *n*th level in terms of the length of the box.)

According to Figure 10.17(a), the wavelength of the particle is given by

$$\lambda = \frac{2L}{n}$$

where $n = 1, 2, 3 \ldots$. The wavelength is also related to the momentum of the particle:

$$\lambda = \frac{h}{p} = \frac{h}{mu}$$

Equating the two expressions for λ and solving for u gives

$$\frac{2L}{n} = \frac{h}{mu}$$

$$u = \frac{nh}{2mL}$$

The particle has only kinetic energy so that

$$E = \frac{1}{2}mu^2 = \frac{1}{2}m\frac{n^2h^2}{4m^2L^2} = \frac{n^2h^2}{8mL^2}$$

10.36 Use Equation 10.57 to calculate the wavelength of the electronic transition in polyenes for $N = 6, 8,$ and 10. Comment on the variation of λ with L, the length of the molecule.

Equation 10.57 gives the wavelength for the electronic transition in polyenes as

$$\lambda = \frac{8m_e L^2 c}{h\left(N + 1\right)}$$

For $N = 6$, L is a sum of 3 double bonds, 2 single bonds and twice the radius of a C atom, that is,

$$L = 3\left(1.35 \text{ Å}\right) + 2\left(1.54 \text{ Å}\right) + 2\left(0.77 \text{ Å}\right) = 8.67 \text{ Å}$$

The wavelength of the electronic transition for $N = 6$ is therefore

$$\lambda = \frac{8\left(9.109 \times 10^{-31} \text{ kg}\right)\left(8.67 \times 10^{-10} \text{ m}\right)^2\left(2.998 \times 10^8 \text{ m s}^{-1}\right)}{\left(6.626 \times 10^{-34} \text{ J s}\right)(7)}$$

$$= 3.54 \times 10^{-7} \text{ m} = 354 \text{ nm}$$

For $N = 8$, L is a sum of 4 double bonds, 3 single bonds and twice the radius of a C atom, that is,

$$L = 4\left(1.35 \text{ Å}\right) + 3\left(1.54 \text{ Å}\right) + 2\left(0.77 \text{ Å}\right) = 11.56 \text{ Å}$$

The wavelength of the electronic transition for $N = 8$ is therefore

$$\lambda = \frac{8\left(9.109 \times 10^{-31}\,\text{kg}\right)\left(11.56 \times 10^{-10}\,\text{m}\right)^2\left(2.998 \times 10^8\,\text{m s}^{-1}\right)}{\left(6.626 \times 10^{-34}\,\text{J s}\right)(9)}$$

$$= 4.896 \times 10^{-7}\,\text{m} = 489.6\,\text{nm}$$

For $N = 10$, L is a sum of 5 double bonds, 4 single bonds and twice the radius of a C atom, that is,

$$L = 5\left(1.35\,\text{Å}\right) + 4\left(1.54\,\text{Å}\right) + 2\left(0.77\,\text{Å}\right) = 14.45\,\text{Å}$$

The wavelength of the electronic transition for $N = 10$ is therefore

$$\lambda = \frac{8\left(9.109 \times 10^{-31}\,\text{kg}\right)\left(14.45 \times 10^{-10}\,\text{m}\right)^2\left(2.998 \times 10^8\,\text{m s}^{-1}\right)}{\left(6.626 \times 10^{-34}\,\text{J s}\right)(11)}$$

$$= 6.259 \times 10^{-7}\,\text{m} = 625.9\,\text{nm}$$

As N increases and the molecule gets longer, λ increases and shifts the light from UV to visible.

10.38 As stated in the chapter, the probability of locating a particle in a one-dimensional box is given by $\int \psi^*\psi\,dx$. Over a small distance, the probability can be calculated without integration as $\psi^*\psi\,\Delta x$. Consider an electron with $n = 1$ in a box of length 2.000 nm. Calculate the probability of locating the electron **(a)** between 0.500 nm and 0.502 nm and **(b)** between 0.999 nm and 1.001 nm. Comment on your results and on the validity of your approximation.

For $n = 1$,

$$\psi^*\psi\,dx = \psi^2\,dx = \frac{2}{L}\sin^2\frac{\pi x}{L}\,dx$$

(a) x can be approximated by the average of 0.500 nm and 0.502 nm. That is, $x = 0.501$ nm. $dx = 0.502\,\text{nm} - 0.500\,\text{nm} = 0.002\,\text{nm}$. Therefore, the probability of locating the electron is

$$\frac{2}{2.000\,\text{nm}}\left[\sin^2\frac{\pi\,(0.501\,\text{nm})}{2.000\,\text{nm}}\right](0.002\,\text{nm}) = (0.002)\sin^2\left(0.2505\pi\right)$$

$$= 1 \times 10^{-3}$$

(b) x can be approximated by the average of 0.999 nm and 1.001 nm. That is, $x = 1.000$ nm. $dx = 1.001\,\text{nm} - 0.999\,\text{nm} = 0.002\,\text{nm}$. Therefore, the probability of locating the electron is

$$\frac{2}{2.000\,\text{nm}}\left[\sin^2\frac{\pi\,(1.000\,\text{nm})}{2.000\,\text{nm}}\right](0.002\,\text{nm}) = (0.002)\sin^2\left(0.5000\pi\right)$$

$$= 0.002$$

The result from part **(b)** is, in fact, for the most probable location to find the particle, the middle of the box. The approximation is valid for small dx, becoming better as dx gets smaller, and is equivalent to assuming that the value of the wave function is approximately constant over the interval.

10.40 Under what conditions will the energy levels of a particle in a two-dimensional box show degeneracies?

The energy levels of a particle in a two-dimensional box will show degeneracies whenever

$$\frac{n_x^2}{a^2} + \frac{n_y^2}{b^2} = \frac{n_x'^2}{a^2} + \frac{n_y'^2}{b^2}$$

for pairs of integers (n_x, n_y) and (n_x', n_y'). In general this requires that a be an integral multiple of b or vice versa.

10.42 Calculate the angular momentum of an electron on a ring of diameter 10 pm with quantum number $m = 3$. In what direction does the angular momentum vector point?

The wave function corresponding to $m = 3$ is $\psi_3(\phi) = \frac{1}{\sqrt{2\pi}}e^{i3\phi}$. It is an eigenfunction of the angular momentum operator, $\hat{L}_z$:

$$\hat{L}_z \psi_3(\phi) = -i\hbar\frac{d}{d\phi}\left(\frac{1}{\sqrt{2\pi}}e^{i3\phi}\right) = 3\hbar\left(\frac{1}{\sqrt{2\pi}}e^{i3\phi}\right) = 3\hbar\psi_3(\phi)$$

The eigenvalue of $\hat{L}_z$, $3\hbar = 3\left(1.055 \times 10^{-34}\ \text{J s}\right) = 3.165 \times 10^{-34}\ \text{J s}$, is the angular momentum of the electron. Since this is a positive number, the angular momentum vector points in the direction of the positive z direction, as determined by application of the right hand rule to the x and y axes.

10.44 To calculate the energy of core electrons in heavy atoms, it is necessary to use Einstein's theory of relativity because the electrons are moving at nearly the speed of light c. As a model illustration, calculate the (nonrelativistic) speed of an electron moving in a 2.0-pm-diameter ring with quantum number $m = 0$, 1, and 2. In each case, how does the speed of the electron compare with c? (Note that 2.0 pm is the order of the size of a $1s$ atomic orbital for a gold atom, and that the color of gold cannot be explained accurately without a relativistic theory. Although larger quantum numbers would predict speeds greater than c, actual electron speeds never exceed c.)

Since the potential energy for the system is 0, the kinetic energy, $\frac{1}{2}m_e u^2$ is simply the total energy, which is $E_m = \frac{\hbar^2}{2I}m^2 = \frac{\hbar^2}{2m_e r^2}m^2$. Here, m_e, u, and I are the mass, the speed, and the moment of inertia of the electron, respectively, and r is the radius of the ring, 1.0 pm. Thus,

$$\frac{1}{2}m_e u^2 = \frac{\hbar^2}{2m_e r^2}m^2$$

$$u = \frac{\hbar m}{m_e r}$$

For $m = 0$, $u = 0$, or 0% of the speed of light.

For $m = 1$,

$$u = \frac{\hbar}{m_e r}$$

$$= \frac{1.055 \times 10^{-34} \, \text{J s}}{\left(9.109 \times 10^{-31} \, \text{kg}\right)\left(1.0 \times 10^{-12} \, \text{m}\right)}$$

$$= 1.16 \times 10^8 \, \text{m s}^{-1} = 1.2 \times 10^8 \, \text{m s}^{-1}$$

This is $\dfrac{1.16 \times 10^8 \, \text{m s}^{-1}}{2.998 \times 10^8 \, \text{m s}^{-1}} \times 100\% = 39\%$ of the speed of light.

For $m = 2$,

$$u = \frac{2\hbar}{m_e r}$$

$$= \frac{2\left(1.055 \times 10^{-34} \, \text{J s}\right)}{\left(9.109 \times 10^{-31} \, \text{kg}\right)\left(1.0 \times 10^{-12} \, \text{m}\right)}$$

$$= 2.32 \times 10^8 \, \text{m s}^{-1} = 2.3 \times 10^8 \, \text{m s}^{-1}$$

This is $\dfrac{2.32 \times 10^8 \, \text{m s}^{-1}}{2.998 \times 10^8 \, \text{m s}^{-1}} \times 100\% = 77\%$ of the speed of light.

10.46   Tungsten has a work function of 4.55 eV. Calculate the longest wavelength that may be used to eject a photoelectron from a clean tungsten surface in a vacuum.

The kinetic energy, K, of the electron is the difference between the photon energy, $E_{\text{photon}} = h\nu$ and the work function, Φ:

$$K = h\nu - \Phi = \frac{hc}{\lambda} - \Phi$$

The longest wavelength that may be used to eject a photoelectron is one that would give it zero kinetic energy. Thus,

$$\frac{hc}{\lambda} - \Phi = 0$$

It follows then

$$\lambda = \frac{hc}{\Phi} = \frac{\left(6.626 \times 10^{-34} \, \text{J s}\right)\left(2.998 \times 10^8 \, \text{m s}^{-1}\right)}{\left(4.55 \, e\text{V}\right)\left(1.602 \times 10^{-19} \, \text{J} \, e\text{V}^{-1}\right)}$$

$$= 2.72 \times 10^{-7} \, \text{m} = 272 \, \text{nm}$$

10.48   Photodissociation of water,

$$H_2O(g) \xrightarrow{h\nu} H_2(g) + \frac{1}{2} O_2(g)$$

has been suggested as a source of molecular hydrogen. The $\Delta_r H^\circ$ for the reaction, calculated from thermochemical data, is 285.8 kJ mol^{-1} of water decomposed. Calculate the maximum wavelength (in nm) that would provide the necessary energy. In principle, is it feasible to use sunlight as a source of energy for this process?

The minimum photon energy is 285.8 kJ mol^{-1}.

$$E = \left(285.8 \times 10^3 \text{ J mol}^{-1}\right) \left(\frac{1 \text{ mol}}{6.022 \times 10^{23}}\right) = 4.7459 \times 10^{-19} \text{ J} = h\nu = \frac{hc}{\lambda}$$

$$\lambda = \frac{\left(6.626 \times 10^{-34} \text{ J s}\right) \left(2.998 \times 10^8 \text{ m s}^{-1}\right)}{4.7459 \times 10^{-19} \text{ J}} = 4.186 \times 10^{-7} \text{ m} = 418.6 \text{ nm}$$

This wavelength is in the visible range of the electromagnetic spectrum. Since water is continuously being struck by visible radiation without decomposition, it seems unlikely that direct photodissociation of water by this method is possible. This is because water does not absorb light at this, or most any, visible wavelength.

10.50   Only a fraction of the electrical energy supplied to a tungsten-filament light bulb is converted to visible light. The rest of the energy shows up as infrared radiation. A 75-W light bulb converts 15.0 percent of the energy supplied to it into visible light. If the wavelength is 550 nm, then how many photons are emitted by the light bulb per second? (1 W = 1 J s^{-1}.)

The energy of visible light emitted by the light bulb per second is

$$E_{\text{bulb}} = \left(75 \text{ J s}^{-1}\right) (1 \text{ s}) (15\%) = 11.3 \text{ J}$$

The energy of a 550 nm photon is

$$E_{\text{photon}} = h\nu = \frac{hc}{\lambda} = \frac{\left(6.626 \times 10^{-34} \text{ J s}\right) \left(2.998 \times 10^8 \text{ m s}^{-1}\right)}{550 \times 10^{-9} \text{ m}} = 3.612 \times 10^{-19} \text{ J}$$

The number of photons emitted by the light bulb per second is

$$\frac{11.3 \text{ J}}{3.612 \times 10^{-19} \text{ J}} = 3.1 \times 10^{19}$$

10.52   The ionization energy of a certain element is 412 kJ mol^{-1}. When the atoms of this element are in the first excited state, however, the ionization energy is only 126 kJ mol^{-1}. Based on this information, calculate the wavelength of light emitted in a transition from the first excited state to the ground state.

The ionization energy of 412 kJ mol^{-1} represents the energy difference between the ground state and the dissociation limit whereas the ionization energy of 126 kJ mol^{-1} represents the energy difference between the first excited state and the dissociation limit. Therefore, the energy difference between the ground state and the first excited state is

$$\Delta E = (412 - 126) \text{ kJ mol}^{-1} = 286 \text{ kJ mol}^{-1}$$

The energy of light emitted in a transition from the first excited state to the ground state is therefore 286 kJ mol^{-1}. The wavelength emitted is calculated as follows.

$$E = \left(286 \times 10^3 \text{ J mol}^{-1}\right) \left(\frac{1 \text{ mol}}{6.022 \times 10^{23}}\right) = 4.749 \times 10^{-19} \text{ J} = h\nu = \frac{hc}{\lambda}$$

$$\lambda = \frac{\left(6.626 \times 10^{-34} \text{ J s}\right)\left(2.998 \times 10^8 \text{ m s}^{-1}\right)}{4.749 \times 10^{-19} \text{ J}} = 4.18 \times 10^{-7} \text{ m} = 418 \text{ nm}$$

10.54   In 1996 physicists created an anti-atom of hydrogen. In such an atom, which is the antimatter equivalent of an ordinary atom, the electrical charges of all the component particles are reversed. Thus, the nucleus of an anti-atom is made of an anti-proton, which has the same mass as a proton but bears a negative charge, while the electron is replaced by an anti-electron (also called positron) with the same mass as an electron but bearing a positive charge. Would you expect the energy levels, emission spectra, and atomic orbitals of an anti-hydrogen atom to be different from those of a hydrogen atom? What would happen if an anti-atom of hydrogen collided with a hydrogen atom?

The anti-atom of hydrogen should show the same characteristics with regard to energy levels, emission spectra, and atomic orbitals as ordinary hydrogen. Should an anti-atom of hydrogen collide with an ordinary hydrogen atom, they would annihilate each other, and energy would be given off.

10.56   Use Equation 2.7 to calculate the de Broglie wavelength of a N_2 molecule at 300 K.

The rms speed of N_2 is

$$v_{\text{rms}} = \sqrt{\frac{3RT}{\mathcal{M}}} = \sqrt{\frac{3\left(8.314 \text{ J K}^{-1} \text{ mol}^{-1}\right)(300 \text{ K})}{28.02 \times 10^{-3} \text{ kg}}} = 516.7 \text{ m s}^{-1}$$

Using $u = v_{\text{rms}}$, the de Broglie wavelength of the molecule is

$$\lambda = \frac{h}{mu} = \frac{6.626 \times 10^{-34} \text{ J s}}{(28.02 \text{ amu})\left(1.6605 \times 10^{-27} \text{ kg amu}^{-1}\right)\left(516.7 \text{ m s}^{-1}\right)} = 2.76 \times 10^{-11} \text{ m}$$

10.58   The sun is surrounded by a white sphere of gaseous material called the corona, which becomes visible during a total eclipse of the sun. The temperature of the corona is in the millions of degrees Celsius, high enough to break up molecules and remove some or all of the electrons from atoms. One way astronomers have been able to estimate the temperature of the corona is by studying the emission lines of ions of certain elements. For example, the emission spectrum of Fe^{14+} ions has been recorded and analyzed. Knowing that it takes 3.5×10^4 kJ mol^{-1} to convert Fe^{13+} to Fe^{14+}, estimate the temperature of the sun's corona. (*Hint*: The average kinetic energy of one mole of a gas is $\frac{3}{2}RT$.)

The energy required to create the Fe^{14+} ion from Fe^{13+} must come from the thermal energy of the corona. That is, collisions with other species in the plasma provide the necessary energy. Thus, estimate the average kinetic energy in the plasma as being equal to the ionization energy of Fe^{13+}.

$$K = IE$$

$$\frac{3}{2}RT = 3.5 \times 10^7 \, \text{J mol}^{-1}$$

$$T = \frac{2\left(3.5 \times 10^7 \, \text{J mol}^{-1}\right)}{3\left(8.314 \, \text{J K}^{-1} \, \text{mol}^{-1}\right)} = 2.8 \times 10^6 \, \text{K}$$

10.60 A clean metal surface in a vacuum is illuminated with 1 W of orange light and 1 W of violet light. Assume that the photon energy from each source exceeds the work function of the metal. **(a)** The number of electrons emitted from the surface illuminated by orange light is (greater than, less than, equal to) the number emitted from the surface illuminated by violet light. **(b)** The kinetic energy of electrons emitted from the surface illuminated by orange light is (greater than, less than, equal to) the kinetic energy of electrons emitted from the surface illuminated by violet light.

(a) The number of electrons emitted from the surface illuminated by orange light is **greater than** the number emitted from the surface illuminated by violet light. Both light sources are delivering the same amount of energy each second, but since orange photons have less energy per photon than violet ones, there must a greater number of orange photons.

(b) The kinetic energy of electrons emitted from the surface illuminated by orange light is **less than** the kinetic energy of electrons emitted from the surface illuminated by violet light. Orange photons contain less energy per photon than violet ones, so the amount of energy exceeding the work function of the surface and that is converted to the kinetic energy of the photoelectrons is less for the surface illuminated by orange light.

10.62 The 2006 Nobel Prize in Physics was awarded to John C. Mather and George F. Smoot "for their discovery of the blackbody form and anisotropy of the cosmic background radiation." Understanding subtle variations in the cosmic background is an active area of research in cosmology, because it has implications for the beginnings of the known universe. The background radiation from the "empty" part of space is well modeled by a 2.7 K blackbody. What is the wavelength of maximum emission for this radiation? In what region of the electromagnetic spectrum does it lie?

The maximum emission is found at the wavelength where the Planck distribution law reaches its maximum value and is found where the derivative with respect to λ is equal to zero. Start with Equation 10.5 and apply both the product and quotient rules for derivatives.

$$\frac{d\rho_\lambda}{d\lambda} = \frac{d}{d\lambda}\left(\frac{8\pi hc}{\lambda^5} \frac{1}{e^{hc/\lambda k_B T} - 1}\right)$$

$$= -5\frac{8\pi hc}{\lambda^6}\frac{1}{e^{hc/\lambda k_B T} - 1} + \frac{8\pi hc}{\lambda^5}\frac{e^{hc/\lambda k_B T}}{\left(e^{hc/\lambda k_B T} - 1\right)^2}\frac{hc}{\lambda^2 k_B T}$$

At $\lambda = \lambda_{max}$, this equals zero.

$$-5\frac{8\pi hc}{\lambda_{max}^6}\frac{1}{e^{hc/\lambda_{max}k_BT}-1}+\frac{8\pi hc}{\lambda_{max}^5}\frac{e^{hc/\lambda_{max}k_BT}}{\left(e^{hc/\lambda_{max}k_BT}-1\right)^2}\frac{hc}{\lambda_{max}^2 k_BT}=0$$

$$-e^{hc/\lambda_{max}k_BT}+1+e^{hc/\lambda_{max}k_BT}\frac{hc}{5\lambda_{max}k_BT}=0$$

$$-1+e^{-hc/\lambda_{max}k_BT}+\frac{hc}{5\lambda_{max}k_BT}=0$$

Now make the substitution, $hc/\left(\lambda_{max}k_BT\right)=x$, to get $e^{-x}+\frac{x}{5}=1$, or $x=5-5e^{-x}$. This equation cannot be solved analytically, but using *Mathematica* a numerical solution is found for $x=4.965$. Recalling the definition of x,

$$\frac{hc}{\lambda_{max}k_BT}=4.965$$

$$\lambda_{max}=\frac{hc}{4.965k_BT}$$

$$=\frac{\left(6.626\times10^{-34}\,\text{J s}\right)\left(2.998\times10^8\,\text{m s}^{-1}\right)}{4.965\left(1.381\times10^{-23}\,\text{J K}^{-1}\right)T}$$

$$=\frac{2.90\times10^{-3}\,\text{m K}}{T},$$

for $T=2.7$ K, $\lambda_{max}=1.1$ mm. This is in the microwave region of the electromagnetic spectrum.

10.64   Consider a particle in a one-dimensional box in the ground state ψ_1 and the first excited state ψ_2 described by the wave functions listed below. For each wave function, calculate the expectation value of the position $\langle x\rangle$, the expectation value of the position squared $\langle x^2\rangle$, the expectation value of the momentum $\langle p\rangle$, and the expectation value of the momentum squared $\langle p^2\rangle$.

(a) $\psi_1(x)=\sqrt{\dfrac{2}{a}}\sin\dfrac{\pi x}{a}$, **(b)** $\psi_2(x)=\sqrt{\dfrac{2}{a}}\sin\dfrac{2\pi x}{a}$ $0\le x\le a$

The expectation value of α is

$$\langle\alpha\rangle=\frac{\int\psi_n^*\hat{A}\psi_n\,d\tau}{\int\psi_n^*\psi_n\,d\tau}$$

where $\hat{A}$ is the operator associated with α. Since both ψ_1 and ψ_2 are normalized (the normalization constant, $\sqrt{\dfrac{2}{a}}$, is determined in the text), the denominator, $\int\psi_n^*\psi_n\,d\tau$ is 1.

(a) For the state described by ψ_1:

$$\langle x \rangle = \int_0^a \psi_1^* \, x \, \psi_1 \, dx = \frac{2}{a} \int_0^a x \sin^2 \frac{\pi x}{a} \, dx = \frac{a}{2}$$

$$\langle x^2 \rangle = \int_0^a \psi_1^* \, x^2 \, \psi_1 \, dx = \frac{2}{a} \int_0^a x^2 \sin^2 \frac{\pi x}{a} \, dx$$

$$= \frac{2}{a} \left[\frac{a^3}{12} \left(2 - \frac{3}{\pi^2} \right) \right]$$

$$= a^2 \left(\frac{1}{3} - \frac{1}{2\pi^2} \right)$$

$$\langle p \rangle = \int_0^a \psi_1^* \left(-i\hbar \frac{d}{dx} \right) \psi_1 \, dx$$

$$= -i\hbar \int_0^a \left(\sqrt{\frac{2}{a}} \sin \frac{\pi x}{a} \right) \frac{d}{dx} \left(\sqrt{\frac{2}{a}} \sin \frac{\pi x}{a} \right) dx$$

$$= -i\hbar \left(\frac{2}{a} \right) \left(\frac{\pi}{a} \right) \int_0^a \sin \frac{\pi x}{a} \cos \frac{\pi x}{a} \, dx$$

$$= -i\hbar \left(\frac{2}{a} \right) \left(\frac{\pi}{a} \right) (0) = 0$$

$$\langle p^2 \rangle = \int_0^a \psi_1^* \left(-i\hbar \frac{d}{dx} \right) \left(-i\hbar \frac{d}{dx} \right) \psi_1 \, dx$$

$$= -\hbar^2 \int_0^a \left(\sqrt{\frac{2}{a}} \sin \frac{\pi x}{a} \right) \frac{d^2}{dx^2} \left(\sqrt{\frac{2}{a}} \sin \frac{\pi x}{a} \right) dx$$

$$= \hbar^2 \left(\frac{2}{a} \right) \left(\frac{\pi^2}{a^2} \right) \int_0^a \sin \frac{\pi x}{a} \sin \frac{\pi x}{a} \, dx$$

$$= \hbar^2 \left(\frac{2}{a} \right) \left(\frac{\pi^2}{a^2} \right) \left(\frac{a}{2} \right)$$

$$= \frac{\pi^2 \hbar^2}{a^2} = \frac{h^2}{4a^2}$$

(b) For the state described by ψ_2:

$$\langle x \rangle = \int_0^a \psi_2^* \, x \, \psi_2 \, dx = \frac{2}{a} \int_0^a x \sin^2 \frac{2\pi x}{a} \, dx = \frac{a}{2}$$

$$\langle x^2 \rangle = \int_0^a \psi_2^* \, x^2 \, \psi_2 \, dx = \frac{2}{a} \int_0^a x^2 \sin^2 \frac{2\pi x}{a} \, dx$$

$$= \frac{2}{a} \left[\frac{a^3}{3} \left(\frac{1}{2} - \frac{3}{16\pi^2} \right) \right]$$

$$= a^2 \left(\frac{1}{3} - \frac{1}{8\pi^2} \right)$$

$$\langle p \rangle = \int_0^a \psi_2^* \left(-i\hbar \frac{d}{dx} \right) \psi_2 \, dx$$

$$= -i\hbar \int_0^a \left(\sqrt{\frac{2}{a}} \sin \frac{2\pi x}{a} \right) \frac{d}{dx} \left(\sqrt{\frac{2}{a}} \sin \frac{2\pi x}{a} \right) dx$$

$$= -i\hbar \left(\frac{2}{a}\right)\left(\frac{2\pi}{a}\right) \int_0^a \sin\frac{2\pi x}{a} \cos\frac{2\pi x}{a} \, dx$$

$$= -i\hbar \left(\frac{2}{a}\right)\left(\frac{2\pi}{a}\right) (0) = 0$$

$$\langle p^2 \rangle = \int_0^a \psi_2^* \left(-i\hbar\frac{d}{dx}\right)\left(-i\hbar\frac{d}{dx}\right) \psi_2 \, dx$$

$$= -\hbar^2 \int_0^a \left(\sqrt{\frac{2}{a}}\sin\frac{2\pi x}{a}\right)\frac{d^2}{dx^2}\left(\sqrt{\frac{2}{a}}\sin\frac{2\pi x}{a}\right) dx$$

$$= \hbar^2 \left(\frac{2}{a}\right)\left(\frac{4\pi^2}{a^2}\right) \int_0^a \sin\frac{2\pi x}{a} \sin\frac{2\pi x}{a} \, dx$$

$$= \hbar^2 \left(\frac{2}{a}\right)\left(\frac{4\pi^2}{a^2}\right)\left(\frac{a}{2}\right)$$

$$= \frac{4\pi^2\hbar^2}{a^2} = \frac{h^2}{a^2}$$

10.66   The following figure represents the emission spectrum of a hydrogenlike ion in the gas phase. All the lines result from the electronic transitions from the excited states to the $n = 2$ state.

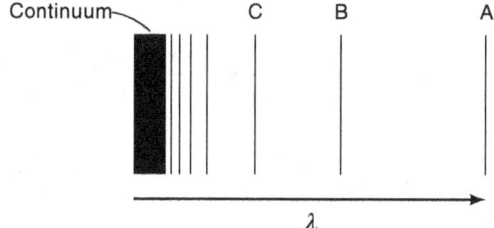

(a) What electronic transitions correspond to lines B and C? **(b)** If the wavelength of line C is 27.1 nm, calculate the wavelengths of lines A and B. **(c)** Calculate the energy needed to remove the electron from the ion in the $n = 4$ state. **(d)** What is the physical significance of the continuum? (*Hint:* Use the Rydberg constant for hydrogen in your calculation.)

(a) Lines B and C correspond to the second and third longest wavelengths (or second and third smallest energy) transitions. Thus, B is a transition from $n = 4$ to $n = 2$ whereas C is a transition from $n = 5$ to $n = 2$.

(b) For an electronic transition for this hydrogenlike atom with atomic number Z,

$$\tilde{\nu} = \frac{1}{\lambda} = Z^2 \tilde{R}_H \left(\frac{1}{n_i^2} - \frac{1}{n_f^2}\right)$$

For line C, $n_i = 5$ and $n_f = 2$,

$$\frac{1}{27.1 \times 10^{-7} \text{ cm}} = Z^2 \left(109737 \text{ cm}^{-1}\right)\left|\frac{1}{25} - \frac{1}{4}\right|$$

$$Z = 4$$

Thus, the wavelength of line A (the longest wavelength transition, $n_i = 3$ and $n_f = 2$) is

$$\frac{1}{\lambda} = Z^2 \tilde{R}_H \left(\frac{1}{n_i^2} - \frac{1}{n_f^2} \right)$$

$$= 16 \left(109737 \text{ cm}^{-1} \right) \left| \frac{1}{9} - \frac{1}{4} \right|$$

$$= 2.4386 \times 10^5 \text{ cm}^{-1}$$

$$\lambda = 41.0 \text{ nm}$$

and the wavelength of line B is

$$\frac{1}{\lambda} = Z^2 \tilde{R}_H \left(\frac{1}{n_i^2} - \frac{1}{n_f^2} \right)$$

$$= 16 \left(109737 \text{ cm}^{-1} \right) \left| \frac{1}{16} - \frac{1}{4} \right|$$

$$= 3.2921 \times 10^5 \text{ cm}^{-1}$$

$$\lambda = 30.4 \text{ nm}$$

(c) When the electron is removed from the $n = 4$ state, $n_f = \infty$:

$$\tilde{\nu} = Z^2 \tilde{R}_H \left(\frac{1}{n_i^2} - \frac{1}{n_f^2} \right)$$

$$= 16 \left(109737 \text{ cm}^{-1} \right) \left| \frac{1}{16} - \frac{1}{\infty} \right|$$

$$= 109737 \text{ cm}^{-1}$$

The corresponding energy is

$$E = h\nu = hc\tilde{\nu}$$

$$= \left(6.626 \times 10^{-34} \text{ J s} \right) \left(2.998 \times 10^{10} \text{ cm s}^{-1} \right) \left(109737 \text{ cm}^{-1} \right)$$

$$= 2.180 \times 10^{-18} \text{ J}$$

(d) The continuum represents transitions from unbound states of the electron to the $n = 2$ state.

10.68 The benzene molecule has a hexagonal symmetry, which may be approximated as a circular ring. Thus, we can apply the particle-on-a-ring model to predict the electronic structure of benzene. (a) From the C–C bond length in benzene, we can show that the radius of the ring is 132 pm. Starting with Equation 10.77, derive an expression for the transition wavenumbers (in cm^{-1} units) of this system. (b) Calculate the transition wavenumbers from the $m = 1$ state to the $m = 2$ state (or equivalently, from the $m = -1$ state to the $m = -2$ state). (c) The experimentally measured value for this transition is 37,900 cm^{-1}. Comment on the discrepancy between the predicted and measured values.

(a) The change in energy associated with a transition from the m_i state to the m_f state is

$$\Delta E = E_\mathrm{f} - E_\mathrm{i} = \frac{\hbar^2}{2I}\left(m_\mathrm{f}^2 - m_\mathrm{i}^2\right)$$

$$= \frac{\hbar^2}{2m_e r^2}\left(m_\mathrm{f}^2 - m_\mathrm{i}^2\right)$$

When the energy of the electromagnetic radiation matches this energy difference, a transition may occur. The transition wavenumbers are

$$hc\tilde{\nu} = \frac{\hbar^2}{2m_e r^2}\left(m_\mathrm{f}^2 - m_\mathrm{i}^2\right)$$

$$\tilde{\nu} = \frac{h}{8\pi^2 cm_e r^2}\left(m_\mathrm{f}^2 - m_\mathrm{i}^2\right)$$

$$= \frac{6.626 \times 10^{-34}\,\mathrm{J\,s}}{8\pi^2\left(2.998 \times 10^{10}\,\mathrm{cm\,s^{-1}}\right)\left(9.109 \times 10^{-31}\,\mathrm{kg}\right)\left(132 \times 10^{-12}\,\mathrm{m}\right)^2}\left(m_\mathrm{f}^2 - m_\mathrm{i}^2\right)$$

$$= 1.764 \times 10^4\,\mathrm{cm^{-1}}\left(m_\mathrm{f}^2 - m_\mathrm{i}^2\right)$$

(b) For the transition from the $m = 1$ state to the $m = 2$ state,

$$\tilde{\nu} = 1.764 \times 10^4\,\mathrm{cm^{-1}}\left(m_\mathrm{f}^2 - m_\mathrm{i}^2\right)$$

$$= 1.764 \times 10^4\,\mathrm{cm^{-1}}\,(4 - 1)$$

$$= 5.29 \times 10^4\,\mathrm{cm^{-1}}$$

(c) The rather large difference between the experimentally measured value and the one predicted using the particle-on-a-ring model suggests that the approximation is not adequate in this case.

10.70 Shown below are contour plots of several wave functions for a particle in a two-dimensional box of dimension $a \times 2a$. Rank these in order of increasing energy and identify which (if any) might be degenerate.

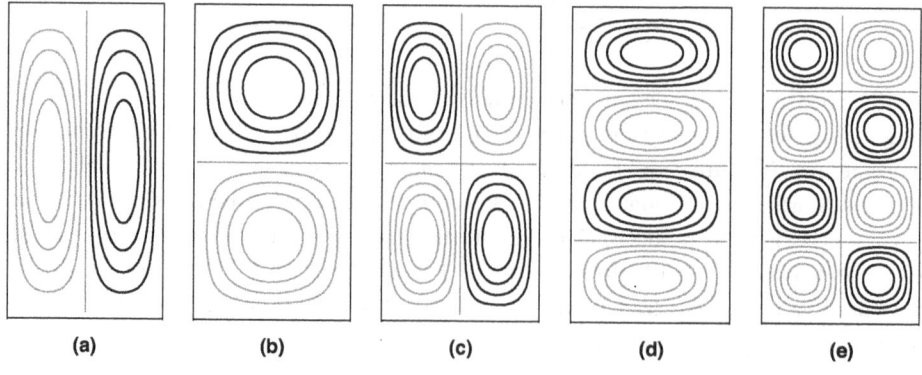

(a)   (b)   (c)   (d)   (e)

The energy for a particle in a two-dimensional box of dimension $a \times 2a$ is

$$E = \frac{n_x^2 h^2}{8ma^2} + \frac{n_y^2 h^2}{8m \, (2a)^2}$$

$$= \frac{h^2}{8ma^2} \left(n_x^2 + \frac{n_y^2}{4} \right)$$

The values of n_x and n_y can be determined by the numbers of nodal lines in the x and y direction, respectively. Specifically, n = number of nodal lines + 1.

(a) $n_x = 2$, $n_y = 1$, $E = \frac{h^2}{8ma^2} \left(4 + \frac{1}{4} \right) = \frac{17}{4} \left(\frac{h^2}{8ma^2} \right)$

(b) $n_x = 1$, $n_y = 2$, $E = \frac{h^2}{8ma^2} \left(1 + \frac{4}{4} \right) = 2 \left(\frac{h^2}{8ma^2} \right)$

(c) $n_x = 2$, $n_y = 2$, $E = \frac{h^2}{8ma^2} \left(4 + \frac{4}{4} \right) = 5 \left(\frac{h^2}{8ma^2} \right)$

(d) $n_x = 1$, $n_y = 4$, $E = \frac{h^2}{8ma^2} \left(1 + \frac{16}{4} \right) = 5 \left(\frac{h^2}{8ma^2} \right)$

(e) $n_x = 2$, $n_y = 4$, $E = \frac{h^2}{8ma^2} \left(4 + \frac{16}{4} \right) = 8 \left(\frac{h^2}{8ma^2} \right)$

In order of increasing energy, $b < a < c = d < e$. The wave functions described by (c) and (d) are degenerate.

10.72 One wavelength in the atomic hydrogen emission spectrum is 1280 nm. What are the initial and final states of the transition responsible for this emission?

The wavenumber for this hydrogen emission is

$$\tilde{\nu} = \frac{1}{1280 \times 10^{-7} \, \text{cm}^{-1}} = 7812.5 \, \text{cm}^{-1}$$

This is related to n_i and n_f by

$$\tilde{\nu} = \tilde{R}_H \left| \frac{1}{n_i^2} - \frac{1}{n_f^2} \right|$$

$$7812.5 \, \text{cm}^{-1} = \left(109737 \, \text{cm}^{-1} \right) \left| \frac{1}{n_i^2} - \frac{1}{n_f^2} \right|$$

$$\left| \frac{1}{n_i^2} - \frac{1}{n_f^2} \right| = 0.0712$$

The longest wavelength associated with the Lyman series (the $n = 2$ to $n = 1$ transition) is 121.5 nm whereas the longest wavelength of the Balmer series (the $n = 3$ to $n = 2$) is 656.1 nm. Thus, the 1280 nm line does not belong in either series.

On the other hand, the longest wavelength transition for the Paschen series (the $n = 4$ to $n = 3$ transition) is 1874.6 nm. Since the observed wavelength is shorter, it is possible that the line comes from the Paschen series (with $n_f = 3$). In fact, when $n_i = 5$,

$$\left| \frac{1}{n_i^2} - \frac{1}{n_f^2} \right| = \left| \frac{1}{9} - \frac{1}{25} \right| = 0.0711$$

Thus, the 1280 nm is due to a transition between $n_i = 5$ and $n_f = 3$.

10.74 A student placed a large unwrapped chocolate bar in a microwave oven without a rotating glass plate. After turning on the oven for less than a minute, she noticed there were evenly spaced dents (due to heating) about 6 cm apart. Based on her observations, calculate the speed of light, given that the microwave frequency is 2.45 GHz.

Assuming that the microwave radiation forms a standing wave inside the oven, its energy will be concentrated at the anti-nodes of the electromagnetic wave, which occur every half wavelength. Thus, $\lambda/2 = 6$ cm, and

$$c = \nu\lambda = (2.45 \times 10^9 \text{ s}^{-1})(2)(6 \times 10^{-2} \text{ m}) = 3 \times 10^8 \text{ m s}^{-1}$$

10.76 In a photoelectric experiment, a student found that a wavelength of 351 nm is needed to just dislodge electrons from a zinc metal surface. Calculate the speed of an ejected electron when she employed light with a wavelength of 313 nm.

At 351 nm, the kinetic energy of any dislodged electron is 0; thus,

$$h\nu = \frac{hc}{\lambda} = \Phi$$

$$\Phi = \frac{(6.626 \times 10^{-34} \text{ J s}) (2.998 \times 10^8 \text{ m s}^{-1})}{351 \times 10^{-9} \text{ m}} = 5.659 \times 10^{-19} \text{ J}$$

At 313 nm,

$$h\nu = \frac{hc}{\lambda} = \Phi + \frac{1}{2}m_e u^2$$

$$u = \left[\frac{2}{m_e} \left(\frac{hc}{\lambda} - \Phi \right) \right]^{1/2}$$

$$= \left\{ \frac{2}{9.109 \times 10^{-31} \text{ kg}} \left[\frac{(6.626 \times 10^{-34} \text{ J s}) (2.998 \times 10^8 \text{ m s}^{-1})}{313 \times 10^{-9} \text{ m}} - 5.659 \times 10^{-19} \text{ J} \right] \right\}^{1/2}$$

$$= \left[\frac{2}{9.109 \times 10^{-31} \text{ kg}} \left(6.88 \times 10^{-20} \text{ J} \right) \right]^{1/2}$$

$$= 3.9 \times 10^5 \text{ m s}^{-1}$$

CHAPTER 11

Applications of Quantum Mechanics to Spectroscopy

PROBLEMS AND SOLUTIONS

11.2 Convert 450 nm into wavenumber (cm^{-1}) and frequency.

$$\lambda = \frac{c}{\nu} = \frac{1}{\tilde{\nu}}$$

$$\tilde{\nu} = \frac{1}{\lambda} = \frac{1}{450 \text{ nm}} \left(\frac{1 \text{ nm}}{1 \times 10^{-7} \text{ cm}} \right) = 2.22 \times 10^4 \text{ cm}^{-1}$$

$$\nu = \frac{c}{\lambda} = \frac{2.998 \times 10^8 \text{ m s}^{-1}}{450 \text{ nm}} \left(\frac{1 \text{ nm}}{1 \times 10^{-9} \text{ m}} \right) = 6.66 \times 10^{14} \text{ s}^{-1}$$

11.4 Convert the following absorbance to percent transmittance: **(a)** 0.0, **(b)** 0.12, and **(c)** 4.6.

$$T = \frac{I}{I_0} = 10^{\log \frac{I}{I_0}} = 10^{-A}$$

(a) $A = 0.0$, and $T = 10^{-0} = 1 = 100\%$.

(b) $A = 0.12$, and $T = 10^{-0.12} = 0.76 = 76\%$.

(c) $A = 4.6$, and $T = 10^{-4.6} = 2.5 \times 10^{-5} = 0.0025\%$.

11.6 The absorption of radiation energy by a molecule results in the formation of an excited molecule. It would seem that given enough time all of the molecules in a sample would have been excited and no more absorption would occur. Yet in practice we find that the absorbance of a sample at any wavelength remains unchanged with time. Why?

Excited molecules lose their excess energy through a variety of relaxation mechanisms and return to the ground state. Under conditions of constant irradiation, a steady state is reached, and the absorbance of the sample remains unchanged in time. If the incident radiation is of sufficient intensity, so that the rate of excitation is faster than all relaxation processes, the sample approaches a state where half of the molecules have been excited, as indicated by Equation (14.10). At this point the populations of the two states involved in the transitions are equal, and the rate of stimulated absorption equals the rate of stimulated emission resulting in no net absorption.

11.8   The frequency of molecular collision in the liquid phase is about 1×10^{13} s^{-1}. Ignoring all other mechanisms contributing to linewidth, calculate the width (in Hz) of vibrational transitions if **(a)** every collision is effective in deactivating the molecule vibrationally, and **(b)** that one collision in 40 is effective.

(a) If every collision is effective in deactivating the molecule, then the lifetime of the excited vibrational state is $\Delta t = 1/1 \times 10^{13}$ s$^{-1} = 1 \times 10^{-13}$ s, and the natural linewidth is

$$\Delta \nu = \frac{1}{4\pi \, \Delta t} = \frac{1}{4\pi \left(1 \times 10^{-13} \text{ s} \right)} = 8 \times 10^{11} \text{ s}^{-1}$$

(b) If one collision in 40 is effective in deactivating the molecule, then the lifetime of the excited vibrational state is $\Delta t = 40/1 \times 10^{13}$ s$^{-1} = 4 \times 10^{-12}$ s, and the natural linewidth is

$$\Delta \nu = \frac{1}{4\pi \, \Delta t} = \frac{1}{4\pi \left(4 \times 10^{-12} \text{ s} \right)} = 2 \times 10^{10} \text{ s}^{-1}$$

11.10   The familiar yellow D line of sodium is actually a doublet at 589.0 nm and 589.6 nm. Calculate the difference in energy (in J) between these two lines. Convert your answer to cm^{-1} units.

$$\Delta E = \frac{hc}{\lambda_1} - \frac{hc}{\lambda_2}$$

$$= hc \left(\frac{1}{\lambda_1} - \frac{1}{\lambda_2} \right)$$

$$= \left(6.626 \times 10^{-34} \text{ J s} \right) \left(2.998 \times 10^{8} \text{ m s}^{-1} \right) \left(\frac{1}{589.0 \times 10^{-9} \text{ m}} - \frac{1}{589.6 \times 10^{-9} \text{ m}} \right)$$

$$= 3.43 \times 10^{-22} \text{ J} = 3.4 \times 10^{-22} \text{ J}$$

This is equivalent to $\frac{\Delta E}{hc} = 17$ cm^{-1}.

11.12   Assuming that the width of a spectral line to be the sole result of lifetime broadening, estimate the lifetime of a state that gives rise to a line of width **(a)** 1.0 cm^{-1}, **(b)** 0.50 Hz.

The lifetime of an excited state is related to its natural linewidth via $\Delta \nu = \dfrac{1}{4\pi \, \Delta t}$.

(a) Since $\nu = c\tilde{\nu}$,

$$\Delta t = \frac{1}{4\pi \Delta \nu}$$

$$= \frac{1}{4\pi c \Delta \tilde{\nu}}$$

$$= \frac{1}{4\pi \left(2.998 \times 10^{10} \text{ cm s}^{-1}\right) \left(1.0 \text{ cm}^{-1}\right)}$$

$$= 2.7 \times 10^{-12} \text{ s}$$

(b)

$$\Delta t = \frac{1}{4\pi \Delta \nu}$$

$$= \frac{1}{4\pi \left(0.50 \text{ s}^{-1}\right)}$$

$$= 0.16 \text{ s}$$

11.14 The molar absorptivity of a benzene solution of an organic compound is 1.3×10^2 L mol^{-1} cm^{-1} at 422 nm. Calculate the percentage reduction in light intensity when light of that wavelength passes through a 1.0-cm cell containing a solution of concentration 0.0033 M.

The Beer–Lambert law gives the absorbance,

$$A = \varepsilon b c$$

$$= \left(1.3 \times 10^2 \text{ L mol}^{-1} \text{ cm}^{-1}\right) (1.0 \text{ cm}) (0.0033 \ M)$$

$$= 0.429$$

The transmittance is found via

$$A = -\log T$$

$$T = 10^{-A} = 10^{-0.429} = 0.372$$

This transmittance corresponds to a reduction in light intensity of $(1 - 0.372) \times 100\% = 63\%$.

11.16 What is the wavelength of a photon that has four times the energy of a photon whose wavelength is 1064 nm?

Since $E = \dfrac{hc}{\lambda}$, the energy of a photon is inversely proportional to its wavelength. For a photon that has four times the energy of that with a wavelength of 1064 nm, its wavelength is

$$\frac{1064 \text{ nm}}{4} = 266 \text{ nm}$$

11.18 Which of the following molecules are microwave active? **(a)** C_2H_2, **(b)** CH_3Cl, **(c)** C_6H_6, **(d)** CO_2, **(e)** H_2O, **(f)** HCN.

Those molecules with a permanent dipole moment are microwave active. These are **(b)**, **(e)**, and **(f)**.

11.20   The $J = 4 \leftarrow 3$ transition for a diatomic molecule occurs at $0.50\,\text{cm}^{-1}$. What is the wavenumber for the $J = 7 \leftarrow 6$ transition for this molecule? Assume a rigid rotor.

Assuming a rigid rotor, the frequency of a rotational transition is given by $\nu = \Delta E_{\text{rot}}/h = 2BJ'$. The wavenumber of the $J = 4 \leftarrow 3$ transition with $J' = 4$ supplies the value of B.

$$B = \frac{\Delta E_{\text{rot}}/h}{2J'}$$

$$= \frac{hc\tilde{\nu}/h}{2J'}$$

$$= \frac{c\tilde{\nu}}{2J'}$$

$$= \frac{\left(2.998 \times 10^{10}\,\text{cm s}^{-1}\right)\left(0.50\,\text{cm}^{-1}\right)}{2(4)}$$

$$= 1.87 \times 10^{9}\,\text{s}^{-1}$$

Then for the $J = 7 \leftarrow 6$ transition with $J' = 7$,

$$\tilde{\nu} = \frac{\nu}{c}$$

$$= \frac{\Delta E_{\text{rot}}/h}{c}$$

$$= \frac{2BJ'}{c}$$

$$= \frac{2\left(1.87 \times 10^{9}\,\text{s}^{-1}\right)(7)}{2.998 \times 10^{10}\,\text{cm s}^{-1}}$$

$$= 0.87\,\text{cm}^{-1}$$

11.22   The equilibrium bond length in nitric oxide ($^{14}\text{N}^{16}\text{O}$) is $1.15\,\text{Å}$. Calculate **(a)** the moment of inertia of NO, and **(b)** the energy for the $J = 1 \leftarrow 0$ transition. **(c)** How many times does the molecule rotate per second in the $J = 1$ level?

In finding the reduced mass of $^{14}\text{N}^{16}\text{O}$ it is important to use masses appropriate for the specific isotopes under consideration and not the average masses found in the periodic table.

$$\mu = \frac{m_{\text{N}}m_{\text{O}}}{m_{\text{N}} + m_{\text{O}}} = \frac{(14.003074\,\text{amu})\,(15.994915\,\text{amu})}{14.003074\,\text{amu} + 15.994915\,\text{amu}}\left(1.6605 \times 10^{-27}\,\text{kg amu}^{-1}\right)$$

$$= 1.23983 \times 10^{-26}\,\text{kg}$$

(a) The moment of inertia is given by

$$I = \mu r^2$$

$$= \left(1.23983 \times 10^{-26}\,\text{kg}\right)\left(1.15 \times 10^{-10}\,\text{m}\right)^2$$

$$= 1.640 \times 10^{-46}\,\text{kg m}^2 = 1.64 \times 10^{-46}\,\text{kg m}^2$$

(b) The rotational constant for the molecule is

$$B = \frac{\hbar}{4\pi I} = \frac{h}{8\pi^2 I} = \frac{6.626 \times 10^{-34}\ \text{J s}}{8\pi^2\ (1.640 \times 10^{-46}\ \text{kg m}^2)} = 5.117 \times 10^{10}\ \text{s}^{-1}$$

and the energy for the $J = 1 \leftarrow 0$ transition is

$$\Delta E_{1\leftarrow 0} = 2BhJ' = 2\ (5.117 \times 10^{10}\ \text{s}^{-1})\ (6.626 \times 10^{-34}\ \text{J s})\ (1) = 6.781 \times 10^{-23}\ \text{J}$$

$$= 6.78 \times 10^{-23}\ \text{J}$$

(c) The frequency of molecular rotation is equal to the frequency of the electromagnetic radiation that causes the transition, which is

$$\nu = \frac{\Delta E_{1\leftarrow 0}}{h} = \frac{6.781 \times 10^{-23}\ \text{J}}{6.626 \times 10^{-34}\ \text{J s}} = 1.02 \times 10^{11}\ \text{s}^{-1}$$

11.24 For a given diatomic molecule, a single absorption line is observed in the microwave region of the electromagnetic spectrum. Is this sufficient information to determine the bond length of the molecule? Explain.

The rotational transitions for diatomic molecules appear as a series of lines at frequencies that are even multiples of the rotational constant ($2B$, $4B$, $6B$, etc.), typically in the microwave region. Observation of a single line does not allow a determination of the multiplying factor. At the minimum, a second line is needed to make a definitive assignment and to determine a value for B. Of course, it is also necessary to know the precise identity of the molecule, including the specific isotopic identity of the two atoms.

11.26 In Example 11.1, we calculated the interatomic distance for CO. A more precise value of the separation between absorption lines is 1.15270×10^{11} Hz for isotopically pure $^{12}\text{C}^{16}\text{O}$. Assume that the bond length does not change with isotopic substitution and predict the separation between absorption lines for $^{13}\text{C}^{16}\text{O}$. Repeat the calculation for $^{12}\text{C}^{18}\text{O}$. Use the relative atomic masses: $^{12}\text{C} = 12.000000$, $^{13}\text{C} = 13.003355$, $^{16}\text{O} = 15.994915$, and $^{18}\text{O} = 17.999160$.

The lines are separated by $2B$, where B is inversely proportional to its moment of inertia

$$B = \frac{\hbar}{4\pi I} = \frac{\hbar}{4\pi \mu r^2}$$

and if bond length does not change with isotopic substitution, then

$$\frac{B_{^{13}\text{C}^{16}\text{O}}}{B_{^{12}\text{C}^{16}\text{O}}} = \frac{\mu_{^{12}\text{C}^{16}\text{O}}}{\mu_{^{13}\text{C}^{16}\text{O}}}$$

$$\frac{B_{^{12}\text{C}^{18}\text{O}}}{B_{^{12}\text{C}^{16}\text{O}}} = \frac{\mu_{^{12}\text{C}^{16}\text{O}}}{\mu_{^{12}\text{C}^{18}\text{O}}}$$

The reduced masses for the three isotopomers are

$$\mu_{^{12}C^{16}O} = \frac{(12.000000 \text{ amu})(15.994915 \text{ amu})}{12.000000 \text{ amu} + 15.994915 \text{ amu}} \left(1.660538921 \times 10^{-27} \frac{\text{kg}}{\text{amu}} \right)$$

$$= 1.13850014 \times 10^{-26} \text{ kg}$$

$$\mu_{^{13}C^{16}O} = \frac{(13.003355 \text{ amu})(15.994915 \text{ amu})}{13.003355 \text{ amu} + 15.994915 \text{ amu}} \left(1.660538921 \times 10^{-27} \frac{\text{kg}}{\text{amu}} \right)$$

$$= 1.19100703 \times 10^{-26} \text{ kg}$$

$$\mu_{^{12}C^{18}O} = \frac{(12.000000 \text{ amu})(17.999160 \text{ amu})}{12.000000 \text{ amu} + 17.999160 \text{ amu}} \left(1.660538921 \times 10^{-27} \frac{\text{kg}}{\text{amu}} \right)$$

$$= 1.19556570 \times 10^{-26} \text{ kg}$$

Thus, the separation between absorption lines, $2B$, for the isotopic species are

$$2B_{^{13}C^{16}O} = \left(\frac{\mu_{^{12}C^{16}O}}{\mu_{^{13}C^{16}O}} \right) (2B_{^{12}C^{16}O})$$

$$= \left(\frac{1.13850014 \times 10^{-26} \text{ kg}}{1.19100703 \times 10^{-26} \text{ kg}} \right) \left(1.15270 \times 10^{11} \text{ Hz} \right)$$

$$= 1.10188 \times 10^{11} \text{ Hz}$$

$$2B_{^{12}C^{18}O} = \left(\frac{\mu_{^{12}C^{16}O}}{\mu_{^{12}C^{18}O}} \right) (2B_{^{12}C^{16}O})$$

$$= \left(\frac{1.13850014 \times 10^{-26} \text{ kg}}{1.19556570 \times 10^{-26} \text{ kg}} \right) \left(1.15270 \times 10^{11} \text{ Hz} \right)$$

$$= 1.09768 \times 10^{11} \text{ Hz}$$

11.28   Give the number of normal vibrational modes of **(a)** O_3, **(b)** C_2H_2, **(c)** CBr_4, **(d)** C_6H_6.

Molecules **(a)**, **(c)**, and **(d)** are non-linear and have $3N - 6$ normal modes, where N is the number of atoms in the molecule. Molecule **(b)** is linear and has $3N - 5$ normal modes.

(a) $3 \times 3 - 6 = 3$

(b) $3 \times 4 - 5 = 7$

(c) $3 \times 5 - 6 = 9$

(d) $3 \times 12 - 6 = 30$

11.30   An object of mass 500 g suspended from the end of a rubber band has a vibrational frequency of 4.2 Hz. Calculate the force constant of the rubber band.

The vibrational frequency is given by $\nu = \frac{1}{2\pi} \sqrt{\frac{k}{m}}$, so that

$$k = 4\pi^2 m v^2$$

$$= 4\pi^2 \left(500 \times 10^{-3}\,\text{kg}\right)\left(4.2\,\text{s}^{-1}\right)^2$$

$$= 3.5 \times 10^2\,\text{kg s}^{-2} = 3.5 \times 10^2\,\text{N m}^{-1}$$

11.32 If molecules did not possess zero-point energy, would they be able to undergo the $v = 1 \leftarrow 0$ transition?

Yes, since the oscillating electric field of the IR radiation would be able to induce motion of the centers of positive and negative charge in much the same way that the rotational motion of a polar molecule is excited. Of course, it is impossible to construct physically meaningful wave functions for an oscillating molecule with no zero-pont energy so that it might be argued that since such a molecule could never exist, neither could the transition take place.

11.34 Show all the fundamental vibration modes of **(a)** carbon disulfide (CS_2) and **(b)** carbonyl sulfide (OCS) and indicate which ones are IR active.

(a) The fundamental vibration modes of CS_2 are identical in form to those of CO_2 (see Figure 11.21). The asymmetric stretch and the (doubly degenerate) bending modes are IR active. The symmetric stretch is not infrared active.

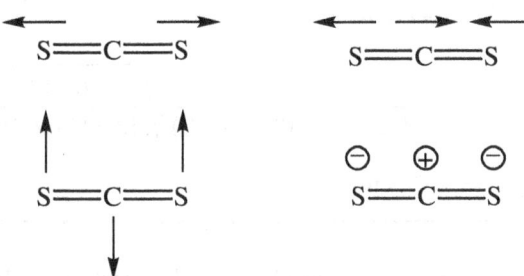

(b) OCS is also a linear molecule with 4 fundamental vibration modes. All four are IR active, leading to 3 IR peaks in the absorption spectrum, because the bend is doubly degenerate.

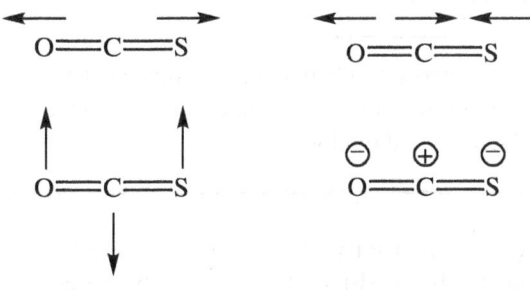

11.36 Which of the following molecules has the highest fundamental frequency of vibration? H_2, D_2, HD. ($D = {}^2H$)

Since the fundamental frequency of vibration is given by $v = \dfrac{1}{2\pi}\sqrt{\dfrac{k}{\mu}}$, and since the force constant for the bond is isotopically invariant, the molecule with the lowest reduced mass will have the highest fundamental frequency of vibration. The reduced masses are

$$\mu_{H_2} = \frac{(1.008 \text{ amu}) (1.008 \text{ amu})}{1.008 \text{ amu} + 1.008 \text{ amu}} = 0.504 \text{ amu}$$

$$\mu_{HD} = \frac{(1.008 \text{ amu}) (2.014 \text{ amu})}{1.008 \text{ amu} + 2.014 \text{ amu}} = 0.672 \text{ amu}$$

$$\mu_{D_2} = \frac{(2.014 \text{ amu}) (2.014 \text{ amu})}{2.014 \text{ amu} + 2.014 \text{ amu}} = 1.007 \text{ amu}$$

Thus, H_2 has the highest fundamental vibration frequency.

11.38   The molecule $H^{79}Br$ has a force constant of 405.7 N m^{-1}. At what wavenumber will $H^{79}Br$ absorb in the infrared?

The reduced mass is

$$\mu = \frac{m_H m_{Br}}{m_H + m_{Br}} = \frac{(1.00783 \text{ amu}) (78.91834 \text{ amu})}{1.00783 \text{ amu} + 78.91834 \text{ amu}} \left(1.6605 \times 10^{-27} \text{ kg amu}^{-1}\right)$$

$$= 1.65244 \times 10^{-27} \text{ kg}$$

The expected wavenumber is

$$\tilde{\nu} = \frac{1}{2\pi c} \sqrt{\frac{k}{\mu}}$$

$$= \frac{1}{2\pi \left(2.998 \times 10^{10} \text{ cm s}^{-1}\right)} \sqrt{\frac{405.7 \text{ N m}^{-1}}{1.65244 \times 10^{-27} \text{ kg}}}$$

$$= 2630 \text{ cm}^{-1}$$

11.40   Consider the 2-propenenitrile molecule whose IR spectrum is shown in Figure 11.25. Which of the following types of molecular motion has the largest number of energy levels appreciably occupied at 300 K? Electronic, C–H stretching vibration, C=C stretching vibration, HCH bending motion, or rotational motion.

The largest number of energy levels populated (equivalent to a large value for the associated partition function) will occur for the molecular motion with the smallest quantized energy level spacing, which is the rotational motion.

11.42   For the molecule BeH_2, identify whether each of the following vibrational modes is Raman active or forbidden: **(a)** bend, **(b)** symmetric stretch, **(c)** asymmetric stretch. Compare and contrast your results with those from the previous problem.

BeH_2 is a linear triatomic molecule that possesses a center of symmetry. Thus, its vibrational modes are either infrared or Raman allowed, but not both. IR-active modes involve a change in dipole moment with vibration, whereas Raman-active modes involve a change in polarizability with vibration.

(a) The bending vibration of BeH_2 is infrared active. This vibration is Raman forbidden.

(b) The symmetric stretch of BeH_2 is infrared forbidden. This vibration is Raman active.

(c) The asymmetric stretch of BeH_2 is infrared active. This vibration is Raman forbidden.

For the bent SO_2 molecule, which does not possess a center of symmetry, all vibrational modes are both infrared and Raman active, as is the case for H_2O discussed in the text. For the linear BeH_2 molecule with a center of symmetry, a vibrational mode cannot be both infrared and Raman allowed.

11.44 Explain why Stokes Raman scattered light is so much more intense than anti-Stokes Raman scattered light.

Anti-Stokes Raman scattered light is caused by scattering from molecules in excited vibrational states. At typical temperatures, only a small fraction of molecules are in such states. The vast majority of molecules are in the ground vibrational state, which can only lead to Rayleigh or Stokes Raman scattering.

11.46 A 633-nm HeNe laser is used to irradiate a liquid sample of benzene, which has a prominent vibrational mode of 992 cm^{-1}. Describe the absolute wavelengths and the relative strengths of the three types of scattered light: Rayleigh, Stokes Raman, and anti-Stokes Raman.

The 633-nm laser line corresponds to $\dfrac{1}{633 \times 10^{-7}\,\text{cm}} = 15798$ cm^{-1}. The strongest scattering will be Rayleigh scattering at this same wavelength, 633 nm. The Stokes and anti-Stokes scattering occur 922 cm^{-1} below and above the Rayleigh line, respectively. The Stokes line is at 14876 cm^{-1}, and the anti-Stokes is at 16720 cm^{-1}.

The wavelengths, $\lambda = 1/\tilde{\nu}$ are 672 nm for the Stokes line and 598 nm for the anti-Stokes line. The anti-Stokes lines is the weakest, the Stokes line much stronger than it, and in turn, the Rayleigh scattering much stronger than the Stokes.

11.48 Why is Raman spectroscopy performed with monochromatic excitation light, and not polychromatic or blackbody radiation (such as a tungsten-filament lightbulb)?

The process responsible for Raman spectroscopy shifts the wavelength of scattered light relative to that of the excitation source. If the excitation source is polychromatic (or continuous), then the Raman spectrum will be "repeated" for each wavelength in the source, leading to a complicated (or continuous) spectrum.

11.50 This problem deals with the amplitude of molecular vibration of a diatomic molecule in its ground vibrational state. (a) When the molecule is stretched by an extent x from the equilibrium position, the increase in the potential energy is given by the integral

$$\int_0^x kx\,dx$$

where k is the force constant. Evaluate this integral. (b) Calculate the amplitude of vibration by equating the potential energy with the vibrational energy in the ground state. Use x_{max} to represent the maximum displacement. (c) Given that the force constant for $H^{35}Cl$ is 4.84×10^2

$N\,m^{-1}$, calculate the amplitude of vibration in the $v = 0$ state. **(d)** What is the percent of the amplitude compared to the bond length (1.27 Å)? **(e)** Repeat the calculations in **(c)** and **(d)** for carbon monoxide, given that the force constant is $1.85 \times 10^3\,N\,m^{-1}$ and the bond length is 1.13 Å. (^{35}Cl: 34.97 amu.)

(a)
$$\int_0^x kx\,dx = \frac{kx^2}{2}$$

(b) Because $E_{\text{vib}} = \left(v + \dfrac{1}{2}\right)h\nu$ and $\nu = \dfrac{1}{2\pi}\sqrt{\dfrac{k}{\mu}}$, for the ground state with $v = 0$,

$$E_{\text{vib}} = \frac{1}{2}h\left(\frac{1}{2\pi}\sqrt{\frac{k}{\mu}}\right) = \frac{1}{2}kx_{\text{max}}^2$$

Solving for x_{max},

$$x_{\text{max}} = \left(\frac{h^2}{4\pi^2 k\mu}\right)^{1/4}$$

(c) For H^{35}Cl the reduced mass is

$$\mu = \left[\frac{(1.008\text{ amu})\,(34.97\text{ amu})}{1.008\text{ amu} + 34.97\text{ amu}}\right]\left(1.6605 \times 10^{-27}\text{ kg amu}^{-1}\right) = 1.6269 \times 10^{-27}\text{ kg}$$

and

$$x_{\text{max}} = \left(\frac{h^2}{4\pi^2 k\mu}\right)^{1/4}$$

$$= \left[\frac{\left(6.626 \times 10^{-34}\text{ J s}\right)^2}{4\pi^2\left(4.84 \times 10^2\text{ N m}^{-1}\right)\left(1.6269 \times 10^{-27}\text{ kg}\right)}\right]^{1/4}$$

$$= 1.090 \times 10^{-11}\text{ m} = 0.109\text{ Å}$$

(d) This amplitude is $\dfrac{0.1090\text{ Å}}{1.27\text{ Å}} \times 100\% = 8.58\%$ of the bond length.

(e) For CO, the reduced mass is

$$\mu = \left[\frac{(12.00\text{ amu})\,(15.99\text{ amu})}{12.00\text{ amu} + 15.99\text{ amu}}\right]\left(1.6605 \times 10^{-27}\text{ kg amu}^{-1}\right) = 1.1384 \times 10^{-26}\text{ kg}$$

and

$$x_{\text{max}} = \left(\frac{h^2}{4\pi^2 k\mu}\right)^{1/4}$$

$$= \left[\frac{\left(6.626 \times 10^{-34}\text{ J s}\right)^2}{4\pi^2\left(1.85 \times 10^3\text{ N m}^{-1}\right)\left(1.1384 \times 10^{-26}\text{ kg}\right)}\right]^{1/4}$$

$$= 4.794 \times 10^{-12}\text{ m} = 0.0479\text{ Å}$$

This amplitude is $\dfrac{0.04794 \text{ Å}}{1.13 \text{ Å}} \times 100\% = 4.24\%$ of the bond length. A triple bond does not stretch as much as a single bond.

11.52 For predicting the electronic absorption spectrum of the C_{60} molecule [the fullerene known as a "Buckyball", named after American inventor R. Buckminster Fuller (1895–1983)], which of the following models would you expect to provide the best prediction of the first three transitions, and why? Particle-on-a-ring, rigid rotor, harmonic oscillator, particle-in-a-box.

The C_{60} molecule approximates a sphere, so that a reasonable model for electron motion would be that of the "particle-on-a-sphere," which is equivalent to the rigid rotor model. Thus, we would expect the best prediction from this model.

11.54 Explain in your own words why molecules vibrate, even at absolute zero of temperature.

The direct proportionality between absolute temperature and energy (the equiparitition principle, requiring $E = \frac{1}{2}k_B$ for each degree of freedom) results from application of classical ideas. Consequently, a quantum system is not required to have energy equal to zero at $T = 0$ K. On the other hand, the Heisenberg uncertainty principle, as discussed in the text, does require that the molecule execute vibrational motion. This is because the atoms have an uncertainty in position comparable to the bond length, which is a finite quantity, not infinite. Therefore, the momentum cannot be known precisely as it would be if vibrational motion were to cease.

11.56 Is the rotational constant B' for an excited vibrational state greater than, less than, or equal to the rotational constant B'' for the ground vibrational state? Explain your reasoning.

The rotational constant, B, is inversely proportional to r^2. For a harmonic oscillator, the bond length of a diatomic molecule would be the same regardless of the vibrational state. Thus, $B' = B''$.

On the other hand, for a real diatomic molecule, because of anharmonicity, the bond length in an excited vibrational state is greater than that in the ground vibrational state. Therefore, $B' < B''$. For polyatomic molecules, a similar effect is noted for stretching vibrations. (They tend to make the molecule "bigger," decreasing the rotational constant.) The situation is a bit more complicated for bending vibrations, since these can lead to the molecule becoming more "compact," which would increase the rotational constant.

11.58 Does the vibrational frequency depend on which rotational state the molecule is in? Explain.

Treating the vibration as a harmonic oscillator, the effect of centrifugal distortion resulting from increasing rotational motion is to shift the origin of the parabolic potential energy function, but not the curvature. Thus, the vibrational frequency will remain the same regardless of the rotational state of the molecule. More generally, because the frequency of the vibrational motion is so much greater than that of the rotational motion (see Problem 11.49), the rotational state is not expected to influence the vibrational frequency.

11.60 Identify each of the following functions as even, odd, or neither even nor odd. (a) $f(x) =$ constant, (b) $f(x) = -x$, (c) $f(x) = \sin(x)$, (d) $f(x) = \cos(x)$, (e) $f(x) = \sin(x)\cos(x)$, (f) $f(x) = 3\cos^2(x) - 1$, (g) $f(x) = \sin^2(x)e^{2ix}$.

If $f(x) = f(-x)$, then f is even and if $f(x) = -f(-x)$, then f is odd.

(a) $f(-x) = \text{constant} = f(x)$; f is even.

(b) $f(-x) = -(-x) = -f(x)$; f is odd.

(c) $f(-x) = \sin(-x) = -\sin(x) = -f(x)$; f is odd.

(d) $f(-x) = \cos(-x) = \cos(x) = f(x)$; f is even.

(e) $f(-x) = \sin(-x)\cos(-x) = -\sin(x)\cos(x) = -f(x)$; f is odd.

(f) $f(-x) = 3\cos^2(-x) - 1 = 3\cos^2(x) - 1 = f(x)$; f is even.

(g) $f(-x) = \sin^2(-x)e^{-2ix} = [-\sin(x)]^2\, e^{-2ix} = \sin^2(x)e^{-2ix}$

Since $f(-x)$ does not equal $f(x)$ or $-f(x)$, f is neither even nor odd.

11.62 For each of the following molecules, identify how many unique rotational quantum numbers are needed to model the rotational behavior of the molecule. (a) NO, (b) CH_4, (c) CH_3Cl, (d) CO_2, (e) SO_2, (f) OCS, (g) SF_6, (h) SF_4.

Diatomic and linear molecules have two rotational degrees of freedom, but they are equivalent. Thus, they require only a single unique rotational quantum number. Non-linear molecules have three rotational degrees of freedom, and would, in general, require three unique rotational quantum numbers, but if the molecule possesses a proper rotation axis, C_n, of order $n \geq 3$, two of the three rotational degrees of freedom are equivalent, and only two unique rotational quantum numbers are required. If the non-linear molecule belongs to a cubic point group, all three rotational degrees of freedom are equivalent, and only one unique rotational quantum number is required.

(a) This is a diatomic molecule and requires one unique rotational quantum number.

(b) This is a non-linear molecule belonging to a cubic point group (T_d). One unique rotational quantum numbers is required.

(c) This is a non-linear molecule that possesses a C_3 proper rotation axis. Two unique rotational quantum numbers are required.

(d) This is a linear molecule and requires one unique rotational quantum number.

(e) This is a non-linear molecule, but does not possess a proper rotation axis, C_n, of order $n \geq 3$. Three unique rotational quantum numbers are required.

(f) This is a linear molecule and requires one unique rotational quantum number.

(g) This is a non-linear molecule belonging to a cubic point group (O_h). One unique rotational quantum numbers is required.

(h) This is a non-linear molecule, but does not possess a proper rotation axis, C_n, of order $n \geq 3$. Three unique rotational quantum numbers are required. (In addition to the four S–F bonds, the central S atom in SF_4 also has one lone pair of electrons. This is a favorite example in the VSEPR method for predicting molecular geometry. The electron pair arrangement is trigonal bipyramidal with the lone pair occupying an equitorial position. This results in the "see-saw" geometry for the molecule. It is described by the C_{2v} point group.)

11.64 For the following series of isotopomers of the hydrogen molecule, rank them in order of increasing zero-point energy. Then rank them in order of increasing dissociation energy. Assume that all of the isotopomers have the same potential energy surface. H_2, HD, D_2, HT, T_2, DT. (Note $H = {}^1H$, $D = {}^2H$, and $T = {}^3H$)

Using the harmonic oscillator model, the zero-point energy is $E_0 = \dfrac{h}{4\pi}\sqrt{\dfrac{k}{\mu}}$. Thus, the smaller the reduced mass of an isotopomer, the greater its zero-point energy, which means the species is easier to dissociate.

The reduced masses of the various isotopomers are

$$\mu_{H_2} = \frac{(m_H)\,(m_H)}{m_H + m_H} = \frac{m_H}{2} = \frac{1.007825 \text{ amu}}{2} = 0.503913 \text{ amu}$$

$$\mu_{HD} = \frac{(m_H)\,(m_D)}{m_H + m_D} = \frac{(1.007825 \text{ amu})\,(2.014102 \text{ amu})}{1.007825 \text{ amu} + 2.014102 \text{ amu}} = 0.671711 \text{ amu}$$

$$\mu_{D_2} = \frac{(m_D)\,(m_D)}{m_D + m_D} = \frac{m_D}{2} = \frac{2.014102 \text{ amu}}{2} = 1.007051 \text{ amu}$$

$$\mu_{HT} = \frac{(m_H)\,(m_T)}{m_H + m_T} = \frac{(1.007825 \text{ amu})\,(3.016049 \text{ amu})}{1.007825 \text{ amu} + 3.016049 \text{ amu}} = 0.755404 \text{ amu}$$

$$\mu_{T_2} = \frac{(m_T)\,(m_T)}{m_T + m_T} = \frac{m_T}{2} = \frac{3.016049 \text{ amu}}{2} = 1.508025 \text{ amu}$$

$$\mu_{DT} = \frac{(m_D)\,(m_T)}{m_D + m_T} = \frac{(2.014102 \text{ amu})\,(3.016049 \text{ amu})}{2.014102 \text{ amu} + 3.016049 \text{ amu}} = 1.207644 \text{ amu}$$

In order of increasing zero-point energy, $T_2 < DT < D_2 < HT < HD < H_2$, and in order of increasing dissociation energy, $H_2 < HD < HT < D_2 < DT < T_2$.

11.66 Isotopic substitution can be used to "tag" vibrational modes in small molecules because the isotopic substitution will result in a shift in the frequency of vibrational modes. In general, which will result in a larger frequency shift, a substitution of ${}^2H$ for ${}^1H$ or a substitution of ${}^{13}C$ for ${}^{12}C$? Why?

It is the change in reduced mass that is responsible for the vibrational frequency shift upon isotopic substitution, because $\tilde{\nu} = \dfrac{1}{2\pi c}\sqrt{\dfrac{k}{\mu}}$. Thus, a larger relative change in reduced mass will cause a larger relative change in vibrational frequency. For a fixed value of m_2, the change in reduced mass accompanying a change in m_1 is

$$\Delta\mu = \frac{\partial\mu}{\partial m_1}\Delta m_1 = \frac{m_2^2}{(m_1 + m_2)^2}\Delta m_1$$

and

$$\frac{\Delta\mu}{\mu} = \frac{m_2}{(m_1 + m_2)}\left(\frac{\Delta m_1}{m_1}\right)$$

For a unit change in m_1, such as that upon substitution of ^{2}H for ^{1}H or substitution of ^{13}C for ^{12}C, the larger change in the reduced mass occurs for the smaller value of m_1, namely ^{2}H for ^{1}H.

11.68   Explain how Raman and IR spectroscopies might be used to distinguish the *ortho-*, *meta-*, and *para-* isomers of dichlorobenzene.

Similar to the situation for the *ortho-*, *meta-*, and *para-* isomers of difluorobenzene discussed in Section 11.5, *para*-dichlorobenzene possesses a center of symmetry and does not have a vibrational mode that is both IR and Raman active. Based on group theory alone, it is not possible to distinguish between *ortho*-dichlorobenzene and *meta*-dichlorobenzene because they both belong to the C_{2v} point group. Like water, both *ortho-* and *meta*-dichlorobenzene lack a center of symmetry and possess vibrational modes that are both IR and Raman active.

11.70   A rigid rotor is described by the wave function

$$\psi(\theta, \phi) = \left(\frac{15}{8\pi}\right)^{1/2} \sin\theta \cos\theta \, e^{-i\phi}$$

What are the quantum numbers ℓ and m_ℓ that correspond to this wave function? What is the energy, the magnitude of angular momentum, and the z-component of angular momentum for a rigid rotor described by this wave function? Express your answers in terms of h and I (moment of inertia).

The wave function is Y_2^{-1} which corresponds $\ell = 2$ and $m_\ell = -1$.

The energy is $E_\ell = \ell(\ell + 1)\dfrac{\hbar^2}{2I} = \dfrac{3\hbar^2}{I}$, the magnitude of the angular momentum is $L = \hbar\sqrt{\ell(\ell + 1)} = \sqrt{6}\hbar$, and the z-component of angular momentum is $L_z = \hbar m_\ell = -\hbar$.

11.72   Apply dimensional analysis to determine the units of ρ_λ and ρ_ν. (*Hint:* See Equations 10.5 and 11.14.)

In the expression $\rho_\lambda = \dfrac{8\pi hc}{\lambda^5}\dfrac{1}{e^{\frac{hc}{\lambda k_B T}} - 1}$, $\dfrac{hc}{\lambda k_B T}$ is unitless as shown below:

$$\frac{hc}{\lambda k_B T} = \frac{(\text{J s})\,(\text{m s}^{-1})}{(\text{m})\,(\text{J K}^{-1})\,(\text{K})}$$

Thus, the units for ρ_λ are the units of $hc\lambda^{-5}$

$$\frac{hc}{\lambda^5} = \frac{(\text{J s})\,(\text{m s}^{-1})}{(\text{m})^5} = \text{J m}^{-4}$$

Similarly, in $\rho_\nu = \dfrac{8\pi h\nu^3}{c^3}\dfrac{1}{e^{\frac{h\nu}{k_B T}} - 1}$, $\dfrac{h\nu}{k_B T}$ is unitless as shown below:

$$\frac{h\nu}{k_B T} = \frac{(\text{J s})\,(\text{s}^{-1})}{(\text{J K}^{-1})\,(\text{K})}$$

Therefore, the units for ρ_v are the units of $h v^3 c^{-3}$

$$\frac{h v^3}{c^3} = \frac{(\text{J s}) \left(\text{s}^{-1}\right)^3}{\left(\text{m s}^{-1}\right)^3} = \text{J s m}^{-3}$$

CHAPTER 12

Electronic Structure of Atoms

PROBLEMS AND SOLUTIONS

12.2   Sketch a 4s hydrogen atom orbital, including an approximate scale, and clearly identify the location of any nodes.

The solid lines in the sketch below represent a contour plot in the xz-plane for the 4s hydrogen atom orbital, which as expected, is spherical with circular contours in this plane. Because the orbital is spherical, the contour plot is the same in any plane containing the origin. The dotted circle at $r = 33.6a_0$ (where a_0 is the Bohr radius) represents the 90% surface. The probability of finding the electron somewhere inside the sphere with this radius is 90%. The dashed circles at $r = 1.9a0$, $6.6a_0$, and $15.5a_0$ represent the three radial nodes for this orbital. Considering that typical chemical bonds are on the order of a few (one or two) Ångstroms, the size of this orbital · ($r = 17.8$ Å for the 90% probability sphere) is quite remarkable.

Hydrogen Atom 4s Orbital

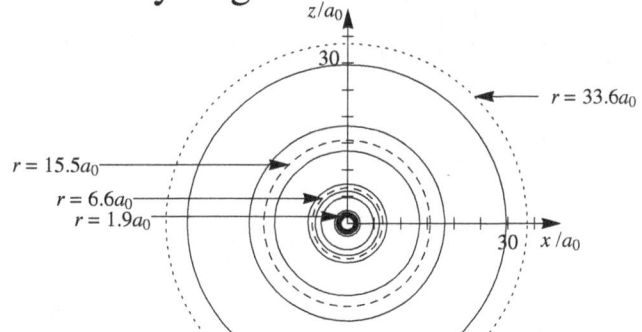

12.4   A hydrogen atom 5d orbital has the radial wave function

$$R_{52}(r) = \frac{1}{150\sqrt{70a_0^3}} \left(42 - 14\rho + \rho^2\right) \rho^2 e^{-\rho/2}$$

How many radial nodes does a 5d orbital have and at what radii do they occur?

Radial nodes occur when $R_{52}(r) = 0$:

$$\frac{1}{150\sqrt{70a_0^3}} \left(42 - 14\rho + \rho^2\right) \rho^2 e^{-\rho/2} = 0$$

$$42 - 14\rho + \rho^2 = 0$$

$$\rho = \frac{14 \pm \sqrt{(14)^2 - 4\,(42)}}{2}$$

$$\rho = \frac{14 \pm \sqrt{28}}{2}$$

$$\rho = 4.354 \text{ and } 9.646$$

Since $r = \rho a_0$, the two radial nodes of a $5d$ orbital are located at $4.354a_0$ and $9.646a_0$ or 230 pm and 510 pm.

12.6   Write the complex hydrogen atom wave functions, $2p_{-1}$, $2p_0$, and $2p_1$ in terms of the real wave functions $2p_x$, $2p_y$, and $2p_z$.

We use the convention of Table 11.3 for the spherical harmonic that describes the angular part of the hydrogen atom wave function. (Although this convention may seem unimportant with respect to the topics covered in this text, the relative signs of the spherical harmonics are important to the proper interpretation of angular momentum in more complex quantum systems.)

$$p_1 = -\sqrt{\frac{3}{8\pi}} \sin\theta e^{i\phi} = -\sqrt{\frac{3}{8\pi}} \sin\theta \,(\cos\phi + i\,\sin\phi)$$

$$= -\frac{1}{\sqrt{2}} \left(\sqrt{\frac{3}{4\pi}} \sin\theta \cos\phi + i\sqrt{\frac{3}{4\pi}} \sin\theta \sin\phi\right)$$

$$p_0 = \sqrt{\frac{3}{4\pi}} \cos\theta$$

$$p_{-1} = \sqrt{\frac{3}{8\pi}} \sin\theta e^{-i\phi} = \sqrt{\frac{3}{8\pi}} \sin\theta \,(\cos\phi - i\,\sin\phi)$$

$$= \frac{1}{\sqrt{2}} \left(\sqrt{\frac{3}{4\pi}} \sin\theta \cos\phi - i\sqrt{\frac{3}{4\pi}} \sin\theta \sin\phi\right)$$

Comparing with Equations 12.9 a, b, and c or Table 12.4,

$$p_1 = -\frac{1}{\sqrt{2}} \left(p_x + ip_y\right)$$

$$p_0 = p_z$$

$$p_{-1} = \frac{1}{\sqrt{2}} \left(p_x - ip_y\right)$$

Combining the radial and angular wave functions,

$$2p_{-1} = \frac{1}{\sqrt{2}}\left(2p_x - i2p_y\right)$$

$$2p_0 = 2p_z$$

$$2p_1 = -\frac{1}{\sqrt{2}}\left(2p_x + i2p_y\right)$$

12.8 Rank the following hydrogen atom orbitals in order of increasing energy, and identify any degeneracies. Assume there are no external fields present. $6s, 5s, 4s, 4p_z, 4p_x, 5d_{xy}, 5d_{xz}$.

The energy of a hydrogen atom orbital depends only on the n quantum number. Thus, in order of increasing energy, $4s = 4p_z = 4p_x < 5s = 5d_{xy} = 5d_{xz} < 6s$.

12.10 Use the hydrogen $2s$ wave function given in Table 12.2 to calculate the value of r (other than $r = \infty$) at which this wave function becomes zero.

The wave function becomes zero when the radial portion is zero:

$$R_{2s} = \frac{1}{\sqrt{2a_0^3}}\left(1 - \frac{\rho}{2}\right)e^{-\rho/2} = 0$$

$$1 - \frac{\rho}{2} = 0$$

$$\rho = 2$$

Since $\rho = r/a_0$, the wave function becomes zero when $r = 2a_0 = 1.058$ Å.

12.12 Explain, in terms of their electron configurations, why Fe^{2+} is more easily oxidized to Fe^{3+} than Mn^{2+} to Mn^{3+}.

The electron configurations of the species being considered are

Fe^{2+}: $[Ar]3d^6$

Fe^{3+}: $[Ar]3d^5$

Mn^{2+}: $[Ar]3d^5$

Mn^{3+}: $[Ar]3d^4$

A half-filled subshell has extra relative stability. On oxidizing Fe^{2+}, the product has a half-filled d subshell. On oxidizing Mn^{2+}, a half-filled d subshell is being lost, which requires more energy.

12.14 The formula for calculating the energies of an electron in a hydrogenlike ion is given in Equation 10.18. This equation cannot be applied to many-electron atoms. One way to modify it for the more complex atoms is to replace Z with $(Z - \sigma)$, where Z is the atomic number and σ is a positive dimensionless quantity called the *shielding constant*. Consider the helium atom as an example. The physical significance of σ is that it represents the extent of shielding that the two $1s$ electrons exert on each other. Thus, the quantity $(Z - \sigma)$ is appropriately called the *effective* *nuclear charge.* Calculate the value of σ if the first ionization energy of helium is 3.94×10^{-18} J per atom. (In your calculations, ignore the minus sign in the given equation.)

The ionization energy for the helium atom ($Z = 2$) can be calculated by using

$$\Delta E = \left[\frac{m_e \, (Z - \sigma)^2 \, e^4}{8h^2 \varepsilon_0^2} \right] \left(\frac{1}{n_i^2} - \frac{1}{n_f^2} \right)$$

$$= \left(2.180 \times 10^{-18} \text{ J} \right) (2 - \sigma)^2 \left(\frac{1}{n_i^2} - \frac{1}{n_f^2} \right)$$

(See Problem 10.13 for the value of the constant)

with $n_f = \infty$. The experimental value of the ionization energy for an electron in He with $n = 1$ is used to determine σ.

$$IE = \left(2.180 \times 10^{-18} \text{ J} \right) (2 - \sigma)^2 \left(\frac{1}{1} \right) = 3.94 \times 10^{-18} \text{ J}$$

$$\sigma = 0.656$$

12.16   A technique called photoelectron spectroscopy (see Section 14.5) is used to measure the ionization energy of atoms. A sample is irradiated with UV light, and electrons are ejected from the valence shell. The kinetic energies of the ejected electrons are measured. Since the energy of the UV photon and the kinetic energy of the ejected electron are known, we can write

$$h\nu = IE + \frac{1}{2} m_e u^2$$

where ν is the frequency of the UV light, and m_e and u are the mass and speed of the electron, respectively. In one experiment the kinetic energy of the ejected electron from potassium is found to be 5.34×10^{-19} J using a UV source of wavelength 162 nm. Calculate the ionization energy of potassium. How can you be sure that this ionization energy corresponds to the electron in the valence shell (i.e., the most loosely held electron)?

The ionization energy for one potassium atom is

$$IE = h\nu - \frac{1}{2} m_e u^2$$

$$= \frac{hc}{\lambda} - E_{\text{kinetic, electron}}$$

$$= \frac{\left(6.626 \times 10^{-34} \text{ J s} \right) \left(2.998 \times 10^8 \text{ m s}^{-1} \right)}{162 \times 10^{-9} \text{ m}} - 5.34 \times 10^{-19} \text{ J}$$

$$= 6.92 \times 10^{-19} \text{ J}$$

The ionization energy for one mole of potassium atoms is

$$IE = \left(6.92 \times 10^{-19} \text{ J} \right) \left(\frac{6.022 \times 10^{23}}{1 \text{ mol}} \right)$$

$$= 4.17 \times 10^5 \text{ J mol}^{-1} = 4.2 \times 10^2 \text{ kJ mol}^{-1}$$

To ensure that the ejected electron is the valence electron, UV light of the longest wavelength (lowest energy) should be used that can still eject electrons.

12.18 Experimentally, the electron affinity of an element can be determined by using a laser light to ionize the anion of the element in the gas phase:

$$X^-(g) + h\nu \rightarrow X(g) + e^-$$

Referring to Table 12.8, calculate the photon wavelength (in nm) corresponding to the electron affinity for chlorine. In what region of the electromagnetic spectrum does this wavelength lie?

The electron affinity for Cl is 349 kJ mol^{-1}. That is,

$$Cl(g) + e^- \rightarrow Cl^-(g) \quad \Delta H = -349 \text{ kJ mol}^{-1}$$

The reverse reaction

$$Cl^-(g) + h\nu \rightarrow Cl(g) + e^-$$

occurs when the photon energy = 349 kJ mol^{-1}. The energy of one photon is

$$E = \left(349 \text{ kJ mol}^{-1}\right) \left(\frac{1 \text{ mol}}{6.022 \times 10^{23}}\right) = 5.795 \times 10^{-22} \text{ kJ} = 5.795 \times 10^{-19} \text{ J}$$

Since

$$E = h\nu = \frac{hc}{\lambda}$$

The wavelength corresponding to this energy is

$$\lambda = \frac{hc}{E}$$

$$= \frac{\left(6.626 \times 10^{-34} \text{ J s}\right) \left(2.998 \times 10^8 \text{ m s}^{-1}\right)}{5.795 \times 10^{-19} \text{ J mol}^{-1}}$$

$$= 3.43 \times 10^{-7} \text{ m} = 343 \text{ nm}$$

This wavelength is in the UV region.

12.20 Explain why the electron affinity of nitrogen is approximately zero, while the elements on either side, carbon and oxygen, have substantial positive electron affinities.

The electron affinity depends on the Z_{eff} for the *empty* orbital into which the additional electron is placed. In general, Z_{eff} increases across a row in the periodic table, so that electrons are held more tightly, which increases electron affinity. Thus, carbon and oxygen have electron affinities on the order of 100 kJ mol^{-1}. In the case of nitrogen, however, the additional electron must go into an orbital that is already half-occupied, and breaks up the half-filled subshell. Consequently, there is little tendency for the atom to accept another electron.

12.22   The first four ionization energies of an element are approximately 738 kJ mol^{-1}, 1450 kJ mol^{-1}, 7.7 × 10^3 kJ mol^{-1}, and 1.1 × 10^4 kJ mol^{-1}. To which periodic group does this element belong? Why?

The large jump between the second and third ionization energies indicates a change in the principal quantum number n. That is, if the first two electrons removed have principal quantum number n, then the next two have principal quantum number $n - 1$. Thus, the element is in the second column of the periodic table, or Group 2A. A comparison with Table 12.7 shows the element is magnesium.

12.24   Use variational theory to calculate the energy of a particle in a one-dimensional box of length L with the following trial function,

$$\phi(x) = Nx^2 \left(L^2 - x^2 \right)$$

where N is a normalization constant to be determined. Recognizing that the true energy is $h^2 / \left(8mL^2 \right)$, calculate the percent error.

The Hamiltonian for a particle in a one-dimensional box is

$$\hat{H} = -\frac{\hbar^2}{2m} \frac{d^2}{dx^2}$$

The trial energy is

$$
\begin{aligned}
E_\phi &= \frac{\int_0^L \phi^* \hat{H} \phi \, dx}{\int_0^L \phi^* \phi \, dx} \\[2mm]
&= \frac{N^2 \left(-\frac{\hbar^2}{2m} \right) \int_0^L \left[x^2 \left(L^2 - x^2 \right) \right] \frac{d^2}{dx^2} \left[x^2 \left(L^2 - x^2 \right) \right] \, dx}{N^2 \int_0^L x^4 \left(L^2 - x^2 \right)^2 \, dx} \\[2mm]
&= \frac{N^2 \left(-\frac{\hbar^2}{2m} \right) \int_0^L \left[x^2 \left(L^2 - x^2 \right) \right] \left(2L^2 - 12x^2 \right) \, dx}{N^2 \int_0^L x^4 \left(L^2 - x^2 \right)^2 \, dx} \\[2mm]
&= \frac{\left(-\frac{\hbar^2}{2m} \right) \left(-\frac{44L^7}{105} \right)}{\frac{8L^9}{315}} \\[2mm]
&= \frac{33\hbar^2}{4mL^2} = \frac{33h^2}{16\pi^2 mL^2} = 0.209 \frac{h^2}{mL^2}
\end{aligned}
$$

The trial energy is 67.2% greater than the true ground state energy of $\dfrac{h^2}{8mL^2} = 0.125 \dfrac{h^2}{mL^2}$.

12.26   Apply the variational method to the harmonic oscillator using the trial wave function

$$\phi(x) = Ne^{-c|x|}$$

where N is a normalization constant, c is a variational parameter, and $|x|$ is the absolute value of the displacement from equilibrium. Because of the discontinuity in the derivative of the trial function at $x = 0$, one would not expect this trial function to give a good solution. Nevertheless, it may be used as a trial wave function. Calculate the variational energy and compare your results to the exact solution. (*Hint*: First calculate $\langle E \rangle$ as a function of c, then take the derivative $d\langle E \rangle / dc$ to find the value of c which minimizes the energy.)

The Hamiltonian for a harmonic oscillator is

$$\hat{H} = -\frac{\hbar^2}{2\mu}\frac{d^2}{dx^2} + \frac{1}{2}kx^2$$

The trial energy is

$$E_\phi = \langle E \rangle = \frac{\int_{-\infty}^{\infty} \phi^* \hat{H}\phi \, dx}{\int_{-\infty}^{\infty} \phi^*\phi \, dx}$$

$$= \frac{N^2 \int_{-\infty}^{\infty} \left(e^{-c|x|}\right)\left(-\frac{\hbar^2}{2\mu}\frac{d^2}{dx^2} + \frac{1}{2}kx^2\right)\left(e^{-c|x|}\right) \, dx}{N^2 \int_{-\infty}^{\infty} \left(e^{-2c|x|}\right) \, dx}$$

Because of the cusp in the trial function, it is best to break up each integral into two. Taking c to be positive, when $-\infty < x < 0$, $|x| = -x$; when $0 < x < \infty$, $|x| = x$.

$$E_\phi = \frac{N^2 \left\{ \int_{-\infty}^{0} \left[e^{-c(-x)}\right]\left(-\frac{\hbar^2}{2\mu}\frac{d^2}{dx^2} + \frac{1}{2}kx^2\right)\left[e^{-c(-x)}\right] \, dx + \int_{0}^{\infty} \left(e^{-cx}\right)\left(-\frac{\hbar^2}{2\mu}\frac{d^2}{dx^2} + \frac{1}{2}kx^2\right)\left(e^{-cx}\right) \, dx \right\}}{N^2 \left[\int_{-\infty}^{\infty} e^{-2c(-x)} \, dx + \int_{-\infty}^{\infty} e^{-2cx} \, dx\right]}$$

$$= \frac{-\frac{\hbar^2}{2\mu}\int_{-\infty}^{0} e^{cx}\frac{d^2}{dx^2}\left(e^{cx}\right) \, dx + \frac{1}{2}k\int_{-\infty}^{0} x^2 e^{2cx} \, dx - \frac{\hbar^2}{2\mu}\int_{0}^{\infty} e^{-cx}\frac{d^2}{dx^2}\left(e^{-cx}\right) \, dx + \frac{1}{2}k\int_{0}^{\infty} x^2 e^{-2cx} \, dx}{\frac{1}{2c} + \frac{1}{2c}}$$

$$= \frac{-\frac{\hbar^2}{2\mu}\left(\frac{c}{2}\right) + \frac{1}{2}k\left(\frac{1}{4c^3}\right) - \frac{\hbar^2}{2\mu}\left(\frac{c}{2}\right) + \frac{1}{2}k\left(\frac{1}{4c^3}\right)}{1/c}$$

$$= -\frac{\hbar^2 c^2}{2\mu} + \frac{k}{4c^2}$$

However, in trying to minimize E_ϕ by setting $\dfrac{dE_\phi}{dc} = 0$, a problem occurs.

$$\frac{d}{dc}\left(-\frac{\hbar^2 c^2}{2\mu} + \frac{k}{4c^2}\right) = 0$$

$$-\frac{\hbar^2 c}{\mu} - \frac{k}{2c^3} = 0$$

$$c^4 = -\frac{k\mu}{\hbar^2}$$

This only has solutions for imaginary values of c, but c must be a positive real number, otherwise the trial function $\phi(x) = Ne^{-cx^2}$ will not vanish as x approaches $\pm\infty$.

Indeed the problem becomes apparent when the trial energy is plotted as a function of c.

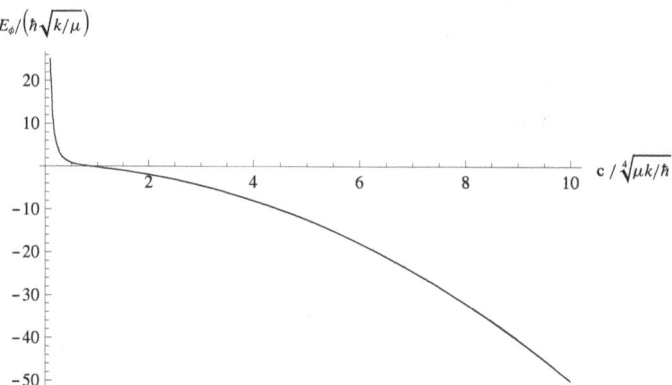

The graph shows that the trial energy decreases without bound as $c \to \infty$. Another way of thinking about this is that the trial energy never reaches a minimum, or alternatively, is only a minimum at $c = \infty$. In this latter case, the wave function shrinks to a infinitesimally thin impulse function at $x = 0$, which would not describe a physical system.

As it turns out, the problem is not the cusp itself in the wave function, but rather that $\phi(x) = Ne^{-c|x|}$ is a function for which the second derivative, or curvature, is positive for all values of x. As a result, the kinetic energy operator, $-\hbar^2 \frac{d^2}{dx^2}$, is negative for all values of x, and consequently, the average value of the kinetic energy for a particle described by this wave function is necessarily negative, which is non-physical. Thus, this wave function cannot represent a harmonic oscillator, even a quantum mechanical one.

Actually, there is one interesting interpretation of this result that is consistent with an admittedly still unphysical quantum mechanical harmonic oscillator. In quantum mechanics, a particle can enter, that is have non-zero probability of being found in, the "classically forbidden" region where the potential energy is greater than the total energy of the particle. Of course, this would imply that the kinetic energy is negative in this region. Thus, the trial function in this problem, which has a negative value for the kinetic energy operator for all values of x, might be considered as describing a system where the classically forbidden region extends over all values of x, or where the potential energy is greater than any possible total energy. Thus, the potential energy must be infinite everywhere, except possibly at a countable number of single, isolated points (which mathematicians call a set of measure zero). For a harmonic oscillator, this would be the case if $k \to \infty$, which would result in the wave function being forced to become an infinitesimally thin impulse function at $x = 0$, which is the same interpretation that the variational method gave.

12.28 Perturbation theory considers small influences on systems with analytical solutions. For each of the following cases, identify the zero-order Hamiltonian operator $\hat{H}^{(0)}$, the perturbation Hamiltonian operator $\hat{H}^{(1)}$, the zero-order wave function $\psi^{(0)}$, and the zero-order energy $E^{(0)}$. It is not necessary to calculate solutions – just use a model quantum mechanical system (such as particle in a box, particle on a ring, rigid rotor, harmonic oscillator, or hydrogen atom) for zero-order items. **(a)** A lithium ion, Li^+. **(b)** A helium atom, He. **(c)** A particle on a line segment of length L, with the potential energy function,

$$V(x) = \infty \qquad x < 0, \text{ or } x > L$$
$$V(x) = bx \qquad 0 \le x \le L$$

where b is a constant. **(d)** An anharmonic oscillator with the potential energy function,

$$V(x) = ax^2 + bx^3 + cx^4$$

where a, b, and c are constants. **(e)** A Morse oscillator with the potential energy function,

$$V(x) = D \left(1 - e^{-\beta x}\right)^2$$

where D and β are constants. (*Hint*: Use a Maclaurin series expansion:

$$e^{-x} = 1 - x + \frac{x^2}{2!} - \frac{x^3}{3!} + \cdots$$

and ignore x^4 and higher terms. The x^3 is the perturbation term.) **(f)** A rigid rotor in an electric field of strength E. The Hamiltonian operator that describes this system is

$$\hat{H} = -\frac{\hbar^2}{2I}\nabla^2 + \mu E \cos\theta$$

where I is the moment of inertia and μ is the dipole moment, and θ is the angle between the electric field and the dipole moment vector. **(g)** A hydrogen atom in a magnetic field of strength B_z in the z direction. The Hamiltonian operator that describes this system is

$$\hat{H} = -\frac{\hbar^2}{2m_e}\nabla^2 - \frac{e^2}{4\pi\varepsilon_0 r} - i\mu_B B_z \left(x\frac{\partial}{\partial y} - y\frac{\partial}{\partial x}\right)$$

where μ_B is the Bohr magneton (a constant), and the other symobols have their usual meaning.

(a) Li^+ has two electrons. The Hamiltonian operator contains the kinetic energy of the electrons (labeled as 1 and 2) and the Coulombic potential energies due to the interaction of each electron with the nucleus ($Z = 3$) and the interaction between the two electrons:

$$\hat{H} = -\frac{\hbar^2}{2m_e}\nabla_1^2 - \frac{\hbar^2}{2m_e}\nabla_2^2 - \frac{3e^2}{4\pi\varepsilon_0 r_1} - \frac{3e^2}{4\pi\varepsilon_0 r_2} + \frac{e^2}{4\pi\varepsilon_0 r_{12}}$$

where r_1 is the distance between the nucleus and electron 1, r_2 is the distance between the nucleus and electron 2, and r_{12} is the distance between the two electrons. The sum of the first and third terms in this Hamiltonian operator,

$$-\frac{\hbar^2}{2m_e}\nabla_1^2 - \frac{3e^2}{4\pi\varepsilon_0 r_1}$$

corresponds to the Hamiltonian operator for a one-electron atom with a nuclear charge of $3e$; so does the sum of the second and fourth terms in the Hamiltonian operator,

$$-\frac{\hbar^2}{2m_e}\nabla_2^2 - \frac{3e^2}{4\pi\varepsilon_0 r_2}$$

The solutions for a one-electron system are the same as those for the hydrogen atom, although the appropriate value for the nuclear charge is $3e$. Thus, the first four terms in the given Hamiltonian operator provide the zero-order Hamiltonian operator:

190   Chapter 12

$$\hat{H}^{(0)} = -\frac{\hbar^2}{2m_e}\nabla_1^2 - \frac{\hbar^2}{2m_e}\nabla_2^2 - \frac{3e^2}{4\pi\varepsilon_0 r_1} - \frac{3e^2}{4\pi\varepsilon_0 r_2}$$

The rest of the Hamiltonian operator is the perturbation:

$$\hat{H}^{(1)} = \frac{e^2}{4\pi\varepsilon_0 r_{12}}$$

The zero-order wave function is the product of the two one-electron atomic wave functions while the zero-order energy is the sum of the two one-electron atomic energies (don't forget that the nuclear charge is $3e$):

$$\psi^{(0)} = \psi_{n_1,\ell_1,m_{\ell_1}}\psi_{n_2,\ell_2,m_{\ell_2}}$$

$$E^{(0)} = -\frac{9m_e e^4}{8h^2\varepsilon_0^2 n_1^2} - \frac{9m_e e^4}{8h^2\varepsilon_0^2 n_2^2}$$

where 1 and 2 denote the quantum numbers for electrons 1 and 2, respectively.

(b) A helium atom has two electrons. Thus, this system has the same Hamiltonian operator, the same zero-order wave function, and the same zero-order energy as Li$^+$ [see part(a)] except that $Z = 2$:

$$\hat{H}^{(0)} = -\frac{\hbar^2}{2m_e}\nabla_1^2 - \frac{\hbar^2}{2m_e}\nabla_2^2 - \frac{2e^2}{4\pi\varepsilon_0 r_1} - \frac{2e^2}{4\pi\varepsilon_0 r_2}$$

$$\hat{H}^{(1)} = \frac{e^2}{4\pi\varepsilon_0 r_{12}}$$

$$\psi^{(0)} = \psi_{n_1,\ell_1,m_{\ell_1}}\psi_{n_2,\ell_2,m_{\ell_2}}$$

$$E^{(0)} = -\frac{4m_e e^4}{8h^2\varepsilon_0^2 n_1^2} - \frac{4m_e e^4}{8h^2\varepsilon_0^2 n_2^2}$$

(c) The zero-order problem is the particle in a box system (where the potential energy is 0 inside the box). The perturbation is the potential energy $V(x)$:

$$\hat{H}^{(0)} = -\frac{\hbar^2}{2m}\frac{d^2}{dx^2}$$

$$\hat{H}^{(1)} = bx \quad 0 \le x \le L$$

$$E^{(0)} = \frac{n^2 h^2}{8mL^2}$$

$$\psi^{(0)} = \sqrt{\frac{2}{L}}\sin\frac{n\pi x}{L}$$

(d) The zero-order system is the harmonic oscillator:

$$\hat{H}^{(0)} = -\frac{\hbar^2}{2\mu}\frac{d^2}{dx^2} + \frac{1}{2}kx^2$$

If we recognize $a = k/2$, then the perturbation to the potential is $V(x) - ax^2 = bx^3 + cx^4$ and $\hat{H} = \hat{H}^{(0)} + \hat{H}^{(1)}$, where

$$\hat{H}^{(1)} = bx^3 + cx^4$$

The zero-order wave function and energy are those of the harmonic oscillator:

$$\psi_v^{(0)} = \frac{1}{(2^v v!)^{1/2}} \left(\frac{\alpha}{\pi}\right)^{1/4} H_v(\alpha^{1/2} x) e^{-\alpha x^2/2}$$

$$E_v^{(0)} = \left(v + \frac{1}{2}\right) h\nu$$

where $\alpha = \sqrt{k\mu/\hbar^2}$.

(e) Expanding the Morse potential,

$$V(x) = D\left(1 - e^{-\beta x}\right)^2$$

$$= D\left[1 - \left(1 - \beta x + \frac{\beta^2 x^2}{2!} - \frac{\beta^3 x^3}{3!} + \cdots\right)\right]^2$$

$$= D\left(\beta x - \frac{\beta^2 x^2}{2} + \frac{\beta^3 x^3}{6} + \cdots\right)^2$$

$$= D\left(\beta^2 x^2 - \beta^3 x^3 + \cdots\right)$$

In fact, as the *Hint* indicates, the x^3 term is the perturbation term. Taking $k = 2D\beta^2$, and ignoring the x^4 and higher order terms,

$$V(x) = \frac{1}{2}kx^2 - D\beta^3 x^3$$

The zero-order system is the harmonic oscillator

$$\hat{H}^{(0)} = -\frac{\hbar^2}{2\mu}\frac{d^2}{dx^2} + \frac{1}{2}kx^2$$

The perturbation is

$$\hat{H}^{(1)} = -D\beta^3 x^3$$

The zero-order wave function and energy are those of the harmonic oscillator:

$$\psi_v^{(0)} = \frac{1}{(2^v v!)^{1/2}} \left(\frac{\alpha}{\pi}\right)^{1/4} H_v(\alpha^{1/2} x) e^{-\alpha x^2/2}$$

$$E_v^{(0)} = \left(v + \frac{1}{2}\right) h\nu$$

where $\alpha = \sqrt{k\mu/\hbar^2}$.

(f) The zero-order Hamiltonian operator is the rigid-rotor Hamiltonian:

$$\hat{H}^{(0)} = -\frac{\hbar^2}{2I}\nabla^2$$

The additional term in the Hamiltonian operator is the perturbation:

$$\hat{H}^{(1)} = \mu E \cos\theta$$

The zero-order wave function and energy are those of the rigid rotor:

$$\psi_{\ell,m_\ell}^{(0)} = Y_\ell^{m_\ell}$$

$$E_\ell^{(0)} = \ell\,(\ell+1)\,\frac{\hbar^2}{2I}$$

(g) The zero-order system is the hydrogen atom:

$$\hat{H}^{(0)} = -\frac{\hbar^2}{2m_e}\nabla^2 - \frac{e^2}{4\pi\varepsilon_0 r}$$

The perturbation is

$$\hat{H}^{(1)} = -i\mu_B B_z\left(x\frac{\partial}{\partial y} - y\frac{\partial}{\partial x}\right)$$

The zero-order wave function and energy are those of the hydrogen atom:

$$E^{(0)} = -\frac{m_e e^4}{8h^2\varepsilon_0^2 n^2}$$

$$\psi^{(0)} = \psi_{n,\ell,m_\ell}$$

12.30   Using a particle in a box as the zero-order system, apply perturbation theory to calculate the energy of a particle of mass m in a box of length a that has the step-potential energy function,

$$V(x) = 0 \qquad 0 \le x \le \frac{a}{2}$$
$$V(x) = c \qquad \frac{a}{2} \le x \le a$$
$$V(x) = \infty \qquad x < 0,\ \text{or}\ x > a$$

where c is a constant.

The perturbation to the Hamiltonian is the (non-zero) potential energy, $H^{(1)} = V(x)$, and the first-order perturbation energy correction to the energy of the nth particle in a box energy level is

$$E_n^{(1)} = \int \psi_n^{(0)*} \hat{H}^{(1)} \psi_n^{(0)} \, d\tau$$

$$= \int_0^{a/2} \left(\sqrt{\frac{2}{a}} \sin \frac{n\pi x}{a} \right) (0) \left(\sqrt{\frac{2}{a}} \sin \frac{n\pi x}{a} \right) dx$$

$$+ \int_{a/2}^a \left(\sqrt{\frac{2}{a}} \sin \frac{n\pi x}{a} \right) (c) \left(\sqrt{\frac{2}{a}} \sin \frac{n\pi x}{a} \right) dx$$

$$= 0 + \frac{2c}{a} \int_{a/2}^a \sin^2 \frac{n\pi x}{a} \, dx$$

$$= \frac{2c}{a} \left(\frac{a}{4} \right)$$

$$= \frac{c}{2}$$

The total energy is

$$E_n = E_n^{(0)} + E_n^{(1)}$$

$$= \frac{n^2 h^2}{8ma^2} + \frac{c}{2}$$

Each energy level differs from the corresponding unperturbed level by a constant amount, $c/2$, independent of the quantum number n.

12.32   Using first-order perturbation theory, calculate the ground-state energy of an anharmonic oscillator of reduced mass μ, with the potential energy function,

$$V(x) = \frac{k}{2} x^2 + bx$$

where k and b are constants. Assume b is sufficiently small that the bx term may be treated as a perturbation.

The zero-order system is the harmonic oscillator. The perturbation is

$$\hat{H}^{(1)} = bx$$

The first-order energy correction to the ground-state energy is

$$E_0^{(1)} = \int \psi_0^{(0)*} \hat{H}^{(1)} \psi_0^{(0)} \, d\tau$$

$$= \int_{-\infty}^{\infty} \left[\left(\frac{\alpha}{\pi} \right)^{1/4} e^{-\alpha x^2/2} \right] (bx) \left[\left(\frac{\alpha}{\pi} \right)^{1/4} e^{-\alpha x^2/2} \right] dx$$

$$= b \left(\frac{\alpha}{\pi} \right)^{1/2} \int_{-\infty}^{\infty} x e^{-\alpha x^2} \, dx$$

$$= 0$$

The ground-state energy of the oscillator is

$$E_0 = E_0^{(0)} + E_0^{(1)}$$

$$= \frac{h\nu}{2} + 0$$

$$= \frac{h\nu}{2}$$

The ground state energy is unperturbed. This is because the perturbation is an odd function while $\psi_0^{(0)*}\psi_0^{(0)}$ is even. Thus, the integral that leads to the first-order energy correction is 0.

12.34   In the hydrogen atom and hydrogenlike ions, the electronic energy levels depend only on the principal quantum number n, whereas in many-electron atoms the energy levels depend on both n and the orbital angular momentum quantum number ℓ. In your own words, explain this difference.

Figure 12.3 shows that electron density is distributed differently with regard to distance from the nucleus for wave functions with the same value of n, but different values of ℓ. In an atom with a single electron these differences are such that they have no effect on the average electrostatic interaction between the nucleus and the electron. (Using the postulates of quantum mechanics, this could be calculated in atomic units as $\int \psi_{n,\ell,m_\ell}^* \left(\frac{1}{r}\right) \psi_{n,\ell,m_\ell} \, d\tau$.) In a many-electron atom, however, electrons with greater density closer to the nucleus (*i.e.,* more penetrating) shield it from those electrons with greater density further away. This causes a lessening in the attraction between the shielded electron and the nucleus. Wave functions with smaller values of the angular momentum quantum number, ℓ, have greater density closer to the nucleus than those with larger values of ℓ. Thus, the electronic energy levels for a given principal quantum number, n, increase in the order $s < p < d < f \cdots$.

12.36   The variational method generally applies to ground-state wave functions. It may be extended to excited states under the condition that the trial wave function is orthogonal to the exact ground-state wave function. Under these circumstances, the trial wave function gives an upper limit for the energy of the first excited state. Given that the wave function,

$$\phi(x) = N\left(x^3 - \frac{3}{2}Lx^2 + \frac{1}{2}L^2x\right)$$

where N is a normalization constant, satisfies these conditions for a particle in a box of length L, determine the formula for the variational energy of the lowest-energy excited state. Next, calculate the energy for an electron in a 0.80-nm box. Then calculate the exact energy for an electron in the first excited state in a 0.80-nm box. How good an approximation is this trial wave function?

The Hamiltonian for a particle in a one-dimensional box is

$$\hat{H} = -\frac{\hbar^2}{2m}\frac{d^2}{dx^2}$$

The trial energy for the lowest-energy excited state is

$$E_\phi = \frac{\int_0^L \phi^* \hat{H} \phi \, dx}{\int_0^L \phi^* \phi \, dx}$$

$$= \frac{N^2 \left(-\frac{\hbar^2}{2m}\right) \int_0^L \left(x^3 - \frac{3}{2}Lx^2 + \frac{1}{2}L^2 x\right) \frac{d^2}{dx^2} \left(x^3 - \frac{3}{2}Lx^2 + \frac{1}{2}L^2 x\right) \, dx}{N^2 \int_0^L \left(x^3 - \frac{3}{2}Lx^2 + \frac{1}{2}L^2 x\right)^2}$$

$$= \frac{\left(-\frac{\hbar^2}{2m}\right) \int_0^L \left(x^3 - \frac{3}{2}Lx^2 + \frac{1}{2}L^2 x\right)(6x - 3L) \, dx}{\int_0^L \left(x^6 - 3Lx^5 + \frac{13}{4}L^2 x^4 - \frac{3}{2}L^3 x^3 + \frac{1}{4}L^4 x^2\right) \, dx}$$

$$= \frac{\left(-\frac{\hbar^2}{2m}\right)\left(-\frac{L^5}{20}\right)}{\frac{L^7}{840}}$$

$$= \frac{21\hbar^2}{mL^2}$$

$$= \frac{21h^2}{4\pi^2 mL^2} = 0.532 \frac{h^2}{mL^2}$$

The exact energy for an electron in the first excited state ($n = 2$) is $\frac{4h^2}{8mL^2} = 0.500 \frac{h^2}{mL^2}$. Thus, the trial energy is 6.4% greater than the exact energy.

For a 0.80-nm box, $L = 0.80$ nm, and the trial energy is

$$\frac{21h^2}{4\pi^2 mL^2} = \frac{21 \left(6.626 \times 10^{-34} \text{ J s}\right)^2}{4\pi^2 \left(9.109 \times 10^{-31} \text{ kg}\right) \left(0.80 \times 10^{-9} \text{ m}\right)^2}$$

$$= \left(4.01 \times 10^{-19} \text{ J}\right) \left(\frac{1 \text{ kJ}}{1000 \text{ J}}\right) \left(\frac{6.022 \times 10^{23}}{1 \text{ mol}}\right)$$

$$= 241 \text{ kJ mol}^{-1} = 2.4 \times 10^2 \text{ kJ mol}^{-1}$$

The exact energy is

$$\frac{4h^2}{8mL^2} = \frac{4 \left(6.626 \times 10^{-34} \text{ J s}\right)^2}{8 \left(9.109 \times 10^{-31} \text{ kg}\right) \left(0.80 \times 10^{-9} \text{ m}\right)^2}$$

$$= \left(3.77 \times 10^{-19} \text{ J}\right) \left(\frac{1 \text{ kJ}}{1000 \text{ J}}\right) \left(\frac{6.022 \times 10^{23}}{1 \text{ mol}}\right)$$

$$= 227 \text{ kJ mol}^{-1} = 2.3 \times 10^2 \text{ kJ mol}^{-1}$$

12.38 Using Equation 10.23, confirm the value of the Rydberg constant for the hydrogen (^{1}H) atom. Next, substitute the reduced mass μ for the mass of the electron m_e, and calculate the value of the Rydberg constant for the hydrogen (^{1}H) atom. Again using the reduced mass, calculate the Rydberg constant for the deuterium (^{2}H) atom. Compare and contrast your results. The mass of the deuterium atom nucleus is $3.34358320 \times 10^{-27}$ kg.

Using Equation 10.23,

$$\tilde{R}_{\mathrm{H}} = \frac{m_e e^4}{8ch^3\varepsilon_0^2}$$

$$= \frac{\left(9.10938291 \times 10^{-31}\ \mathrm{kg}\right)\left(1.602176565 \times 10^{-19}\ \mathrm{C}\right)^4}{8\left(2.99792458 \times 10^8\ \mathrm{m\,s^{-1}}\right)\left(6.62606957 \times 10^{-34}\ \mathrm{J\,s}\right)^3\left(8.854187817 \times 10^{-12}\ \mathrm{C^2\,N^{-1}\,m^{-2}}\right)^2}$$

$$= 10973731.6\ \mathrm{m}^{-1} = 109737.316\ \mathrm{cm}^{-1}$$

For the hydrogen atom, the reduced mass is

$$\mu_{\mathrm{H}} = \frac{m_{\mathrm{proton}} m_{\mathrm{electron}}}{m_{\mathrm{proton}} + m_{\mathrm{electron}}}$$

$$= \frac{\left(1.672621777 \times 10^{-27}\ \mathrm{kg}\right)\left(9.10938291 \times 10^{-31}\ \mathrm{kg}\right)}{1.672621777 \times 10^{-27}\ \mathrm{kg} + 9.10938291 \times 10^{-31}\ \mathrm{kg}}$$

$$= 9.104424485 \times 10^{-31}\ \mathrm{kg}$$

Note that the hydrogen nucleus is just a proton. The Rydberg constant is then

$$\tilde{R}_{\mathrm{H}} = \frac{\mu_{\mathrm{H}} e^4}{8ch^3\varepsilon_0^2}$$

$$= \frac{\left(9.104424485 \times 10^{-31}\ \mathrm{kg}\right)\left(1.602176565 \times 10^{-19}\ \mathrm{C}\right)^4}{8\left(2.99792458 \times 10^8\ \mathrm{m\,s^{-1}}\right)\left(6.62606957 \times 10^{-34}\ \mathrm{J\,s}\right)^3\left(8.854187817 \times 10^{-12}\ \mathrm{C^2\,N^{-1}\,m^{-2}}\right)^2}$$

$$= 10967758.4\ \mathrm{m}^{-1} = 109677.584\ \mathrm{cm}^{-1}$$

For the deuterium atom, the reduced mass is

$$\mu_{\mathrm{D}} = \frac{m_{\mathrm{D\ nucleus}} m_{\mathrm{electron}}}{m_{\mathrm{D\ nucleus}} + m_{\mathrm{electron}}}$$

$$= \frac{\left(3.34358320 \times 10^{-27}\ \mathrm{kg}\right)\left(9.10938291 \times 10^{-31}\ \mathrm{kg}\right)}{3.34358320 \times 10^{-27}\ \mathrm{kg} + 9.10938291 \times 10^{-31}\ \mathrm{kg}}$$

$$= 9.106901792 \times 10^{-31}\ \mathrm{kg}$$

The Rydberg constant is then

$$\tilde{R}_{\mathrm{H}} = \frac{\mu_{\mathrm{D}} e^4}{8ch^3\varepsilon_0^2}$$

$$= \frac{\left(9.106901792 \times 10^{-31}\ \mathrm{kg}\right)\left(1.602176565 \times 10^{-19}\ \mathrm{C}\right)^4}{8\left(2.99792458 \times 10^8\ \mathrm{m\,s^{-1}}\right)\left(6.62606957 \times 10^{-34}\ \mathrm{J\,s}\right)^2\left(8.854187817 \times 10^{-12}\ \mathrm{C^2\,N^{-1}\,m^{-2}}\right)^2}$$

$$= 10970742.7\ \mathrm{m}^{-1} = 109707.427\ \mathrm{cm}^{-1}$$

We see that the value of the Rydberg constant for the hydrogen atom (^{1}H) is roughly 30 cm^{-1} smaller than that for the deuterium atom, (^{2}H) which is roughly 30 cm^{-1} smaller than the value obtained using the mass of the electron m_e, which is equivalent to assuming an infinite nuclear mass. These differences are on the order of 0.03% of the Rydberg constant itself. Looking at this another way, the wavelengths of the first Balmer line are predicted to be 656.47 nm, 656.29 nm, 656.11 nm, for ^{1}H, ^{2}H, and a hypothetical hydrogen atom with infinite nuclear mass, respectively.

12.40   The ionization energy of a certain element is 412 kJ mol^{-1}. When the atoms of the element are in the first excited electronic state, however, the ionization energy is only 126 kJ mol^{-1}. Based on this information, calculate the wavelength of light emitted in a transition from the first excited state to the ground state.

Ionization energy is the energy required to remove an electron. Typically, the term refers to the removal of an electron from the ground state of an atom, but it is also possible to start with the atom in an excited state. In this case, it will take less energy to remove the electron, as indicated in the problem. The difference in the values of ionization energies reflects the difference in energies of the initial states. In this case, it is the difference between the ground and the first excited states:

$$\Delta E = 412 \text{ kJ mol}^{-1} - 126 \text{ kJ mol}^{-1} = 286 \text{ kJ mol}^{-1}$$

The electromagnetic radiation energy emitted must match this energy difference,

$$\frac{hc}{\lambda} = \left(286 \times 10^3 \text{ J mol}^{-1}\right)\left(\frac{1 \text{ mol}}{6.022 \times 10^{23}}\right)$$

$$\lambda = \frac{hc}{4.749 \times 10^{-19} \text{ J}}$$

$$= \frac{\left(6.626 \times 10^{-34} \text{ J s}\right)\left(2.998 \times 10^8 \text{ m s}^{-1}\right)}{4.749 \times 10^{-19} \text{ J}}$$

$$= 4.183 \times 10^{-7} \text{ m} = 418 \text{ nm}$$

CHAPTER 13

Molecular Electronic Structure
and the Chemical Bond

PROBLEMS AND SOLUTIONS

13.2 In solving the Schrodinger equation (Equation 13.5), qualitatively describe how the bond energies and bond lengths would vary amongst a series of one-electron cations: H_2^+, HHe^{2+}, He_2^{3+}, and LiH^{3+}.

Moving across the series, the number of protons in the nuclei increases. This has two effects. First, it increases the coulombic attraction between the nuclei and the single electron, but as noted in Section 13.1, this not the most important contribution to bonding. The most important contribution is the lowering of kinetic energy that results from giving the electron a larger space in which to move. However, with greater nuclear charge, the electron is more localized near a nucleus, and this major contribution to bonding is diminished. This is exacerbated by the second effect of increasing nuclear charge, namely greater coulombic repulsion between the positively charged nuclei. In fact, only the first molecule, H_2^+, is predicted to form a bond. The others do not.

13.4 A sigma bonding orbital is taken as the sum of two hydrogen atom $1s$ orbitals,

$$\sigma(1) = N\left[s_A(1) + s_B(1)\right]$$

where "1" represents the coordinates of an electron numbered one. Assuming that the hydrogen atomic orbitals are normalized, calculate the normalization constant, N, for this sigma orbital. Express your answer in terms of the overlap integral, S, defined in Equation 13.15.

To normalize $\sigma(1)$,

$$\int \left\{N\left[s_A^*(1) + s_B^*(1)\right]\right\}\left\{N\left[s_A(1) + s_B(1)\right]\right\}\,d\tau = 1$$

$$N^2\left(\int s_A^*(1)s_A(1)\,d\tau + \int s_A^*(1)s_B(1)\,d\tau + \int s_B^*(1)s_A(1)\,d\tau + \int s_B^*(1)s_B(1)\,d\tau\right) = 1$$

$$N^2\left(1 + S + S + 1\right) = 1$$

$$N^2 = \frac{1}{2 + 2S}$$

$$N = \frac{1}{\sqrt{2 + 2S}}$$

Note that the overlap integral can be expressed as $\int s_A^*(1)s_B(1)\,d\tau$ or $\int s_B^*(1)s_A(1)\,d\tau$. These integrals are the same because S is real.

13.6 Sigma bonding and antibonding orbitals (σ and σ^*) are defined in Problems 13.4 and 13.5. Show that these two orbitals are orthogonal.

Multiply the complex conjugate of the σ orbital with the σ^* orbital and integrate over all space. (Unfortunately, we use the star, "*", in this problem for two different purposes: (1) to denote complex conjugation, and (2) to distinguish between N and N^* from Problems 13.4 and 13.5, respectively.)

$$\int \left\{ N\left[s_A^*(1) + s_B^*(1)\right] \right\} \left\{ N^*\left[s_A(1) - s_B(1)\right] \right\}\,d\tau$$

$$= NN^* \left(\int s_A^*(1)s_A(1)\,d\tau - \int s_A^*(1)s_B(1)\,d\tau + \int s_B^*(1)s_A(1)\,d\tau - \int s_B^*(1)s_B(1)\,d\tau \right)$$

$$= NN^*\,(1 - S + S - 1)$$

$$= 0$$

13.8 The valence bond and molecular orbital pictures are two ways of describing a chemical bond. In a sense, the valence bond approach contains too little ionic character and the molecular orbital approach contains too much ionic character. One technique that might be applied is that of perturbation theory, in which ionic bonding is treated as a perturbation of valence bonding. Outline how to solve for the energetics of a chemical bonding system using perturbation theory.

Typically, one thinks of applying perturbation theory by identifying terms in the Hamiltonian that are ignored in a first approximation to lead to a known or easily found solution (the zero-order solution) and then accounting for the "perturbation" terms with increasing levels of approximation. (See Section 12.11.) However, valence bond theory is, in fact, an application of perturbation theory using the complete Hamiltonian for the molecule, such as the following for H_2.

$$\hat{H} = \underbrace{-\frac{1}{2}\nabla_1^2 - \frac{1}{r_{1A}} - \frac{1}{2}\nabla_2^2 - \frac{1}{r_{2B}}}_{\hat{H}^{(0)}} \underbrace{- \frac{1}{r_{1B}} - \frac{1}{r_{2A}} + \frac{1}{R} + \frac{1}{r_{12}}}_{\hat{H}^{(1)}}$$

The first four terms are the zero-order Hamiltonian, $\hat{H}^{(0)}$, and represent the Hamiltonians for two independent hydrogen atoms. The last four terms are the perturbation, $\hat{H}^{(1)}$. (Note that the sign before the $\frac{1}{R}$ term in Equation 13.16 should be a plus sign.) Valence bond theory then assumes a zero-order solution (wave function) that contains only covalent terms, but does satisfy the Pauli principle (Equation 13.10).

$$\psi_{VB+} = N_+\left[1s_A(1)\,1s_B(2) + 1s_A(2)\,1s_B(1)\right]\left[\alpha(1)\beta(2) - \beta(1)\alpha(2)\right]$$

The energy associated with this choice of wave function is evaluated as outlined in Section 13.3. (In doing this, the association of the nucleus-electron Coulomb interaction terms with zero-order and perturbation Hamiltonian switches when integrating over the second term in the spatial part of the wave function.) Effectively, one must evaluate the quantity

$$\langle E \rangle = 2E_{1s} + \frac{\int \psi_{VB+}^* \hat{H}^{(1)} \psi_{VB+} \, d\tau}{\int \psi_{VB+}^* \psi_{VB+} \, d\tau}$$

$$= 2E_{1s} + \frac{J + K}{1 + S^2}$$

where the integrals J, K, and S are defined in Equations 13.15, 13.18 and 13.19. (The sign before the $\frac{1}{R}$ term in Equations 13.18 and 13.19, as well as those in 13.23 and 13.24 used below, should be a plus sign.)

If one wanted to "correct" the valence bond method with the addition of ionic terms, they would be added to the zero-order wave function, as in Equation 13.25.

$$\psi_{MO} = N_{MO} \left[\overbrace{1s_A(1)1s_B(2) + 1s_A(2)1s_B(1)}^{\psi_{VB}} \right.$$

$$\left. + \underbrace{1s_A(1)1s_A(2) + 1s_B(1)1s_B(2)}_{\psi_{ionic}} \right] \left[\alpha(1)\beta(2) - \beta(1)\alpha(2) \right]$$

Perturbation theory would then be applied in approximating the energy associated with this wave function by evaluating

$$\langle E \rangle = 2E_{1s} + \frac{\int \psi_{MO}^* \hat{H}^{(1)} \psi_{MO} \, d\tau}{\int \psi_{MO}^* \psi_{MO} \, d\tau}$$

$$= 2E_{1s} + \frac{J' + K'}{1 + S}$$

where the integrals J' and K' are defined in Equations 13.23 and 13.24 (as corrected) and S is as before.

Of course, as discussed in Section 13.4, this approach would greatly overestimate the ionic character of the bond. One way of proceeding would be to use a wave function with an adjustable amount of ionic contribution.

$$\psi_{MO} = N_{MO} \left[\overbrace{1s_A(1)1s_B(2) + 1s_A(2)1s_B(1)}^{\psi_{VB}} \right.$$

$$\left. + \lambda \left(\underbrace{1s_A(1)1s_A(2) + 1s_B(1)1s_B(2)}_{\psi_{ionic}} \right) \right] \left[\alpha(1)\beta(2) - \beta(1)\alpha(2) \right]$$

The perturbation integrals could then be evaluated as a function of λ and variation theory applied to find the value of λ that gives the lowest energy. Although the calculation is beyond the scope of this treatment, the best value is found to be approximately $\lambda = \frac{1}{6}$, which roughly corresponds to 3% ionic character $((1/6)^2 = 1/36 = 0.03)$.

13.10   Explain why the overlap integral S (Equation 13.15) is never zero for two hydrogen atom $1s$ orbitals.

The wave function for the hydrogen atom $1s$ orbital is greater than zero for all finite values of r, and consequently extends over all space. Thus, the product of two such wave functions, centered on different atoms, and forming the integrand for the overlap integral is likewise greater than zero throughout all space. Consequently, the integral itself likewise takes on a positive, non-zero value. In fact, it can be shown that

$$S = e^{-ZR}\left(1 + ZR + \frac{Z^2 R^2}{3}\right)$$

where Z is the atomic number ($Z = 1$ for hydrogen) and R is the distance between the two hydrogen atoms.

13.12   The four-electron species H_2^{2-} is not covalently bound in its ground electronic state, but has a bond order of one in a low-lying excited electronic state. Explain this phenomenon by drawing a molecular orbital diagram constructed from hydrogen $1s$ and $2s$ orbitals, and assigning electrons to molecular orbitals.

In the low-lying electronic state formed by exciting one of the electrons from the $1s\sigma_u^*$ orbital into the $2s\sigma_s$ orbital, there are 3 bonding electrons and 1 antibonding electron, giving a bond order of $(3 - 1)/2 = 1$.

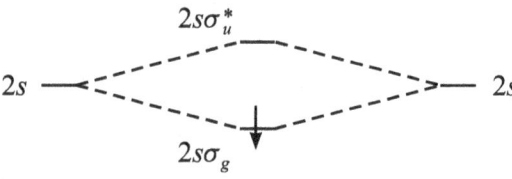

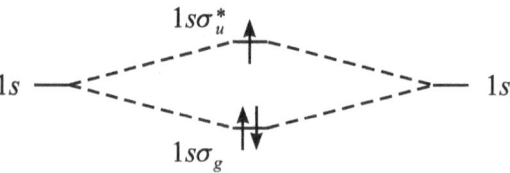

13.14   Explain the difference between the (often interchanged) terms "molecular orbital" and "molecular wave function."

The molecular wave function, strictly speaking, is the (single) wave function that describes all the electrons in a molecule. It is a function of the coordinates of all the electrons and returns a single value, the magnitude squared of which can be considered the probability density for finding the electrons arranged one each at the coordinates specified. In molecular orbital theory, the molecular wave function is written as the (antisymmetrized) product of molecular orbitals. The molecular orbitals are functions of just three spatial coordinates, corresponding to the coordinates of a single electron. To satisfy the Pauli Principle, each molecular orbital can describe at most two, spin-paired electrons. Molecular orbitals are most often expressed as linear combinations of atomic orbitals.

13.16   Diatomic neon (Ne_2) has a dissociative ground state and simple molecular orbital theory predicts bond order zero. Excited states of diatomic neon may have nonzero bond order. Sketch the molecular orbital diagram for diatomic neon, and place electrons in orbitals to produce an excited state with a nonzero bond order. Identify the bond order of your excited state.

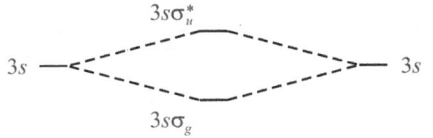

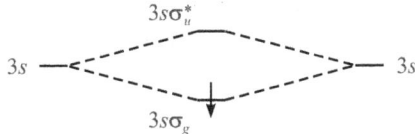

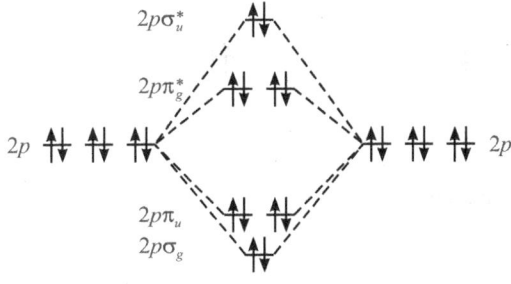

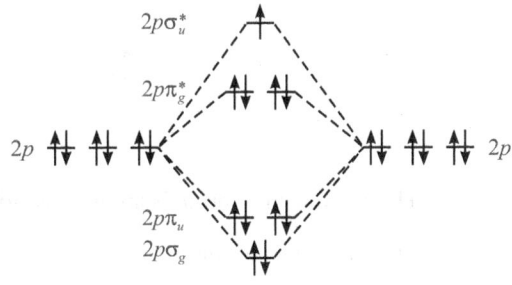

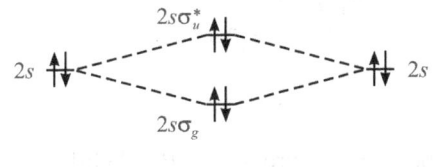

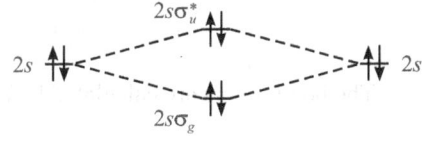

Ground State   An Excited State

The $1s$ molecular orbitals are not included in this sketch.

The electron configuration for Ne_2 in its ground state is

$$(\sigma_g 1s)^2 (\sigma_u^* 1s)^2 (\sigma_g 2s)^2 (\sigma_u^* 2s)^2 (\sigma_g 2p_z)^2 (\pi_u 2p_x)^2 (\pi_u 2p_y)^2 (\pi_g^* 2p_x)^2 (\pi_g^* 2p_y)^2 (\sigma_u^* 2p_z)^2$$

There are 10 bonding electrons and 10 antibonding electrons, giving a bond order of $(10 - 10)/2 = 0$.

One excited state of Ne_2 has the electron configuration

$$(\sigma_g 1s)^2 (\sigma_u^* 1s)^2 (\sigma_g 2s)^2 (\sigma_u^* 2s)^2 (\sigma_g 2p_z)^2 (\pi_u 2p_x)^2 (\pi_u 2p_y)^2 (\pi_g^* 2p_x)^2 (\pi_g^* 2p_y)^2 (\sigma_u^* 2p_z)^1 (\sigma_g 3s)^1$$

There are 11 bonding electrons and 9 antibonding electrons, giving a bond order of $(11 - 9)/2 = 1$. There are, of course, other possibilities.

13.18   For the homonuclear diatomic ions in their ground electronic state – Be_2^+, B_2^+, C_2^+ – identify the bond order and whether the species is paramagnetic or diamagnetic. Which species would have the longest bond length? Which would have the greatest bond strength? Use molecular orbital diagrams to support your conclusions.

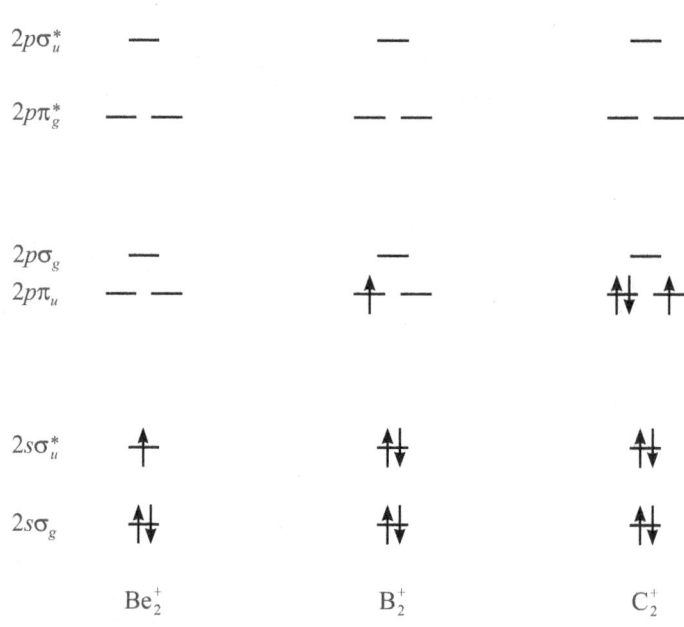

The $1s$ molecular orbitals are not included in this sketch.

The electron configurations are:

$$Be_2^+ \quad (\sigma_g 1s)^2 (\sigma_u^* 1s)^2 (\sigma_g 2s)^2 (\sigma_u^* 2s)^1$$
$$B_2^+ \quad (\sigma_g 1s)^2 (\sigma_u^* 1s)^2 (\sigma_g 2s)^2 (\sigma_u^* 2s)^2 (\pi_u 2p_x)^1$$
$$C_2^+ \quad (\sigma_g 1s)^2 (\sigma_u^* 1s)^2 (\sigma_g 2s)^2 (\sigma_u^* 2s)^2 (\pi_u 2p_x)^2 (\pi_u 2p_y)^1$$

The bond orders are calculated below:

Species	# of Bonding Electrons	# of Antibonding Electrons	Bond Order
Be_2^+	4	3	$(4-3)/2 = 0.5$
B_2^+	5	4	$(5-4)/2 = 0.5$
C_2^+	7	4	$(7-4)/2 = 1.5$

All three species have an unpaired electron, they are paramagnetic.

Be_2^+ and B_2^+ both have the smallest bond order (0.5); thus, one of these two would be expected to have the longest bond length. Due to the smaller nuclear charge, Be_2^+ would be expected to have a longer bond length than B_2^+. Because C_2^+ has the highest bond order, it would have the greatest bond strength.

13.20   Among the simplest heteronuclear diatomic species is HeH^+. Sketch a correlation diagram for the molecular orbitals of this species. What is the predicted bond order for the ground electronic state? If this species is thermally dissociated, remaining on the ground state potential energy surface, which atomic species would be formed, He and H^+ or He^+ and H? Next draw the electron configuration for when an electron is promoted from the HOMO (highest-occupied molecular orbital) to the LUMO (lowest-unoccupied molecular orbital). Is this excited electronic state bound or dissociative? If HeH^+ dissociates from this excited state, will it form He and H^+ or He^+ and H?

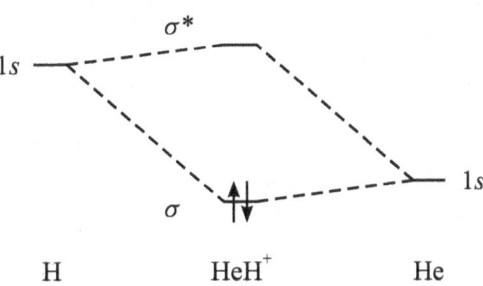

The predicted bond order is $(2 - 0)/2 = 1$.

The σ bonding orbital in HeH^+ has its electron density more concentrated on the helium atom. Thus, when the molecule is dissociated from its ground state, we would expect both electrons to remain located there and giving He and H^+. Additionally, because the electron affinity of H^+ has a smaller magnitude than the first ionization energy of He, this would be the lower energy state for the separated atoms.

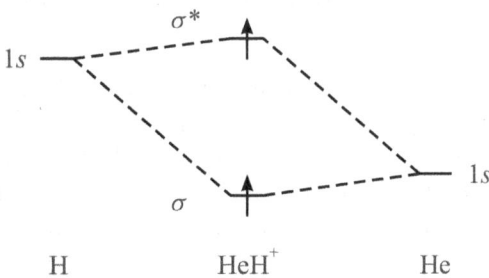

The lowest energy excited state for HeH^+ is a triplet state with the electron configuration $(\sigma)^1(\sigma^*)^1$. The antibonding orbital has its electron density more concentrated on the hydrogen atom. Thus, the molecule has one electron located primarily near the helium atom and the other near the hydrogen atom. Upon dissociation, we would expect the electrons to remain with their respective atoms giving He^+ and H. Furthermore, because the electron spins are not paired, the Pauli Exclusion Principle would prevent both of them from being accommodated by the $1s$ orbital on helium without a spin flip occurring.

13.22   First consider a homonuclear diatomic molecule, then consider a heteronuclear diatomic molecule. For each molecule, is it possible to form a molecular orbital from the combination of a $1s$ atomic orbital and a $2s$ atomic orbital? If so, would it be σ or π or neither? g(gerade) or u(ungerade) or neither? Bonding, antibonding or nonbonding?

Molecular orbitals gain their greatest stability when the atomic orbitals from which they are formed are close in energy. For a homonuclear diatomic molecule, the $1s$ and $2s$ atomic orbitals will be of different energies, and more importantly, the two $1s$ atomic orbitals and the two $2s$ atomic orbitals on the two atoms will have exactly the same energy. Consequently, although the $1s$ orbital and the $2s$ orbital have the proper symmetries to combine, the most important contributions to any molecular orbital will be from the combination of $1s$ with $1s$ and $2s$ with $2s$. Of course, a better description of the bonding in the molecule will be obtained by using a more flexible basis set that includes contributions from all orbitals of the proper symmetry, but for a homonuclear diatomic molecule, a molecular orbital would have major contributions primarily from $1s$ and very little $2s$ or vice versa.

For a heteronuclear diatomic molecule, it is possible that the $1s$ orbital on one atom is close in energy to the $2s$ orbital on the other. In this case, one would form two σ molecular orbitals, which are neither gerade or ungerade. One would be bonding and the other would be antibonding.

13.24 Quantitatively estimate the largest value of the dipole moment that could be obtained for a diatomic species. Explicitly state any assumptions that you make. Using the scientific literature, look up any information that you might need to make your estimate more accurate.

The factors that contribute to a large value for the dipole moment of a diatomic molecule, $\mu = QR$, are significant ionic character (large Q) and long bond length (large R). This suggests looking for a molecule formed from atoms on opposite sides of the periodic table. One question is whether an alkaline earth salt, with doubly charged cation and anion, should be chosen, since a molecule such as BaO would have a transfer of two electrons if 100% ionic. In fact, as seen below, these molecules tend to have lower dipole moments than alkali halides, presumably because they have significant covalent character. This is not surprising in the gas phase, which lacks the lattice energy stabilization of the ionic solid, or the hydration energy of the aqueous solution, to provide the stabilization necessary to overcome the large second ionization energy of the alkaline earth atom.

Setting aside francium, which is radioactive and present only in trace amounts in the Earth's crust, cesium is a natural choice for the electropositive element. It has an ionic radius of 161 pm. The choice for the anionic partner is not so obvious. Fluorine is more electronegative, but F^- has a smaller ionic radius (136 pm) than I^- (216 pm). Assuming 100% ionic character for the molecule, CsI, with a predicted ionic contact distance of 377 pm, would be predicted to have the larger dipole moment.

$$\mu_{\text{pred}} = QR$$

$$= \left(1.602 \times 10^{-19}\,\text{C}\right)\left(377 \times 10^{-12}\,\text{m}\right)\left(\frac{1\,\text{D}}{3.336 \times 10^{-30}\,\text{C m}}\right)$$

$$= 18.1\,\text{D}$$

Experimental dipole moments are available from the National Institute of Standards and Technology web site, http://www.nist.gov/data/nsrds/NSRDS-NBS-10.pdf. CsI is not included there because the values available in 1967 were deemed untrustworthy. However, a more recent reference [T. L. Story, Jr. and A. J. Hebert, *J. Chem. Phys.* **64**, 855 (1976)] provides a value of 11.69 D. This can be compared to the NIST values of 10.42 D and 7.88 D for CsCl and CsF, respectively, suggesting that the longer bond length is more important than electronegativity difference. Finally, the value in the NIST table for BaO is 7.95 D, which indicates that greater ionic character is indeed found in the alkali halides.

13.26 Based on the following table, describe the periodic trend in percent ionic character of these bonds. Which neutral diatomic species has the most covalent bond? Does this agree with your expectation based on electronegativity differences?

Species	μ/debye	eR/debye	R/pm
BH	1.733	5.936	123.6
CH	1.570	5.398	112.4

Species	μ/debye	eR/debye	R/pm
NH	1.627	4.985	103.8
OH	1.780	4.661	97.05
FH	1.942	4.405	91.71

The percent ionic character for each species is calculated using $\dfrac{\mu}{eR} \times 100\%$, and listed below:

Species	% ionic character
BH	29.19
CH	29.08
NH	32.64
OH	38.19
FH	44.09

The general trend is one of increasing ionic character going across the period from BH to FH. This is in accord with expectations based on electronegativity differences. There is a small discontinuity between BH and CH, which have very nearly the same ionic character. Nevertheless, the most species with the (barely) most covalent bond is CH, whereas electronegativity differences would predict BH to be the most covalent. On the other hand, boron has an electronegativity (2.0) below that of hydrogen (2.1), while all the other binding partners in the list have greater electronegativities, so that the direction of charge transfer changes between BH and CH.

13.28 Consider the series of hydride molecules LiH, BeH_2, BH_3, CH_4, NH_3, H_2O, and HF. For each molecule, would the covalent single bond to the hydrogen atom be better described using simple VB or MO theory? For this series, identify molecular properties that might best be described by VB theory. Recall that VB theory underestimates and MO theory overestimates the percent ionic character in a covalent bond. Is there a periodic trend in ionic character?

As noted in the answer to Problem 13.26, the value of the electronegativity for hydrogen (2.1) falls between those of boron (2.0) and carbon (2.5). Thus, the percent ionic character of each bond is expected to decrease along the series from LiH to BH_3 or CH_4 and then turn around and increase continuing on to HF. This suggests that simple VB theory will better describe the single bonds to hydrogen for BH_3 and CH_4 while simple MO theory will work better at either end of the series, LiH, BeH_2, NH_3, H_2O, and HF. Simple VB theory does an excellent job describing molecular geometry for these species.

13.30 Table 13.2 lists most of the more common situations where VSEPR theory is applied to small molecules. Omitted from this table is the situation where four groups of electrons surround a central atom and three of these groups of electrons are lone pairs. (a) What might the hybridization of the central atom be under these circumstances? (b) What would the molecular geometry be? (c) Give an example of a molecule that might be described by this electronic structure. (d) Discuss the advantages and disadvantages of applying VSEPR theory in this situation.

(a) With four electron groups surrounding the central atom, it would be predicted to adopt sp^3 hybridization.

(b) Three of the four groups are lone pairs, leaving only one bonding pair. Assuming that the bond is to a single atom, and not a more complicated chemical grouping, the molecule is diatomic and is necessarily linear.

(c) An example of such a molecule would be HF.

(d) VSEPR has the advantage of specifying locations for all four electron groups, and it does indicate that the three lone pairs are in equivalent orbitals. It has the disadvantage of being unnecessary for predicting the geometry of the molecule. All diatomic molecules have the "same" geometry, differing only in bond length, which VSEPR does not address. Furthermore, the orbitals it does predict for the lone pairs do not have the correct symmetry for the molecule.

13.32   An extension of Table 13.2 is a molecule or ion in which six groups of electrons surround a central atom and three of these groups of electrons are lone pairs. **(a)** What might the hybridization of the central atom be under these circumstances? **(b)** What would the molecular geometry be?

(a) As in the cases presented in Table 13.2 for six groups of electrons surrounding a central atom, the central atom would adopt sp^3d^2 hybridization.

(b) The molecule would have a (likely distorted) T-shaped geometry obtained, for example, by replacing one of the four fluorine atoms shown for XeF_4 in Table 13.2 with a lone pair. This choice gives two neighboring lone pair interactions at 90° (one between the leg of the T and each of the two arms), and a 180° interaction between the two arms. This is less repulsion than would result from the only possible alternative, replacing two equatorial F atoms in the structure shown for ClF_5 with lone pairs. That choice would give three neighboring lone pair interactions at 90°.

13.34   How would the substitution of a fluorine atom for a single hydrogen atom influence the results of a simple Hückel MO theory calculation on an aromatic hydrocarbon molecule? What if all hydrogen atoms were replaced with fluorine atoms?

In its simplest form, Hückel MO theory treats all carbons equally, so that the parameters α and β are not influenced by substitution for hydrogen. (The parameters are different if carbon itself is replaced by a heteroatom.) Nevertheless, a slight extension to the simple theory could be developed that would account for the difference in electron density expected at the carbon bonded to the electron-withdrawing fluorine atom. This would replace the Coulomb integral α by α' for the affected carbon atom in the secular determinant and β by β' for the resonance integral with its nearest neighbors. This, of course, would affect the simple results often found using Hückel MO theory.

If all of the hydrogen atoms were replaced with fluorine atoms, then once again all the carbon atoms would be equivalent, and although the actual values of α and β might be changed, the same qualitative pattern of orbital energies would be obtained and the orbitals themselves would have identical shapes as in the all hydrogen case.

13.36   Use Hückel MO theory to compare two ethylene molecules (two C_2H_4) with butadiene (C_4H_6) and with cyclobutadiene (C_4H_4). Express the energies of these three systems in terms of the Hückel parameters α and β. In the simple Hückel approximation, which system has the lowest π energy?

The π binding energies for ethylene, butadiene, and cyclobutadiene have been calculated in the textbook and are listed below:

Species	π binding energy
ethylene	$2\alpha + 2\beta$
butadiene	$4\alpha + 4.48\beta$
cyclobutadiene	$4\alpha + 4\beta$

Butadiene has the lowest π energy according to the simple Hückel approximation. Specifically, the binding energy of butadiene is lower than that of two ethylene molecules by $4\alpha + 4.48\beta - 2(2\alpha + 2\beta) = 0.48\beta$, indicating that delocalization of π electrons stabilizes butadiene. On the other hand, the binding energy of cyclobutadiene is the same as that for two ethylene molecules; cyclobutadiene is not resonance stabilized.

13.38   Set up (but do not solve by hand) the Hückel determinants for the following series of C_8H_8 molecules. Predict which of these molecules will be aromatic.

According to Hückel's rule, aromatic molecules have $(4n + 2)$ delocalized π electrons in a planar ring or rings, where $n = 1, 2, 3, \cdots$.

The Hückel determinant for each species is written using the numbering system indicated.

$$
\begin{array}{cccc}
 & 2 \quad 1 & \\
3 & & 8 \\
4 & & 7 \\
 & 5 \quad 6 &
\end{array}
$$

The secular determinant is

$$
\begin{vmatrix}
\alpha - E & \beta & 0 & 0 & 0 & 0 & 0 & \beta \\
\beta & \alpha - E & \beta & 0 & 0 & 0 & 0 & 0 \\
0 & \beta & \alpha - E & \beta & 0 & 0 & 0 & 0 \\
0 & 0 & \beta & \alpha - E & \beta & 0 & 0 & 0 \\
0 & 0 & 0 & \beta & \alpha - E & \beta & 0 & 0 \\
0 & 0 & 0 & 0 & \beta & \alpha - E & \beta & 0 \\
0 & 0 & 0 & 0 & 0 & \beta & \alpha - E & \beta \\
\beta & 0 & 0 & 0 & 0 & 0 & \beta & \alpha - E
\end{vmatrix} = 0
$$

After dividing each element by β and letting $x = (\alpha - E)/\beta$, the secular determinant becomes

$$
\begin{vmatrix}
x & 1 & 0 & 0 & 0 & 0 & 0 & 1 \\
1 & x & 1 & 0 & 0 & 0 & 0 & 0 \\
0 & 1 & x & 1 & 0 & 0 & 0 & 0 \\
0 & 0 & 1 & x & 1 & 0 & 0 & 0 \\
0 & 0 & 0 & 1 & x & 1 & 0 & 0 \\
0 & 0 & 0 & 0 & 1 & x & 1 & 0 \\
0 & 0 & 0 & 0 & 0 & 1 & x & 1 \\
1 & 0 & 0 & 0 & 0 & 0 & 1 & x
\end{vmatrix} = 0
$$

The secular determinant is

$$
\begin{vmatrix}
\alpha - E & \beta & 0 & 0 & 0 & 0 & 0 & 0 \\
\beta & \alpha - E & \beta & 0 & 0 & 0 & 0 & \beta \\
0 & \beta & \alpha - E & \beta & 0 & 0 & 0 & 0 \\
0 & 0 & \beta & \alpha - E & \beta & 0 & 0 & 0 \\
0 & 0 & 0 & \beta & \alpha - E & \beta & 0 & 0 \\
0 & 0 & 0 & 0 & \beta & \alpha - E & \beta & 0 \\
0 & 0 & 0 & 0 & 0 & \beta & \alpha - E & \beta \\
0 & \beta & 0 & 0 & 0 & 0 & \beta & \alpha - E
\end{vmatrix} = 0
$$

After dividing each element by β and letting $x = (\alpha - E)/\beta$, the secular determinant becomes

$$
\begin{vmatrix}
x & 1 & 0 & 0 & 0 & 0 & 0 & 0 \\
1 & x & 1 & 0 & 0 & 0 & 0 & 1 \\
0 & 1 & x & 1 & 0 & 0 & 0 & 0 \\
0 & 0 & 1 & x & 1 & 0 & 0 & 0 \\
0 & 0 & 0 & 1 & x & 1 & 0 & 0 \\
0 & 0 & 0 & 0 & 1 & x & 1 & 0 \\
0 & 0 & 0 & 0 & 0 & 1 & x & 1 \\
0 & 1 & 0 & 0 & 0 & 0 & 1 & x
\end{vmatrix} = 0
$$

The secular determinant is

$$
\begin{vmatrix}
\alpha - E & \beta & 0 & 0 & 0 & 0 & 0 & 0 \\
\beta & \alpha - E & \beta & 0 & 0 & 0 & 0 & 0 \\
0 & \beta & \alpha - E & \beta & 0 & 0 & 0 & \beta \\
0 & 0 & \beta & \alpha - E & \beta & 0 & 0 & 0 \\
0 & 0 & 0 & \beta & \alpha - E & \beta & 0 & 0 \\
0 & 0 & 0 & 0 & \beta & \alpha - E & \beta & 0 \\
0 & 0 & 0 & 0 & 0 & \beta & \alpha - E & \beta \\
0 & 0 & \beta & 0 & 0 & 0 & \beta & \alpha - E
\end{vmatrix} = 0
$$

After dividing each element by β and letting $x = (\alpha - E)/\beta$, the secular determinant becomes

$$
\begin{vmatrix}
x & 1 & 0 & 0 & 0 & 0 & 0 & 0 \\
1 & x & 1 & 0 & 0 & 0 & 0 & 0 \\
0 & 1 & x & 1 & 0 & 0 & 0 & 1 \\
0 & 0 & 1 & x & 1 & 0 & 0 & 0 \\
0 & 0 & 0 & 1 & x & 1 & 0 & 0 \\
0 & 0 & 0 & 0 & 1 & x & 1 & 0 \\
0 & 0 & 0 & 0 & 0 & 1 & x & 1 \\
0 & 0 & 1 & 0 & 0 & 0 & 1 & x
\end{vmatrix} = 0
$$

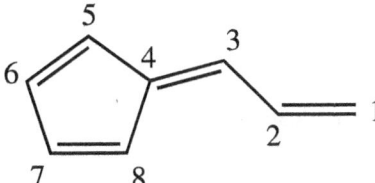

The secular determinant is

$$
\begin{vmatrix}
\alpha - E & \beta & 0 & 0 & 0 & 0 & 0 & 0 \\
\beta & \alpha - E & \beta & 0 & 0 & 0 & 0 & 0 \\
0 & \beta & \alpha - E & \beta & 0 & 0 & 0 & 0 \\
0 & 0 & \beta & \alpha - E & \beta & 0 & 0 & \beta \\
0 & 0 & 0 & \beta & \alpha - E & \beta & 0 & 0 \\
0 & 0 & 0 & 0 & \beta & \alpha - E & \beta & 0 \\
0 & 0 & 0 & 0 & 0 & \beta & \alpha - E & \beta \\
0 & 0 & 0 & \beta & 0 & 0 & \beta & \alpha - E
\end{vmatrix} = 0
$$

After dividing each element by β and letting $x = (\alpha - E)/\beta$, the secular determinant becomes

$$
\begin{vmatrix}
x & 1 & 0 & 0 & 0 & 0 & 0 & 0 \\
1 & x & 1 & 0 & 0 & 0 & 0 & 0 \\
0 & 1 & x & 1 & 0 & 0 & 0 & 0 \\
0 & 0 & 1 & x & 1 & 0 & 0 & 1 \\
0 & 0 & 0 & 1 & x & 1 & 0 & 0 \\
0 & 0 & 0 & 0 & 1 & x & 1 & 0 \\
0 & 0 & 0 & 0 & 0 & 1 & x & 1 \\
0 & 0 & 0 & 1 & 0 & 0 & 1 & x
\end{vmatrix} = 0
$$

According to Hückel's rule, only the seven-membered ring and the six-membered ring have $4n + 2$ (with $n = 1$) electrons and thus are aromatic.

13.40 Obtain a computer program that is able to perform simple Hückel MO theory electronic structure calculations (there are many such programs available). Perform calculations on the bicyclic $C_{10}H_8$ constitutional isomers naphthalene and azulene shown here and comment on the difference. Are each of these molecules aromatic?

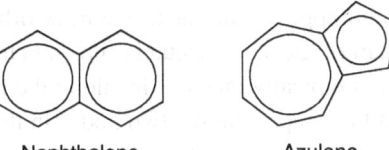

Naphthalene   Azulene

The Hückel MO theory energy diagrams shown below were obtained using *Mathematica* to set up and evaluate the secular determinants for the two molecules. The molecular symmetry of naphthalene is reflected in the symmetric pattern of orbital energies with paired bonding and antibonding orbitals equally displaced above and below the Coulomb integral value, α. Azulene, with rings of two different sizes, does not show this symmetric pattern. Each molecule has 10 π electrons, which are accommodated in bonding MO's for each molecule. The π binding energies of the two molecules are nearly equal, $10\alpha + 13.684\beta$ for naphthalene and $10\alpha + 13.364\beta$ for azulene. In each case this is lower than that of five ethylene molecules ($10\alpha + 10\beta$), making both molecules aromatic.

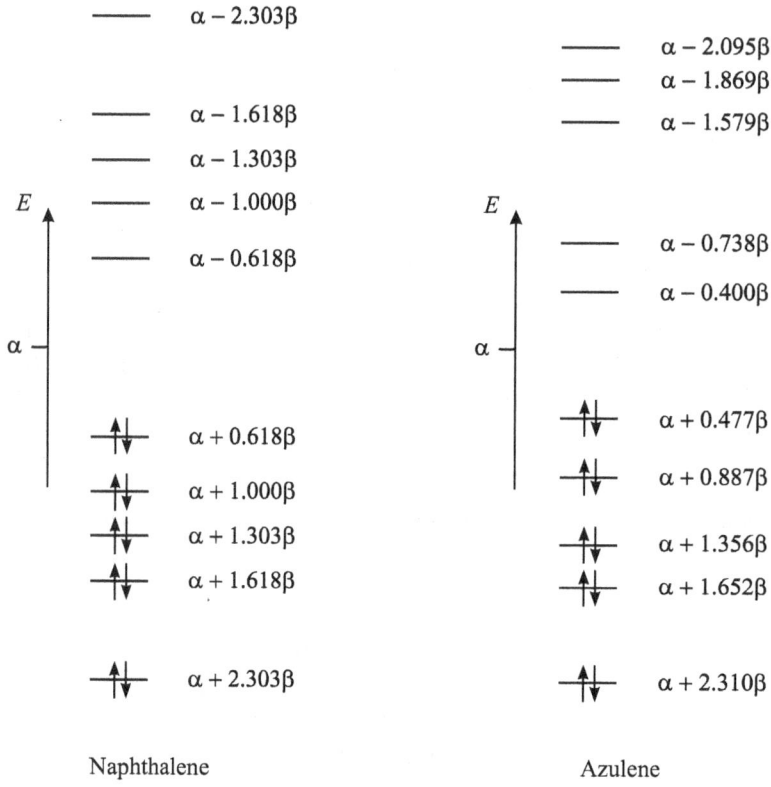

13.42   Noting the appearance of the $\sigma_g 1s$ and $\sigma_g 2s$ orbitals for the H_2^+ molecular ion, sketch the appearance of the $\sigma_g 3s$ molecular orbital. How many nodes does the $\sigma_g 3s$ orbital contain? Describe their shape.

The $\sigma_g 3s$ orbital for the H_2^+ molecular ion is formed by the overlap of two hydrogen $3s$ atomic orbitals, one centered at each of the two hydrogen nuclei. As such, it has the cylindrical symmetry of a σ molecular orbital and appears similar to the $\sigma_g 1s$ orbital shown in Figure 13.4 of the textbook. A contour diagram of the $\sigma_g 3s$ orbital is shown below with coordinate axes drawn to indicate scale. (The z axis is typically taken to lie along the bond.) The contours are given by solid lines, and the dashed lines represent the two nodes in the orbital. The inner node appears more clearly ellipsoidal, while the outer node is nearly, but not exactly, spherical.

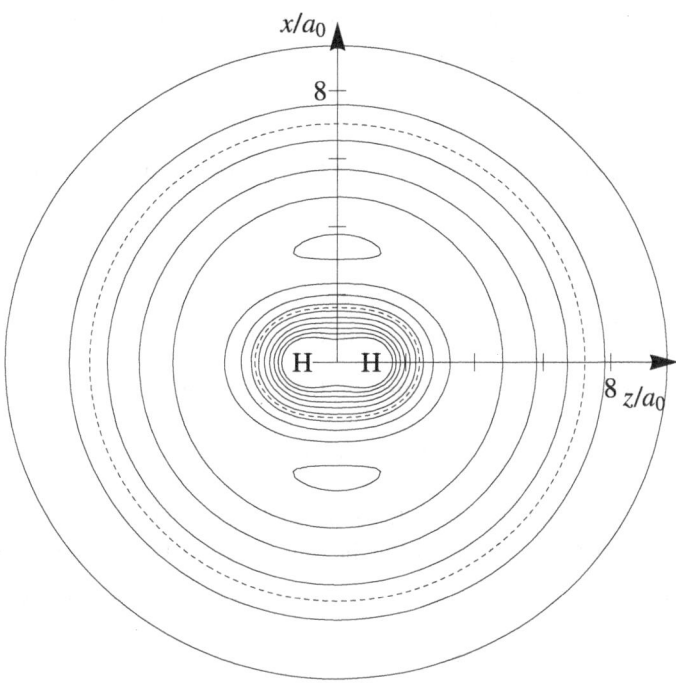

13.44 For which of the following chemical species is it possible to find an analytical solution to the time-independent electronic Schrödinger equation using the Born–Oppenheimer approximation? Explain. $_1^2\text{H}$, $_1^2\text{H}_2^+$, H_3^{2+}, $_2^3\text{He}$, $_2^4\text{He}^+$, Li_2^{5+}, C_2, N^{6+}, an α particle ($_2^4\text{He}^{2+}$).

An analytical solution to the time-independent electronic Schrödinger equation is possible for one-electron atoms or one-electron diatomic molecules. The Born-Oppenheimer approximation is the neglect of the kinetic energy due to the relative motion of nuclei, and is therefore only applicable to (and necessary for) molecular systems. Thus, the chemical species in this list for which it is possible to find an analytical solution to the time-independent electronic Schrödinger equation using the Born–Oppenheimer approximation are the one-electron diatomic molecules: $_1^2\text{H}_2^+$ and Li_2^{5+}. The others species include atoms (monoelectronic, polyelectronic, and even one with no electrons, the α particle) and molecules, either with more than one electron or more than two atoms.

13.46 The resonance concept is sometimes described by analogy to a mule, which is a cross between a horse and a donkey. Compare this analogy with the description of a rhinoceros as a cross between a griffin and a unicorn. Which description is more appropriate? Explain.

A mule, which shares characteristics of both, is a hybrid between two actual animals, the horse and the donkey. Describing a rhinoceros, which is a real animal, as a hybrid sharing the characteristics of a griffin and a unicorn, is to do so in terms of two mythical creatures that exist only in imagination. This latter description is the more appropriate analogy to resonance, since none of the resonance structures that are drawn "exist" as a real electron arrangement in the molecule. Rather, the molecule has an actual electronic structure that shares some of the characteristics of each resonance form that is drawn. Both analogies are accurate, however, in that in neither case is a mule "resonating" between being a horse part of the time and a donkey at other times, nor a rhinoceros switching back and forth being a griffin or a unicorn.

13.48   A single bond is almost always a sigma bond, and a double bond is almost always made up of a sigma bond and a pi bond. There are very few exceptions to this rule. Show that the B_2 and C_2 molecules are examples of the exceptions.

The molecular orbital electron configuration for B_2 with six electrons (see Figure 13.13a) is

$$(2s\sigma_g)^2(2s\sigma_u^*)^2(2p\pi_u)^2$$

The bond order is one, and the molecule has a single bond, but the net bonding is due to the electrons in π MOs. This molecule has a single π bond.

The molecular orbital electron configuration for C_2 with eight electrons (see Figure 13.13a) is

$$(2s\sigma_g)^2(2s\sigma_u^*)^2(2p\pi_u)^4$$

The bond order is two, and the molecule has a double bond, but the net bonding is once again due to the electrons in π MOs. This molecule has a double bond as a consequence of two π bonds.

13.50   The Lewis structure for O_2 is

$$\ddot{\text{O}}=\ddot{\text{O}}$$

Use the molecular orbital theory to show that the structure actually corresponds to an excited state of the oxygen molecule.

The Lewis structure indicates that all electrons are paired. This would correspond to an electron configuration of

$$\left(\sigma_g 1s\right)^2 \left(\sigma_u^* 1s\right)^2 \left(\sigma_g 2s\right)^2 \left(\sigma_u^* 2s\right)^2 \left(\sigma_g 2p_z\right)^2 \left(\pi_u 2p_x\right)^2 \left(\pi_u 2p_y\right)^2 \left(\pi_g^* 2p_x\right)^2$$

Compared to the electron configuration of O_2 in the ground state:

$$\left(\sigma_g 1s\right)^2 \left(\sigma_u^* 1s\right)^2 \left(\sigma_g 2s\right)^2 \left(\sigma_u^* 2s\right)^2 \left(\sigma_g 2p_z\right)^2 \left(\pi_u 2p_x\right)^2 \left(\pi_u 2p_y\right)^2 \left(\pi_g^* 2p_x\right)^1 \left(\pi_g^* 2p_y\right)^1$$

the configuration in the figure has paired the electrons in the π^* orbitals. Because it takes energy to pair electrons in the same orbital, the configuration implied by the Lewis structure is at a higher energy than the ground state, and is thus an excited state.

13.52   Does the molecule HBrC=C=CBrH have a dipole moment? Explain.

The two CBrH groups are in different (perpendicular) planes (see figure), and the two C–Br bond moments will not cancel. Thus, HBrC=C=CBrH is a polar molecule.

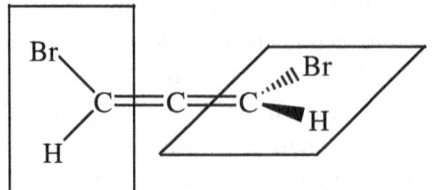

CHAPTER 14

Electronic Spectroscopy and Magnetic Resonance Spectroscopy

PROBLEMS AND SOLUTIONS

14.2 Using the Franck–Condon principle, qualitatively explain how the electronic spectra originating from the following two potential-energy diagrams will differ.

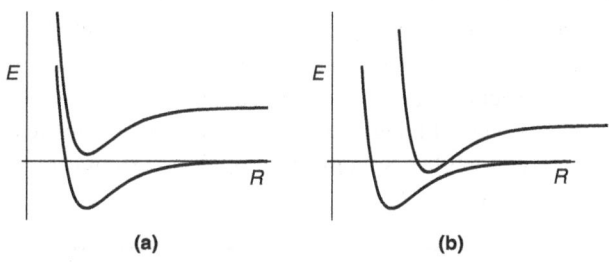

When the two states involved in an electronic transition have their minima at approximately the same internuclear separation, as is the case in diagram **(a)**, the most probable transitions are those where $\Delta v = 0$, since the best overlap of vibrational wave functions in the two states occurs between states with the same vibrational quantum number. At the same time, vibrational wave functions with different values for v' and v'' remain nearly orthogonal. Thus, the electronic spectrum is dominated by a series of "diagonal" vibrational transitions, *i.e.* $0 - 0$, $1 - 1$, etc., that occur at practically the same wavelength close to the electronic band origin. Since most diatomic molecules have significant population only in the ground vibrational state, only the $0 - 0$ transition would be expected.

When the two states have potential minima at significantly different internuclear separations, as in diagram **(b)**, the best overlap of vibrational wave functions occurs when the vibrational quantum numbers in the two states have different values, such as $0 - 5$, or $5 - 0$ to pick just two illustrative examples. Often a sequence of vibrational levels in one state will have good overlap with any given level in the other, leading to a sequence of lines in the electronic spectrum spaced by the appropriate vibrational energy interval and displaced to either side of the electronic band origin in a polyatomic molecule with many populated vibrational levels. In a diatomic molecule, a single sequence originating in the ground vibrational state is expected and extends to shorter wavelengths from the band origin.

14.4 Rank the following in order of increasing rate: absorption, fluorescence, phosphorescence, vibrational energy transfer, internal conversion, intersystem crossing. (Some of these may occur at approximately the same rate.)

The slowest process is phosphorescence (10^{-3} s or longer) and the fastest is (electronic) absorption (10^{-15} s). In the chapter, we find that typical fluorescence lifetimes are 10^{-9} s and that vibrational energy transfer occurs with a time scale of approximately 10^{-13} s. Furthermore, we know that internal conversion and intersystem crossing have similar rates and can compete with fluorescence, but not so effectively with vibrational energy transfer. Thus, in order of increasing rate,

> phosphorescence
> fluorescence
> internal conversion $\approx$ intersystem crossing
> vibrational energy transfer
> absorption

14.6   Explain why fluorescence is generally a more sensitive detection technique than absorption or phosphorescence.

The high sensitivity of fluorescence detection is primarily a consequence of it being a "zero-background" technique. That is, the signal is detected as light arising from an otherwise dark sample. Absorption, on the other hand, measures a decrease in intensity of the light illuminating the sample. It is much easier to detect a small increase from a baseline of no signal than to detect a small decrease in a large one. Although phosphorescence shares this feature, the small transition probability and consequent ability of competing processes to interfere leads to smaller signals and a decrease in sensitivity.

14.8   List some important differences between fluorescence and phosphorescence.

Fluorescence has much shorter lifetimes than phosphorescence. Because most molecular ground states are singlet states, fluorescence is generally decay from an excited singlet state, and phosphorescence is decay from an excited triplet state. As Problems 14.14 and 14.15 indicate, the multiplicity of the ground state will have an impact on this correlation. For a given molecule, fluorescence occurs at a longer wavelength (lower frequency) than phosphorescence.

14.10   The luminescent first-order decay of a certain organic molecule yields the following data:

t/s	0	1	2	3	4	5	10
I	100	43.5	18.9	8.2	3.6	1.6	0.02

where I is the relative intensity. Calculate the mean lifetime, τ, for the process. Is the decay fluorescent or phosphorescent?

Because $I = I_0 e^{-t/\tau}$, where τ is the mean lifetime of the state, then $\ln\left(I/I_0\right) = -t/\tau$, and a graph of $\ln\left(I/I_0\right)$ vs. t will result in a straight line with slope equal to $-1/\tau$. The necessary data are

t/s	0	1	2	3	4	5	10
$\ln\left(I/I_0\right)$	0	-0.83	-1.67	-2.50	-3.32	-4.14	-8.52

The slope of the best fit line to the data is -0.851 s^{-1}, which implies that $\tau = -1/\left(-0.851 \text{ s}^{-1}\right) = 1.2$ s. This is a long lifetime corresponding to phosphorescence.

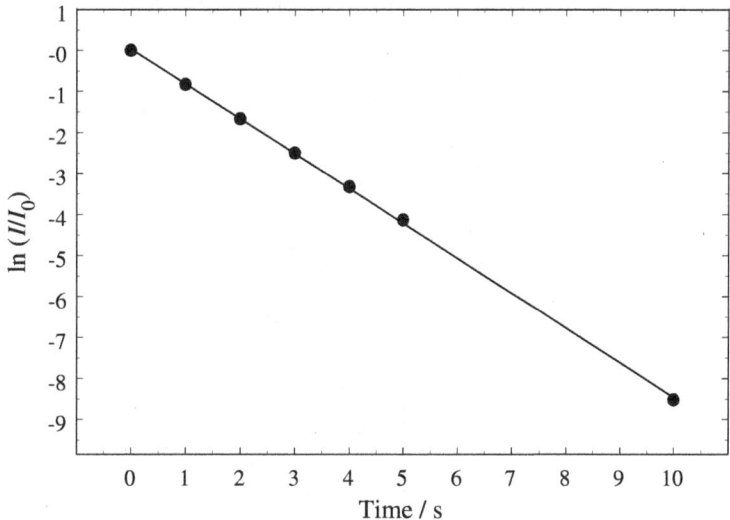

14.12 The fluorescence of a protein is due to tryptophan, tyrosine, and phenylalanine (assuming that the protein does not contain a prosthetic group that is fluorescent). Iodide ions are known to quench the fluorescence of tryptophan. If a protein is known to contain only one tryptophan residue and iodide fails to quench its fluorescence, what can you conclude about the location of the tryptophan residue?

The tryptophan is most likely located in the interior of the protein in a region inaccessible to the iodide ion.

14.14 Molecules with a net electron spin of $S = \frac{1}{2}$ have a ground electronic state called a doublet state because the multiplicity $2S + 1 = 2$. For molecules with a doublet ground state, what is the multiplicity of the electronic state from which phosphorescence originates? (*Note*: Multiplicity = singlet, doublet, triplet, quartet, $\cdots$)

Phosphorescence is emission of radiation accompanied by a change in electron spin. If the ground state is a doublet state, the excited state from which phosphorescence originates has three unpaired electrons, that is, a quartet state.

14.16 Name four characteristic properties of lasers.

The four characteristic properties of laser light are high intensity, coherence, monochromaticity, and spatial collimation.

14.18 Consider a three-level laser system. The wavelength for an absorption from level A to level C is 466 nm and that the wavelength for a transition between levels B and C is 752 nm. What is the wavelength for a transition between levels A and B?

The wavelengths indicate that level B lies between levels A and C. It is then easy to see that

$$E_C - E_A = (E_C - E_B) + (E_B - E_A)$$

$$\frac{hc}{\lambda_{CA}} = \frac{hc}{\lambda_{CB}} + \frac{hc}{\lambda_{BA}}$$

$$\frac{1}{\lambda_{CA}} = \frac{1}{\lambda_{CB}} + \frac{1}{\lambda_{BA}}$$

$$\frac{1}{466 \text{ nm}} = \frac{1}{752 \text{ nm}} + \frac{1}{\lambda_{BA}}$$

$$\lambda_{BA} = \lambda_{AB} = 1225 \text{ nm}$$

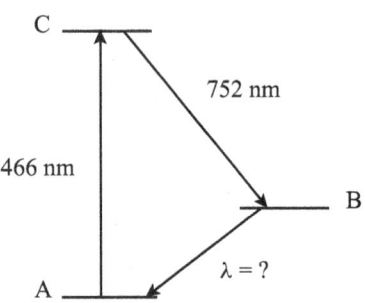

14.20 The Raman effect was first discovered prior to the invention of the laser. Why is essentially all contemporary Raman spectroscopy performed using a laser? (Section 11.5 describes Raman spectroscopy.)

As noted in Section 11.5, only a small portion of the incident light is inelastically scattered and gives rise to the Raman effect. Increasing the intensity of the incident light will lead to a stronger and more easily observed Raman signal. Because lasers are the most intense light sources available, they are well-suited for Raman spectroscopy. Furthermore, lasers are highly monochromatic which is advantageous in separating the excitation light from the scattered light and also for calculating the energy difference between the excitation and scattered light.

14.22 To extend the wavelength range of lasers, frequency doubling, tripling, and quadrupling may be used when suitable crystalline materials are available for the conversion process. A Nd:YAG laser, operating at 1064 nm is frequency doubled to make 532 nm (green) light. Calculate the wavelengths obtained by frequency tripling and quadrupling a Nd:YAG laser. In what region of the electromagnetic spectrum do these lie?

Since frequency and wavelength are inversely proportional, frequency tripling and quadrupling a Nd:YAG laser result in wavelengths of $(1064 \text{ nm})\left(\dfrac{1}{3}\right) = 355 \text{ nm}$ and $(1064 \text{ nm})\left(\dfrac{1}{4}\right) = 266 \text{ nm}$, respectively. These two wavelengths lie in the ultraviolet region.

14.24 A laser operating at 10 Hz produces pulses with a wavelength of 600 nm and a pulse duration of 8 ns. The average power of the laser is measured to be 400 mW. Calculate the energy per laser pulse, the peak power of a laser pulse, and the number of photons per laser pulse.

Recalling that power is a measure of energy delivered per second, the laser provides $400\ \text{mJ} = 4.00 \times 10^{-1}$ J each second. In a second, there are 10 laser pulses. Thus, the energy per pulse is $40.0\ \text{mJ} = 4.00 \times 10^{-2}$ J.

The peak power associated with each pulse is the energy per pulse divided by the pulse duration.

$$P = \frac{4.00 \times 10^{-2}\ \text{J}}{8 \times 10^{-9}\ \text{s}} = 5.0 \times 10^6\ \text{J s}^{-1} = 5\ \text{MW}$$

The number of photons in each pulse is the energy of the pulse divided by the energy per photon, $E = \frac{hc}{\lambda}$.

$$n_{\text{photons}} = \frac{4.00 \times 10^{-2}\ \text{J}}{\frac{(6.626 \times 10^{-34}\ \text{J s})(2.998 \times 10^8\ \text{m s}^{-1})}{600 \times 10^{-9}\ \text{m}}} = 1.2 \times 10^{17}$$

14.26 Compare and contrast ultraviolet photoelectron spectroscopy (UPS) with X-ray photoelectron spectroscopy (XPS).

Both forms of photoelectron spectroscopy are based on the photoelectric effect, and in both, a gaseous sample is illuminated with light of a given, fixed frequency, ν, which causes an electron to be ejected. The kinetic energy of the ejected electron is measured. UPS uses vacuum ultraviolet photons obtained from a He(I) or He(II) source and is used to probe electrons in valence orbitals. XPS uses X-ray photons from a variety of sources and the shorter wavelength, higher energy, photons can probe more tightly bound core electrons. Synchrotron sources are able to provide fixed frequencies in both the vacuum ultraviolet and X-ray regions of the spectrum, and can be used in both UPS and XPS.

14.28 Explain in your own words why methane (CH_4) exhibits two (only) different peaks corresponding to C–H bonds in its photoelectron spectrum.

Although the valence bond picture for the C–H bonding in methane envisions four equivalent sp^3 hybrid orbitals on carbon that each overlap with the $1s$ orbital on a hydrogen atom, leading to four equivalent sigma bonds, the molecular orbital picture suggests that the bonding is better described by one singly degenerate molecular orbital extending over all four bonds and three equivalent, triply degenerate orbitals comprising various combinations of the four.

14.30 A high-level molecular electronic structure calculation predicts oxygen valence molecular orbitals with binding energies of 0.5389, 0.60354, and 0.7171 hartree. With a 58.46-nm photon source, what will the corresponding electron kinetic energies (in eV) be in a photoelectron spectrum? (1 hartree = 27.211 eV)

The energy of the excitation photon is

$$E_{\text{photon}} = \frac{hc}{\lambda} = \frac{(6.626 \times 10^{-34}\ \text{J s})(2.998 \times 10^8\ \text{m s}^{-1})}{(58.46 \times 10^{-9}\ \text{m})(1.602 \times 10^{-19}\ \text{J } eV^{-1})} = 21.21\ eV$$

Note that the precision in the photon energy only extends to $\pm 0.01\ eV$ and will similarly limit the precision of the predicted electron kinetic energies. From Equation 14.13,

$$E_{photon} = E_{ionization,\ electron} + E_{kinetic,\ electron}$$

Converting the binding energies from hartree to eV and taking the difference with the photon energy gives the following predicted electron kinetic energies.

Binding energy/hartree	Binding energy/eV	Electron kinetic energy/eV
0.5389	14.66	6.55
0.60354	16.423	4.79
0.7171	19.51	1.70

14.32   One way to measure the kinetic energy of ejected photoelectrons is to measure their stopping voltage, that is, the voltage required to stop the transmission of photoelectrons and thus result in zero current. The stopping voltages for the photoelectrons from a particular sample are 952.9, 1083.9, and 1201.2 eV when using an aluminum K_α X-ray source. What are the binding energies of the corresponding molecular orbitals? Based on these binding energies, from what type of atomic or molecular orbital do you anticipate that these photoelectrons originate, core or valence?

The energy of an aluminum K_α X-ray photon is 1486.6 eV (Table 14.4). From Equation 14.13,

$$E_{photon} = E_{ionization,\ electron} + E_{kinetic,\ electron}$$

Taking the difference between the measured stopping voltages, which corresponds to the kinetic energies of the ejected photoelectrons, and the photon energy gives the following binding energies.

Electron kinetic energy/eV	Binding energy/eV
952.9	533.7
1083.9	402.7
1201.2	285.4

These are very large binding energies that correspond to core orbitals.

14.34   The NMR signal of a compound is found to be 240 Hz downfield from the TMS peak using a spectrometer operating at 60 MHz. Calculate its chemical shift in ppm relative to TMS.

According to Equation 14.22,

$$\delta = \frac{\nu - \nu_{ref}}{\nu_{spec}} \times 10^6 \text{ ppm} = \frac{240 \text{ Hz}}{60 \times 10^6 \text{ Hz}} \times 10^6 \text{ ppm} = 4.0 \text{ ppm}$$

14.36   Suppose the NMR spectrum of acetaldehyde (see Figure 14.32) is recorded at 200 MHz and 400 MHz. State whether each of the following quantities remains unchanged or is different from 200 MHz to 400 MHz: **(a)** sensitivity of detection, **(b)** $|\delta_{CH_3} - \delta_H|$, **(c)** $|\nu_{CH_3} - \nu_H|$, **(d)** J.

(a) Increased (due to a more favorable Boltzmann distribution),

(b) unchanged (chemical shifts are independent of magnetic field strength),

(c) increased (frequency shifts are proportional to magnetic field strength),

(d) unchanged (coupling constants are independent of magnetic field strength).

14.38 For each of the following molecules, state how many proton NMR peaks occur, and whether each peak is a singlet, doublet, triplet, etc. (a) CH_3OCH_3, (b) $C_2H_5OC_2H_5$, (c) C_2H_6, (d) CH_3F, and (e) $CH_3COOC_2H_5$.

(a) One singlet,

(b) two peaks, one triplet (1:2:1) and one quartet (1:3:3:1),

(c) one singlet,

(d) one doublet (1:1) due to the coupling of the protons to the fluorine nucleus which has $I = 1/2$,

(e) three peaks, one singlet, one triplet (1:2:1), one quartet (1:3:3:1).

14.40 The toluene proton NMR spectrum, consisting of two peaks due to the methyl and aromatic protons, has been recorded at 60 MHz and 1.41 T. (a) What would the magnetic field be at 300 MHz? (b) At 60 MHz, the resonance frequencies are: methyl, 140 Hz; aromatic, 430 Hz. What would the frequencies be if recorded by a 300 MHz spectrometer? (c) Calculate the chemical shifts (δ) of the two signals, using both the 60 MHz and 300 MHz data. (d) What are the positions of the methyl and aromatic signals in Hz at 300 MHz?

(a) Equation 14.21 shows that the NMR frequency is linearly proportional to the magnetic field strength. Thus a 300 MHz proton NMR spectrum requires $B_0 = \dfrac{300 \text{ MHz}}{60 \text{ MHz}} \times 1.41 \text{ T} = 7.05$ T.

(b) The frequencies of both the reference proton and the proton of interest are linearly proportional to the magnetic field strength. Thus,

$$\Delta\nu_{300} = \frac{300 \text{ MHz}}{60 \text{ MHz}}\nu_{60} - \frac{300 \text{ MHz}}{60 \text{ MHz}}\nu_{60}^{ref}$$

$$= \frac{300 \text{ MHz}}{60 \text{ MHz}}\left(\nu_{60} - \nu_{60}^{ref}\right)$$

$$= \frac{300 \text{ MHz}}{60 \text{ MHz}}\Delta\nu_{60}$$

For the methyl protons the 300 MHz frequency is 5×140 Hz $= 700$ Hz relative to reference.

For the aromatic protons the 300 MHz frequency is 5×430 Hz $= 2150$ Hz relative to reference.

(c) In each case $\delta = \dfrac{\nu - \nu_{ref}}{\nu_{ref}} \times 10^6$ ppm.

At 60 MHz:

$$\delta_{methyl} = \frac{140 \text{ Hz}}{60 \times 10^6 \text{ Hz}} \times 10^6 \text{ ppm} = 2.33 \text{ ppm}$$

$$\delta_{aromatic} = \frac{430 \text{ Hz}}{60 \times 10^6 \text{ Hz}} \times 10^6 \text{ ppm} = 7.17 \text{ ppm}$$

At 300 MHz:

$$\delta_{\text{methyl}} = \frac{700 \text{ Hz}}{300 \times 10^6 \text{ Hz}} \times 10^6 \text{ ppm} = 2.33 \text{ ppm}$$

$$\delta_{\text{aromatic}} = \frac{2150 \text{ Hz}}{300 \times 10^6 \text{ Hz}} \times 10^6 \text{ ppm} = 7.17 \text{ ppm}$$

The chemical shift is independent of frequency.

14.42 Account for the number of lines observed in the ESR spectra of benzene and naphthalene anion radicals shown in Figure 14.37. How would you use isotopic substitution to assign the two hyperfine splitting constants in naphthalene?

In the benzene anion radical the splitting arises from 6 equivalent protons with $I = 1/2$, leading to $2nI + 1 = 2(6)(1/2) + 1 = 7$ lines.

In the naphthalene radical anion, there are two sets of 4 equivalent protons (those α to the common side of the two rings and those β to it). This gives rise to $(2n_1 I + 1)(2n_2 I + 1) = [2(4)(1/2) + 1][2(4)(1/2) + 1] = 25$ lines.

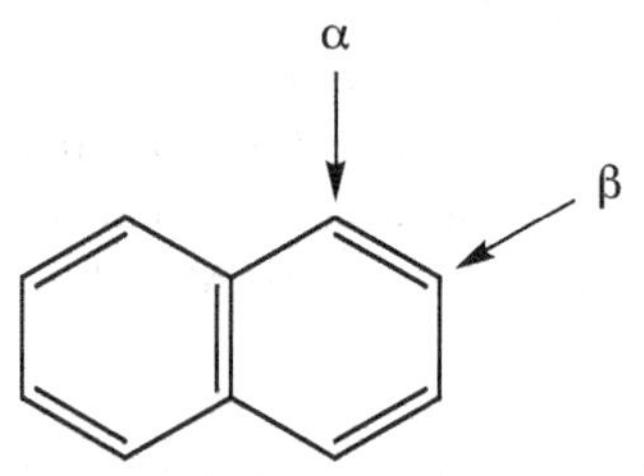

The isotopically substituted naphthalene shown below, for example, would have an ESR spectrum in which only the α hyperfine splitting would be affected, allowing the assignment of the observed hyperfine splitting constants in the normal isotopic species to the α and β protons.

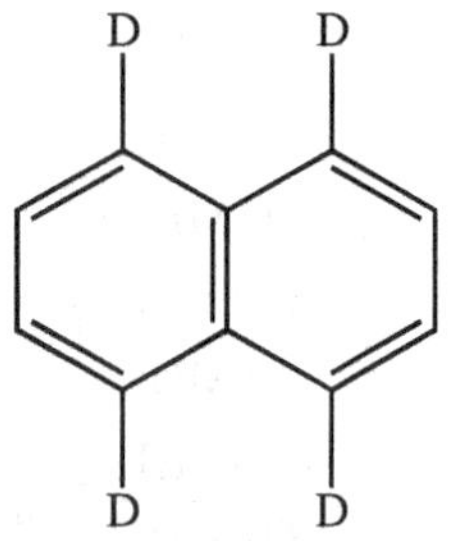

14.44 Both NMR and ESR spectroscopy differ from other branches of spectroscopy discussed in this chapter in one important respect. Explain.

Both NMR and ESR spectroscopy utilize the interaction of the oscillating magnetic field of the electromagnetic radiation with a molecular property. The others forms of spectroscopy discussed involve the oscillating electric field of the electromagnetic radiation. Additionally, in both NMR and ESR spectroscopy, the energy levels are degenerate in the absence of an external magnetic field, and the separation of the energy levels depends on the strength of the external field. This is not a feature of the other spectroscopies discussed in the chapter, although a similar effect of an external electric or magnetic field may be observed on the degenerate rotational levels of molecules.

14.46 A laser dye is dissolved in a polymer to a concentration of 5.5×10^{-9} *M*. A 1.0-μm-thick layer of the polymer is deposited on a microscope slide. Calculate the radius of a probe laser needed to contain a single laser dye molecule.

The number of molecules per area is

$$\left(5.5 \times 10^{-9} \text{ mol L}^{-1}\right)\left(\frac{6.022 \times 10^{23}}{1 \text{ mol}}\right)\left(1.0 \times 10^{-6} \text{ m}\right)\left(\frac{1000 \text{ L}}{1 \text{ m}^3}\right) = 3.31 \times 10^{12} \text{ m}^{-2}$$

The area is related to the radius of a probe laser by πr^2. Therefore, to contain a single molecule,

$$\pi r^2 = \frac{1}{3.31 \times 10^{12} \text{ m}^{-2}}$$

$$r = \left[\frac{1}{\pi}\left(3.02 \times 10^{-13} \text{ m}^2\right)\right]^{1/2} = 3.1 \times 10^{-7} \text{ m} = 3.1 \times 10^2 \text{ nm}$$

This is close to the diffration limit for UV–visible light.

14.48 The molar absorptivity of a solute at 664 nm is 895 $\text{L mol}^{-1} \text{cm}^{-1}$. When light at that wavelength is passed through a 2.0-cm cell containing a solution of the solute, 74.6% of the light is absorbed. Calculate the concentration of the solution.

With 74.6% of the light absorbed, the transmittance is $T = 1.000 - 0.746 = 0.254$, and the absorbance is $A = -\log T = -\log 0.254 = 0.5952$. Using the Beer–Lambert law, $A = \varepsilon b c$,

$$c = \frac{A}{\varepsilon b}$$

$$= \frac{0.5952}{\left(895 \text{ L mol}^{-1} \text{cm}^{-1}\right)(2.0 \text{ cm})}$$

$$= 3.3 \times 10^{-4} \text{ M}$$

14.50 The frequency of molecular collision in the liquid phase is about $1 \times 10^{13} \text{ s}^{-1}$. Ignoring all other mechanisms contributing to linewidth, calculate the width (in Hz) of vibrational transitions if **(a)** every collision is effective in deactivating the molecule vibrationally and **(b)** one collision in 40 is effective.

(a) If every collision is effective in deactivating the molecule, then the lifetime of the excited vibrational state is $\Delta t = 1/1 \times 10^{13} \text{ s}^{-1} = 1 \times 10^{-13} \text{ s}$, and the natural linewidth is

$$\Delta \nu = \frac{1}{4\pi \Delta t} = \frac{1}{4\pi \left(1 \times 10^{-13} \text{ s}\right)} = 8 \times 10^{11} \text{ s}^{-1}$$

(b) If one collision in 40 is effective in deactivating the molecule, then the lifetime of the excited vibrational state is $\Delta t = 40/1 \times 10^{13} \text{ s}^{-1} = 4 \times 10^{-12} \text{ s}$, and the natural linewidth is

$$\Delta \nu = \frac{1}{4\pi \Delta t} = \frac{1}{4\pi \left(4 \times 10^{-12} \text{ s}\right)} = 2 \times 10^{10} \text{ s}^{-1}$$

14.52 The molecule fluorescein is a red solid known as the color additive "D&C Yellow no. 7" because it appears yellow in aqueous solution. When excited, fluorescein emits green light (~520 nm). Draw a schematic energy level diagram that explains these colors.

The yellow color of the solution arises from the absorption of the complementary color, blue (Table 14.2), in the 435–480 nm wavelength region. This leads to excitation of an excited state that relaxes via a radiationless transition to a lower energy excited state. This lower energy state then emits light with the longer wavelength (~520 nm) in returning to the ground electronic state. The schematic energy level diagram below illustrates this.

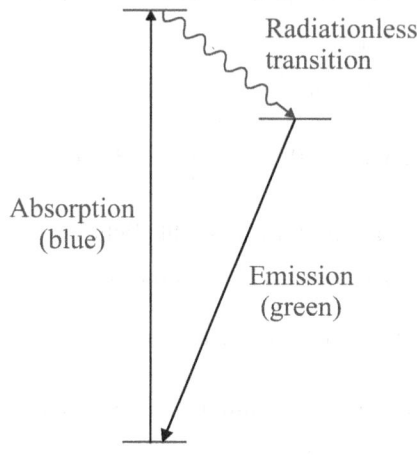

14.54 The Beer-Lambert law predicts a linear dependence of absorbance on concentration. The law often breaks down at very high concentrations. Why?

There are several reasons, and some are given here. Among them are consequences of changes to the nominal absorbing species at high concentrations such as (i) ion-pair formation (if ionic species are present), or (ii) aggregation or polymerization. Others are due to effects on the interaction of light with the absorbing species such as (iii) interactions between the transition dipoles of different chromophores (similar to the hypochromism effect in DNA), (iv) scattering of light (mostly due to large biological molecules), and (v) fluorescence of the absorbing species.

14.56 (a) Calculate the energy difference between the two spin states of ^{1}H and of ^{13}C in a magnetic field of 4.7 T. (b) What is the precession frequency of a ^{1}H nucleus at this magnetic field? Of a ^{13}C nucleus? (c) At what magnetic field do protons precess at a frequency of 500 MHz?

(a) The energy difference can be calculated from Equation 14.17:

$$\Delta E = B_0 \gamma \hbar$$

where B_0 = 4.70 T in this case. Thus,

$$\Delta E = (4.70 \text{ T}) \gamma \left(1.055 \times 10^{-34} \text{ J s}\right) = \left(4.959 \times 10^{-34} \text{ T J s}\right) \gamma$$

For the ^{1}H nucleus,

$$\Delta E = \left(4.959 \times 10^{-34} \text{ T J s}\right) \left(26.75 \times 10^7 \text{ T}^{-1} \text{s}^{-1}\right) = 1.327 \times 10^{-25} \text{ J} = 1.33 \times 10^{-25} \text{ J}$$

For the ^{13}C nucleus,

$$\Delta E = \left(4.959 \times 10^{-34}\ \mathrm{T\,J\,s}\right)\left(6.73 \times 10^{7}\ \mathrm{T^{-1}\,s^{-1}}\right) = 3.337 \times 10^{-26}\ \mathrm{J} = 3.34 \times 10^{-26}\ \mathrm{J}$$

(b) The precession frequency is given by Equation 14.19:

$$\nu_{precession} = \frac{B_0 \gamma}{2\pi} = \frac{\Delta E}{h}$$

For the ^{1}H nuclues,

$$\nu_{precession} = \frac{1.327 \times 10^{-25}\ \mathrm{J}}{6.626 \times 10^{-34}\ \mathrm{J\,s}} = 2.00 \times 10^{8}\ \mathrm{Hz} = 200\ \mathrm{MHz}$$

For the ^{13}C nuclues,

$$\nu_{precession} = \frac{3.337 \times 10^{-26}\ \mathrm{J}}{6.626 \times 10^{-34}\ \mathrm{J\,s}} = 5.03 \times 10^{7}\ \mathrm{Hz} = 50.3\ \mathrm{MHz}$$

(c)

$$\nu_{precession} = \frac{B_0 \gamma}{2\pi}$$

$$B_0 = \frac{2\pi \nu_{precession}}{\gamma} = \frac{2\pi \left(500 \times 10^{6}\ \mathrm{Hz}\right)}{26.75 \times 10^{7}\ \mathrm{T^{-1}\,s^{-1}}} = 11.7\ \mathrm{T}$$

14.58 Oxygen is an effective quencher of fluorescence because it is a triplet in its electronic ground state. The unpaired spins of O_2 can induce the excited state of the fluorescent molecule to undergo intersystem crossing from the singlet state to the triplet state, that is, $S_1 \rightarrow T_1$ (see Figure 14.9). **(a)** How would you verify the mechanism experimentally? **(b)** Assume that the quenching rate constant (i.e., the rate constant for the collision between O_2 and fluorescent molecules) to be $1 \times 10^{10}\ M^{-1}\,s^{-1}$ at 25°C. How many collisions s^{-1} on average does each fluorescent molecule in solution experience? The concentrations are: $[O_2] = 3.4 \times 10^{-4}\ M$ and $[F] = 0.50\ M$, where F is the fluorescent molecule. **(c)** The fluorescence lifetime of pyrene, a molecule that is often used to probe biological systems, is 500 ns, while that of tryptophan is about 5 ns. Explain why under normal atmospheric conditions O_2 can interfere only with the fluorescence of pyrene but not that of tryptophan. **(d)** A quantitative relationship of fluorescence quenching is the Stern–Volmer equation,

$$\frac{I_0}{I} = 1 + k_Q \tau_0\, [Q]$$

where I_0 and I are the fluorescence intensities in the absence and presence of the quencher, k_Q is the quenching rate constant, τ_0 is the mean lifetime of the fluorescent state in the absence of the quencher, and [Q] is the concentration of the quencher. Use this equation to support your conclusion in (c). **(e)** What air pressure is needed to get 50% quenching of tryptophan in solution?

(a) Measure the fluorescence lifetime in the presence and absence of O_2. If the mechanism is correct, the lifetime will be shorter in the presence of O_2.

(b) The collision rate is

$$\text{rate} = k_Q \left[O_2 \right] \left[F \right]$$

$$= \left(1.0 \times 10^{10} \ M^{-1} \text{s}^{-1} \right) \left(3.4 \times 10^{-4} \ M \right) (0.5 \ M)$$

$$= 1.7 \times 10^6 \ M \ \text{s}^{-1} = 2 \times 10^6 \ M \ \text{s}^{-1}$$

(c) From part (b), in each second, there are 1.7×10^6 moles of collisions in each liter between fluorescent molecules and the quencher (O_2), and there are $0.5 \ M$ of fluorescent molecules in each liter ([F] = 0.50 M). Thus, an individual F molecule experiences a collision rate of $\frac{1.7 \times 10^6 \ M \ \text{s}^{-1}}{0.5 \ M} = 3.4 \times 10^6 \ \text{s}^{-1}$. Consequently, the average time between collisions is $\frac{1}{3.4 \times 10^6 \ \text{s}^{-1}} = 3 \times 10^{-7} \ \text{s} = 300$ ns. This is comparable to pyrene's fluorescence lifetime, but much longer than that of tryptophan. The excited state of a tryptophan molecule will relax via fluorescence long before it encounters an O_2 molecule.

(d) Using the Stern-Volmer equation to calculate the ratio of the fluorescence intensities for pyrene in the absence and presence of quencher

$$\frac{I_0}{I} = 1 + k_Q \tau_0 \left[Q \right]$$

$$= 1 + \left(1.0 \times 10^{10} \ M^{-1} \text{s}^{-1} \right) \left(500 \times 10^{-9} \ \text{s} \right) \left(3.4 \times 10^{-4} \ M \right)$$

$$= 1 + 1.7 = 2.7$$

Nearly two thirds of pyrene's fluorescence is quenched by dissolved atmospheric O_2.

Repeating for tryptophan,

$$\frac{I_0}{I} = 1 + \left(1.0 \times 10^{10} \ M^{-1} \text{s}^{-1} \right) \left(5 \times 10^{-9} \ \text{s} \right) \left(3.4 \times 10^{-4} \ M \right)$$

$$= 1 + 0.017 = 1.017 \approx 1$$

shows that the quenching is negligible in this case.

(e) At 50% quenching, $I_0/I = 2$, or

$$2 = 1 + \left(1.0 \times 10^{10} \ M^{-1} \text{s}^{-1} \right) \left(5 \times 10^{-9} \ \text{s} \right) \left[O_2 \right]$$

$$\left[O_2 \right] = \frac{1}{5.0 \times 10^1 \ M^{-1}} = 0.02 \ M$$

According to Henry's Law (Section 6.4), the concentration of dissolved oxygen is directly proportional to the pressure. Because at 1 atm, $\left[O_2 \right] = 3.4 \times 10^{-4} \ M$, and

$$\frac{0.02 \ M}{3.4 \times 10^{-4} \ M} = 60$$

it would require that the air pressure be 60 atm to get 50% quenching of tryptophan in solution. Alternatively, since air is approximately $\frac{1}{5}$ oxygen, 12 atm of pure oxygen could be used.

CHAPTER 15

Chemical Kinetics

PROBLEMS AND SOLUTIONS

15.2 The rate law for the reaction

$$NH_4^+(aq) + NO_2^-(aq) \rightarrow N_2(g) + 2H_2O(l)$$

is given by rate = k [NH_4^+] [NO_2^-]. At 25°C, the rate constant is 3.0×10^{-4} $M^{-1}\,s^{-1}$. Calculate the rate of the reaction at this temperature if [NH_4^+] = 0.26 M and [NO_2^-] = 0.080 M.

$$\text{Rate} = k\,[NH_4^+]\,[NO_2^-] = \left(3.0 \times 10^{-4}\,M^{-1}\,s^{-1}\right)(0.26\,M)(0.080\,M) = 6.2 \times 10^{-6}\,M\,s^{-1}$$

15.4 The following reaction is found to be first order in A:

$$A \rightarrow B + C$$

If half of the starting quantity of A is used up after 56 s, calculate the fraction that will be used up after 6.0 min.

From the half-life (56 s), k can be determined:

$$k = \frac{\ln 2}{t_{1/2}} = \frac{\ln 2}{56\,s} = 1.24 \times 10^{-2}\,s^{-1}$$

The fraction of A that will remain after 6.0 min (3.6×10^2 s) is

$$\frac{[A]}{[A]_0} = e^{-kt} = e^{-\left(1.24 \times 10^{-2}\,s^{-1}\right)\left(3.6 \times 10^2\,s\right)} = 0.0115$$

Thus, the fraction that will be used up after 6.0 min is $1 - 0.0115 = 0.989$.

15.6 **(a)** The half-life of the first-order decay of radioactive ^{14}C is about 5720 years. Calculate the rate constant for the reaction. **(b)** The natural abundance of ^{14}C isotope is 1.1×10^{-13} mol % in living matter. Radiochemical analysis of an object obtained in an archaeological excavation shows that the ^{14}C isotope content is 0.89×10^{-14} mol %. Calculate the age of the object. State any assumptions.

(a)
$$k = \frac{\ln 2}{t_{1/2}} = \frac{\ln 2}{5720 \text{ yr}} = 1.212 \times 10^{-4} \text{ yr}^{-1} = 1.21 \times 10^{-4} \text{ yr}^{-1}$$

(b)
$$\frac{[^{14}\text{C}]}{[^{14}\text{C}]_0} = e^{-kt}$$

$$t = -\frac{1}{k} \ln \frac{[^{14}\text{C}]}{[^{14}\text{C}]_0}$$

Due to constant exchange of material with the surroundings, the mol % of ^{14}C of all living matter is assumed to be the same. However, when the object ceases to live, it no longer exchanges material with the environment and the mol % of ^{14}C will decrease according to first-order decay kinetics. Therefore, the ratio between $[^{14}\text{C}]$ and $[^{14}\text{C}]_0$ depends on the time elapsed since the object's "death." Thus, t in the equation above gives the age of the object.

$$t = -\frac{1}{1.212 \times 10^{-4} \text{ yr}^{-1}} \ln \frac{0.89 \times 10^{-14}}{1.1 \times 10^{-13}} = 2.1 \times 10^4 \text{ yr}$$

A key assumption in radiocarbon dating is that the natural abundance of ^{14}C has remained constant throughout the ages. Since the production of terrestrial ^{14}C is due to bombardment of ^{14}N by cosmic rays, variations in cosmic ray flux have in fact led to variations in the natural abundance of ^{14}C.

15.8 When the concentration of A in the reaction A $\rightarrow$ B was changed from 1.20 M to 0.60 M, the half-life increased from 2.0 min to 4.0 min at 25°C. Calculate the order of the reaction and the rate constant.

The half-life is related to the initial concentration of A by

$$t_{1/2} \propto \frac{1}{[A]_0^{n-1}}$$

According to the data, the half-life doubled when $[A]_0$ was halved. This is only possible if the half-life is inversely proportional to $[A]_0$, or the reaction order, $n = 2$, indicating a second-order reaction.

The rate constant can be calculated using either $[A]_0 = 1.20$ M or 0.60 M and the corresponding half-life.

$$k = \frac{1}{[A]_0 t_{1/2}} = \frac{1}{(1.20 \text{ } M)(2.0 \text{ min})} = 0.42 \text{ } M^{-1} \text{ min}^{-1}$$

15.10 Cyclobutane decomposes to ethylene according to the equation

$$C_4H_8(g) \rightarrow 2C_2H_4(g)$$

Determine the order of the reaction and the rate constant based on the following pressures, which were recorded when the reaction was carried out at 430°C in a constant-volume vessel:

Time/s	$P_{C_4H_8}$/mmHg
0	400
2000	316
4000	248
6000	196
8000	155
10000	122

At constant temperature and volume, the pressure of cyclobutane is proportional to its concentration. A plot of $\ln P$ vs t shows a straight line with an equation of $y = -1.19 \times 10^{-4}x + 5.99$. Thus, the reaction is first order.

The slope of the line is $-k$. In other words, $k = 1.19 \times 10^{-4}$ s^{-1}.

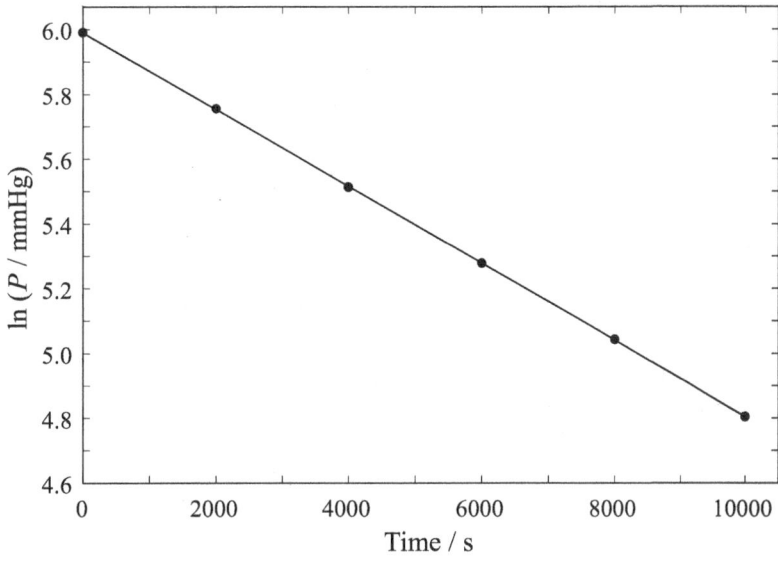

15.12   The rate constant for the second-order reaction

$$2NO_2(g) \rightarrow 2NO(g) + O_2(g)$$

is 0.54 M^{-1} s^{-1} at 300°C. How long (in seconds) would it take for the concentration of NO$_2$ to decrease from 0.62 M to 0.28 M?

$$\frac{1}{[A]} - \frac{1}{[A]_0} = kt$$

$$t = \frac{1}{k}\left(\frac{1}{[A]} - \frac{1}{[A]_0}\right) = \frac{1}{0.54 \ M^{-1}\,s^{-1}}\left(\frac{1}{0.28 \ M} - \frac{1}{0.62 \ M}\right) = 3.6 \text{ s}$$

15.14   The integrated rate law for the zero-order reaction A $\rightarrow$ B is $[A] = [A]_0 - kt$. **(a)** Sketch the following plots: (**(i)** rate versus [A] and **(ii)** [A] versus t. **(b)** Derive an expression for the half-life of the reaction. **(c)** Calculate the time in half-lives when the integrated rate law is no longer valid, that is, when [A] = 0.

(a) (i) The rate law is

$$\text{Rate} = k$$

Therefore, the rate of reaction is independent of the concentration of A.

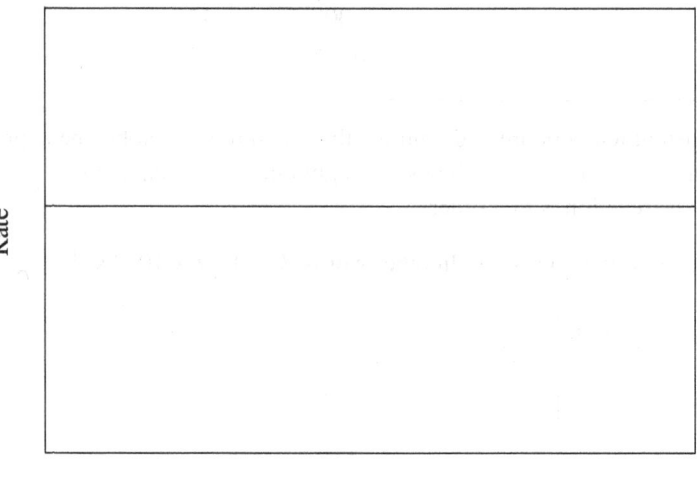

(ii) Because

$$[A] = [A]_0 - kt$$

a plot of [A] vs t is a straight line with a slope of $-k$.

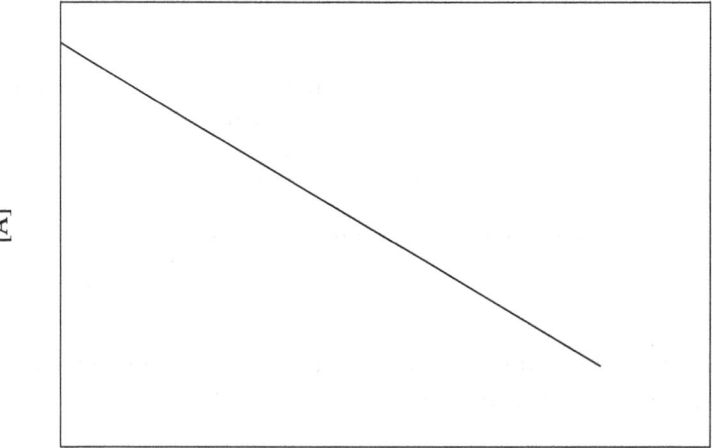

(b) At $t = t_{1/2}$, [A] = $[A]_0/2$. Therefore,

$$\frac{[A]_0}{2} = [A]_0 - kt_{1/2}$$

$$t_{1/2} = \frac{1}{2k}[A]_0$$

(c) When [A] = 0,

$$[A] = 0 = [A]_0 - kt$$

According to part **(b)**,

$$k = \frac{1}{2t_{1/2}}[A]_0$$

Therefore, the time it takes to consume all the reactant is

$$t = \frac{[A]_0}{k} = \frac{[A]_0}{\frac{1}{2t_{1/2}}[A]_0} = 2t_{1/2}$$

The integrated rate law is no longer valid after 2 half-lives.

15.16   Many reactions involving heterogeneous catalysis are zero order; that is, rate $= k$. An example is the decomposition of phosphine (PH_3) over tungsten (W):

$$4PH_3(g) \rightarrow P_4(g) + 6H_2(g)$$

The rate for this reaction is independent of $[PH_3]$ as long as phosphine's pressure is sufficiently high (≥ 1 atm). Explain.

With sufficient PH_3, all the catalytic sites on the tungsten surface are occupied. Further increases in the amount of phosphine cannot affect the reaction, and the rate is independent of $[PH_3]$.

15.18   Consider the following nuclear decay

$$^{64}Cu \rightarrow {}^{64}Zn + {}_{-1}^{0}\beta \quad t_{1/2} = 12.8 \text{ h}$$

Starting with one mole of ^{64}Cu, calculate the number of grams of ^{64}Zn formed after 25.6 hours.

Two half-lives would have elapsed after 25.6 hours. After the first half-life, 1/2 mole of ^{64}Cu would remain. After the second half-life, 1/4 mol of ^{64}Cu would remain. Thus, a total of 3/4 mole of ^{64}Cu decays to form 3/4 mole of ^{64}Zn. The mass of ^{64}Zn produced is

$$m = \left(\frac{3}{4} \text{ mol}\right)\left(63.93 \text{ g mol}^{-1}\right) = 47.95 \text{ g}$$

15.20   Derive Equation 15.35 using the steady-state approximation for both the H and Br atoms.

Apply the steady state approximation to both the H and Br atoms:

$$\frac{d\,[H]}{dt} = k_2[Br][H_2] - k_3[H][Br_2] - k_4[H][HBr] = 0 \qquad (15.20.1)$$

$$\frac{d\,[Br]}{dt} = 2k_1[Br_2] - k_2[Br][H_2] + k_3[H][Br_2] + k_4[H][HBr] - k_5[Br]^2 = 0 \;(15.20.2)$$

Summing Eqs. 15.20.1 and 15.20.2,

$$2k_1[Br_2] - k_5[Br]^2 = 0$$

$$[Br] = \left(\frac{2k_1[Br_2]}{k_5}\right)^{1/2} \qquad (15.20.3)$$

Solving for [H] using Eq. 15.20.1,

$$[H] = \frac{k_2[Br][H_2]}{k_3[Br_2] + k_4[HBr]} \tag{15.20.4}$$

Substitute Eq. 15.20.3 into Eq. 15.20.4 to obtain an expression for [H] in terms of the concentrations of the reactants and products, and of the rate constants.

$$[H] = \frac{k_2 \left(\frac{2k_1}{k_5}\right)^{1/2} [Br_2]^{1/2} [H_2]}{k_3[Br_2] + k_4[HBr]} \tag{15.20.5}$$

According to the reaction mechanism,

$$\frac{d[HBr]}{dt} = k_2[Br][H_2] + k_3[H][Br_2] - k_4[H][HBr]$$

$$= k_2[Br][H_2] + [H]\left(k_3[Br_2] - k_4[HBr]\right) \tag{15.20.6}$$

Substitute Eqs. 15.20.3 and 15.20.5 into Eq. 15.20.6,

$$\frac{d[HBr]}{dt} = k_2 \left(\frac{2k_1}{k_5}\right)^{1/2} [Br_2]^{1/2} [H_2] + \frac{k_2 \left(\frac{2k_1}{k_5}\right)^{1/2} [Br_2]^{1/2} [H_2]}{k_3[Br_2] + k_4[HBr]} \left(k_3[Br_2] - k_4[HBr]\right)$$

$$= k_2 \left(\frac{2k_1}{k_5}\right)^{1/2} [Br_2]^{1/2} [H_2] \left\{1 + \frac{k_3[Br_2] - k_4[HBr]}{k_3[Br_2] + k_4[HBr]}\right\}$$

$$= k_2 \left(\frac{2k_1}{k_5}\right)^{1/2} [Br_2]^{1/2} [H_2] \left\{\frac{2k_3[Br_2]}{k_3[Br_2] + k_4[HBr]}\right\}$$

$$= k_2 \left(\frac{2k_1}{k_5}\right)^{1/2} [Br_2]^{1/2} [H_2] \left\{\frac{2}{1 + \frac{k_4}{k_3}\frac{[HBr]}{[Br_2]}}\right\} \tag{15.20.7}$$

Setting

$$\alpha = 2k_2 \left(\frac{2k_1}{k_5}\right)^{1/2}$$

$$\beta = \frac{k_4}{k_3}$$

Eq. 15.20.7 becomes

$$\frac{d[HBr]}{dt} = \frac{\alpha[H_2][Br_2]^{1/2}}{1 + \beta[HBr]/[Br_2]}$$

15.22 The following data were collected for the reaction between hydrogen and nitric oxide at 700° C:

$$2H_2(g) + 2NO(g) \rightarrow 2H_2O(g) + N_2(g)$$

Experiment	[H_2]/M	[NO]/M	Initial rate/M s^{-1}
1	0.010	0.025	2.4×10^{-6}
2	0.0050	0.025	1.2×10^{-6}
3	0.010	0.0125	0.60×10^{-6}

(a) What is the rate law for the reaction? **(b)** Calculate the rate constant for the reaction. **(c)** Suggest a plausible reaction mechanism that is consistent with the rate law. (*Hint:* Assume that the oxygen atom is the intermediate.) **(d)** More careful studies of the reaction show that the rate law over a wide range of concentrations of reactants should be

$$\text{rate} = \frac{k_1[NO]^2[H_2]}{1 + k_2[H_2]}$$

What happens to the rate law at very high and very low hydrogen concentrations?

(a) Comparing Experiment 1 and Experiment 2, the concentration of NO is constant and the concentration of H_2 has decreased by one-half. The initial rate has also decreased by one-half. Therefore, the initial rate is directly proportional to the concentration of H_2. That is, the reaction is first order in H_2.

Comparing Experiment 1 and Experiment 3, the concentration of H_2 is constant and the concentration of NO has decreased by one-half. The initial rate has decreased by one-fourth. Therefore, the initial rate is proportional to the squared concentration of NO. That is, the reaction is second order in NO.

Therefore, the rate law is

$$\text{Rate} = k[NO]^2[H_2]$$

(b) Using Experiment 1 to calculate the rate constant,

$$k = \frac{\text{Rate}}{[NO]^2[H_2]} = \frac{2.4 \times 10^{-6} \, M \, s^{-1}}{(0.025 \, M)^2 \, (0.010 \, M)} = 0.38 \, M^{-2} \, s^{-1}$$

(c) The rate law suggests that the slow step in the reaction mechanism will probably involve one H_2 molecule and two NO molecules. Additionally, the hint suggests that the O atom is an intermediate. A plausible mechanism is

$$H_2 + 2NO \longrightarrow N_2 + H_2O + O \quad \text{slow step}$$

$$O + H_2 \longrightarrow H_2O \quad\quad\quad\quad \text{fast step}$$

(d) At very high hydrogen concentrations, $k_2 [H_2] \gg 1$. Therefore, the rate law becomes second order:

$$\text{rate} = \frac{k_1[NO]^2[H_2]}{k_2[H_2]} = \frac{k_1}{k_2}[NO]^2$$

At very low hydrogen concentrations, $k_2 [H_2] \ll 1$. Therefore, the rate law becomes third order:

$$\text{rate} = k_1[NO]^2[H_2]$$

15.24 The gas-phase reaction between H_2 and I_2 to form HI involves a two-step mechanism:

$$I_2 \underset{k_{-1}}{\overset{k_1}{\rightleftharpoons}} 2I$$

$$H_2 + 2I \overset{k_2}{\longrightarrow} 2HI$$

(a) Assume the first step is a rapid equilibrium, and derive the rate law for the reaction. (b) The rate of formation of HI increases with the intensity of visible light. How does this fact support the two-step mechanism given.

(a) Because step 2 is the slow step,

$$\text{Rate} = k_2[H_2][I]^2$$

The intermediate, I, can be written in terms of $[I_2]$ using the first fast equilibrium step.

$$\text{Forward rate of step 1} = \text{Reverse rate of step 1}$$

$$k_1[I_2] = k_{-1}[I]^2$$

$$[I]^2 = \frac{k_1}{k_{-1}}[I_2]$$

The rate law becomes

$$\text{Rate} = k_2[H_2][I]^2 = \frac{k_1 k_2}{k_{-1}}[H_2][I_2]$$

(b) In this two-step mechanism, the rate determining step is the second one in which a hydrogen molecule collides with two iodine atoms. The absorption of visible light by the purple molecular iodine vapor weakens the I_2 bond and increases the number of I atoms present, which, according to this mechanism, should increase the reaction rate. This is indeed observed and supports the proposed mechanism. (Hydrogen gas is colorless and does not absorb visible light. Ultraviolet light is required to photodissociate H_2 molecules.)

15.26 Use Equation 15.36 to calculate the rate constant at 300 K for E_a = 0, 2, and 50 kJ mol^{-1}. Assume that $A = 10^{11}$ s^{-1} in each case.

Equation 15.36 is $k = Ae^{-E_a/RT}$. For E_a = 0 kJ mol^{-1},

$$k = \left(10^{11}\ \text{s}^{-1}\right) e^{-\left(0\ \text{J mol}^{-1}\right)/\left[\left(8.314\ \text{J K}^{-1}\ \text{mol}^{-1}\right)(300\ \text{K})\right]} = 10^{11}\ \text{s}^{-1}$$

For E_a = 2 kJ mol^{-1},

$$k = \left(10^{11}\ \text{s}^{-1}\right) e^{-\left(2\times10^3\ \text{J mol}^{-1}\right)/\left[\left(8.314\ \text{J K}^{-1}\ \text{mol}^{-1}\right)(300\ \text{K})\right]} = 4.5 \times 10^{10}\ \text{s}^{-1}$$

For E_a = 50 kJ mol^{-1},

$$k = \left(10^{11}\ \text{s}^{-1}\right) e^{-\left(50\times10^3\ \text{J mol}^{-1}\right)/\left[\left(8.314\ \text{J K}^{-1}\ \text{mol}^{-1}\right)(300\ \text{K})\right]} = 2.0 \times 10^2\ \text{s}^{-1}$$

15.28 Over a range of about $\pm3°C$ from normal body temperature the metabolic rate, M_T, is given by $M_T = M_{37}(1.1)^{\Delta T}$, where M_{37} is the normal rate and ΔT is the change in T. Discuss this equation in terms of a possible molecular interpretation. [Source: "Eco-Chem," J. A. Campbell, *J. Chem. Educ.* **52**, 327 (1975).]

Converting to kelvin, and using the Arrhenius equation,

$$\ln \frac{M_T}{M_{37}} = -\frac{E_a}{R}\left(\frac{1}{T} - \frac{1}{310.15 \text{ K}}\right)$$

Since the temperature range is so small, $f(T) = \frac{1}{T} - \frac{1}{310.15\text{ K}}$ may be expanded in a Taylor series about $T_0 = 310.15$ K. Keeping only the first non-zero term results in $f(T) \approx -\frac{\Delta T}{T_0^2}$, where $\Delta T = 310.15 \text{ K} - T$. Thus,

$$\ln \frac{M_T}{M_{37}} = \frac{E_a}{R}\frac{\Delta T}{T_0^2}$$

or

$$M_T = M_{37}e^{\frac{E_a}{RT_0^2}\Delta T} = M_{37}\left(e^{\frac{E_a}{RT_0^2}}\right)^{\Delta T} = M_{37}(\text{constant})^{\Delta T}$$

which is of the observed form, providing an implicit factor of 1 K^{-1} is incorporated into the argument of the exponential function and the ΔT is interpreted as a unitless number. Specifically, it must be true that

$$e^{\frac{E_a}{RT_0^2}} = 1.1$$

$$\frac{E_a}{RT_0^2} = \ln 1.1 = 0.0953$$

$$E_a = \left(1\,\text{K}^{-1}\right)\left(8.314\,\text{J K}^{-1}\,\text{mol}^{-1}\right)(310.15\,\text{K})^2\,(0.0953)$$

$$= 7.6 \times 10^4\,\text{J mol}^{-1} = 76\,\text{kJ mol}^{-1}$$

This activation energy is consistent with a single rate determining step controlling the metabolic rate within this temperature range.

15.30 The rate constants for the first-order decomposition of an organic compound in solution are measured at several temperatures:

k/s^{-1}	4.92×10^{-3}	0.0216	0.0950	0.326	1.15
$t/^\circ\text{C}$	5.0	15	25	35	45

Determine graphically the pre-exponential factor and the energy of activation for the reaction.

Since

$$\ln k = \ln A - \frac{E_a}{RT}$$

A plot of $\ln k$ vs $1/T$ gives a slope of $-E_a/R$ and an intercept of $\ln A$. The following data are used for the plot:

10^3 K/T	3.595	3.470	3.354	3.245	3.143
$\ln(k/\text{s}^{-1})$	-5.314	-3.835	-2.354	-1.121	0.140

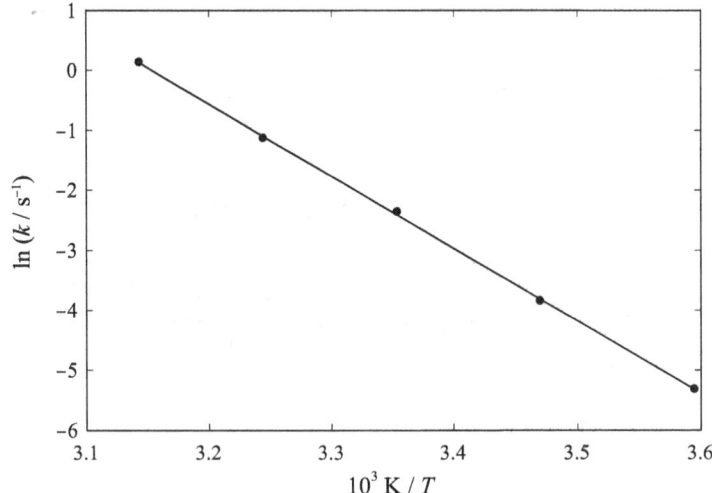

The equation for the line that best fits these points is $y = -1.207 \times 10^4 x + 38.06$. Therefore,

$$E_a = -\left(-1.207 \times 10^4\ \text{K}^{-1}\right)\left(8.314\ \text{J K}^{-1}\,\text{mol}^{-1}\right)$$

$$= 1.00 \times 10^5\ \text{J mol}^{-1} = 100.\ \text{kJ mol}^{-1}$$

$$A = e^{38.06} = 3.38 \times 10^{16}\ \text{s}^{-1}$$

15.32 The rate constant of a first-order reaction is $4.60 \times 10^{-4}\ \text{s}^{-1}$ at 350°C. If the activation energy is $104\ \text{kJ mol}^{-1}$, calculate the temperature at which its rate constant is $8.80 \times 10^{-4}\ \text{s}^{-1}$.

Assume Arrhenius behavior,

$$\ln \frac{k_2}{k_1} = -\frac{E_a}{R}\left(\frac{1}{T_2} - \frac{1}{T_1}\right)$$

$$\ln \frac{8.80 \times 10^{-4}}{4.60 \times 10^{-4}} = -\frac{104 \times 10^3\ \text{J mol}^{-1}}{8.314\ \text{J K}^{-1}\,\text{mol}^{-1}}\left(\frac{1}{T_2} - \frac{1}{623.15\ \text{K}}\right)$$

$$T_2 = 644.0\ \text{K} = 371°\text{C}$$

15.34 Consider the following parallel reactions

$$A \begin{array}{c} \xrightarrow{k_1} B \\ \xrightarrow{k_2} C \end{array}$$

The activation energies are $45.3\ \text{kJ mol}^{-1}$ for k_1 and $69.8\ \text{kJ mol}^{-1}$ for k_2. If the rate constants are equal at 320 K, at what temperature will $k_1/k_2 = 2.00$?

Assume Arrhenius behavior. The ratio of the rate constants is

$$\frac{k_1}{k_2} = \frac{A_1 e^{-E_{a1}/RT}}{A_2 e^{-E_{a2}/RT}}$$

$$= \frac{A_1}{A_2} e^{(E_{a2}-E_{a1})/RT} = \frac{A_1}{A_2} e^{\left(69.8\times10^3 \text{ J mol}^{-1} - 45.3\times10^3 \text{ J mol}^{-1}\right)/\left[\left(8.314 \text{ J K}^{-1}\text{mol}^{-1}\right)T\right]}$$

$$= \frac{A_1}{A_2} e^{2.947\times10^3 \text{ K}/T}$$

First use data at 320 K to calculate A_1/A_2:

$$\frac{k_1}{k_2} = 1.00 = \frac{A_1}{A_2} e^{2.947\times10^3 \text{ K}/320 \text{ K}}$$

$$\frac{A_1}{A_2} = 1.001 \times 10^{-4}$$

When $k_1/k_2 = 2.00$,

$$2.00 = \frac{A_1}{A_2} e^{2.947\times10^3 \text{ K}/T} = \left(1.001 \times 10^{-4}\right) e^{2.947\times10^3 \text{ K}/T}$$

$$\frac{1}{T} = \frac{1}{2.947 \times 10^3 \text{ K}} \ln \frac{2.00}{1.001 \times 10^{-4}} = 3.360 \times 10^{-3} \text{ K}^{-1}$$

$$T = 298 \text{ K}$$

15.36   The rate of the electron-exchange reaction between naphthalene ($C_{10}H_8$) and its anion radical ($C_{10}H_8^-$) in an organic solvent is diffusion-controlled:

$$C_{10}H_8^- + C_{10}H_8 \rightleftharpoons C_{10}H_8 + C_{10}H_8^-$$

The reaction is bimolecular and second order. The rate constants are

T/K	307	299	289	273
$k/10^9 \, M^{-1}\text{s}^{-1}$	2.71	2.40	1.96	1.43

Calculate the values of E_a, $\Delta H^{o\ddagger}$, $\Delta S^{o\ddagger}$ and $\Delta G^{o\ddagger}$ at 307 K for the reaction. [*Hint*: Rearrange Equation 15.49 and plot ($\ln k/T$) versus $1/T$.]

Equation 15.49 gives

$$k = \frac{k_B T}{h} e^{\Delta S^{o\ddagger}/R} e^{-\Delta H^{o\ddagger}/RT} M^{-1}$$

or

$$\ln \frac{k/(M^{-1}\text{s}^{-1})}{T/\text{K}} = \ln \frac{k_B K}{h \text{ s}^{-1}} + \frac{\Delta S^{o\ddagger}}{R} - \frac{\Delta H^{o\ddagger}}{RT}$$

where implied units are explicitly included in the logarithm terms. Although these are often omitted, as indeed they are in the next sentence, a good reason for keeping them will be seen when calculating $\Delta S^{o\ddagger}$ below. A plot of $\ln (k/T)$ vs $1/T$ gives a slope of $-\Delta H^{o\ddagger}/R$ and an intercept of $\ln k_B/h + \Delta S^{o\ddagger}/R$. The data used for the plot are

10^3 K/T	3.257	3.344	3.460	3.663
$\ln \frac{k/(M^{-1}\,s^{-1})}{T/K}$	15.9934	15.8983	15.7298	15.4715

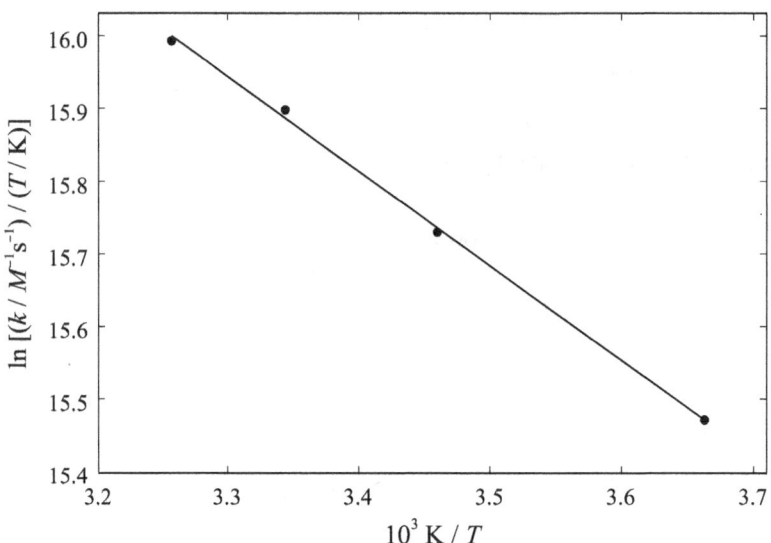

The best fit line has a formula of $y = -1302.0x + 20.24$. Therefore,

$$\Delta H^{o\ddagger} = -(-1302.0\ \text{K})\left(8.314\ \text{J K}^{-1}\,\text{mol}^{-1}\right)$$

$$= 1.082 \times 10^4\ \text{J mol}^{-1}$$

$$= 1.08 \times 10^4\ \text{J mol}^{-1}$$

and

$$\Delta S^{o\ddagger} = R\left(20.24 - \ln \frac{k_B \text{K}}{h\ \text{s}^{-1}}\right)$$

$$= \left(8.314\ \text{J K}^{-1}\,\text{mol}^{-1}\right)\left(20.24 - \ln \frac{1.381 \times 10^{-23}\ \text{J K}^{-1}\ \text{K}}{6.626 \times 10^{-34}\ \text{J s s}^{-1}}\right)$$

$$= -29.3\ \text{J K}^{-1}\,\text{mol}^{-1}$$

Notice how the units would have appeared inconsistent without the explicit consideration of the implied units.

From Equation 15.51 and the discussion following it (see also the answer to Problem 15.37), the activation energy for this reaction, which occurs in solution (condensed phase) is

$$E_a = \Delta H^{o\ddagger} + RT$$

$$= 1.082 \times 10^4\ \text{J mol}^{-1} + \left(8.314\ \text{J K}^{-1}\,\text{mol}^{-1}\right)(307\ \text{K})$$

$$= 1.34 \times 10^4\ \text{J mol}^{-1} = 13.4\ \text{kJ mol}^{-1}$$

This is a relatively small activation energy.

From $\Delta H^{o\ddagger}$ and $\Delta S^{o\ddagger}$, $\Delta G^{o\ddagger}$ at 307 K is calculated.

$$\Delta G^{o\ddagger} = \Delta H^{o\ddagger} - T \Delta S^{o\ddagger}$$

$$= 1.082 \times 10^4 \text{ J mol}^{-1} - (307 \text{ K}) \left(-29.3 \text{ J K}^{-1} \text{mol}^{-1} \right)$$

$$= 1.98 \times 10^4 \text{ J mol}^{-1} = 19.8 \text{ kJ mol}^{-1}$$

15.38   A person may die after drinking D_2O instead of H_2O for a prolonged period (on the order of days). Explain. Since D_2O has practically the same properties as H_2O, how would you test the presence of large quantities of the former in a victim's body?

Because of the lower zero-point energy for bonds in which D is substituted for H, there is a higher activation energy required for reactions in which this bond breaks. Thus, the rate of H^+ ion exchange is faster than that for the D^+ ion. Additionally, the dissociation constants of deuterated acids are smaller than the corresponding acid with the normal H^+ ion. These differences will affect the delicate acid-base balance in the body as well as the kinetics of biological processes, and could lead to death. A mass spectrum of a body fluid sample should reveal the presence of a larger than natural abundance of the heavier isotope of hydrogen. Alternatively, the infrared absorption spectrum of water vapor from a body fluid sample should show a characteristic peak associated with an O–D stretch in D_2O.

15.40   Lubricating oils for watches or other mechanical objects are made of long-chain hydrocarbons. Over long periods of time they undergo auto-oxidation to form solid polymers. The initial step in this process involves hydrogen abstraction. Suggest a chemical means for prolonging the life of these oils.

Deuterating the oils, that is, replacing the H atoms with D atoms, will slow down the rate of hydrogen abstraction. The previous question suggests that deuterated oils might last an order of magnitude longer.

15.42   Measurements of a certain enzyme-catalyzed reaction give $k_1 = 8 \times 10^6 \ M^{-1} s^{-1}$, $k_{-1} = 7 \times 10^4 \ s^{-1}$, and $k_2 = 3 \times 10^3 \ s^{-1}$. Does the enzyme–substrate binding follow the equilibrium or steady-state scheme?

The dissociation constant, K_S, and the Michaelis constant, K_M must be compared.

$$K_S = \frac{k_{-1}}{k_1}$$

$$= \frac{7 \times 10^4 \ s^{-1}}{8 \times 10^6 \ M^{-1} s^{-1}}$$

$$= 8.8 \times 10^{-3} \ M = 9 \times 10^{-3} \ M$$

and

$$K_M = \frac{k_{-1} + k_2}{k_1}$$

$$= \frac{7 \times 10^4 \text{ s}^{-1} + 3 \times 10^3 \text{ s}^{-1}}{8 \times 10^6 \text{ } M^{-1} \text{s}^{-1}}$$

$$= 9.1 \times 10^{-3} M = 9 \times 10^{-3} M$$

Within the precision of the measurements, the two constants are equal. Thus, the binding follows the equilibrium scheme. That is, k_{-1} is sufficiently greater than k_2 so that the binding reaches equilibrium.

15.44 Derive the following equation from Equation 15.70

$$\frac{v_0}{[S]} = \frac{V_{max}}{K_M} - \frac{v_0}{K_M}$$

and show how you would obtain values of K_M and V_{max} graphically from this equation.

Starting with Equation 15.70, multiply both sides by $K_M + [S]$, then divide by $K_M[S]$ and rearrange.

$$v_0 = \frac{V_{max}[S]}{K_M + [S]}$$

$$v_0 K_M + v_0[S] = V_{max}[S]$$

$$v_0 K_M = V_{max}[S] - v_0[S]$$

$$\frac{v_0}{[S]} = \frac{V_{max}}{K_M} - \frac{v_0}{K_M}$$

Thus, a plot of $v_0/[S]$ vs. v_0 will have a slope of $-1/K_M$ and a y-intercept of V_{max}/K_M. The same data, however, when plotted in a Eadie–Hofstee plot, v_0 vs. $v_0/[S]$, gives more straightforward results (see Problem 15.46).

15.46 The hydrolysis of N-glutaryl-L-phenylalanine-p-nitroanilide (GPNA) to p-nitroaniline and N-glutaryl-L-phenylalanine is catalyzed by α-chymotrypsin. The following data are obtained:

$[S]/10^{-4}$ M	2.5	5.0	10.0	15.0
$v_0/10^{-6}$ M min^{-1}	2.2	3.8	5.9	7.1

where [S] = [GPNA]. Assuming Michaelis–Menten kinetics, calculate the values of V_{max}, K_M, and k_2 using the Lineweaver–Burk plot. Another way to treat the data is to plot v_0 versus $v_0/[S]$, which is the Eadie–Hofstee plot. Calculate the values of V_{max}, K_M, and k_2 from the Eadie–Hofstee treatment, given that $[E]_0 = 4.0 \times 10^{-6}$ M. [*Source*: J. A. Hurlbut, T. N. Ball, H. C. Pound, and J. L. Graves, *J. Chem. Educ.* **50,** 149 (1973).]

For the Lineweaver–Burk plot, the following data are needed.

$(1/[S])/10^3\ M^{-1}$	4.00	2.00	1.00	0.667
$(1/v_0)/10^5\ M^{-1}$ min	4.55	2.63	1.69	1.41

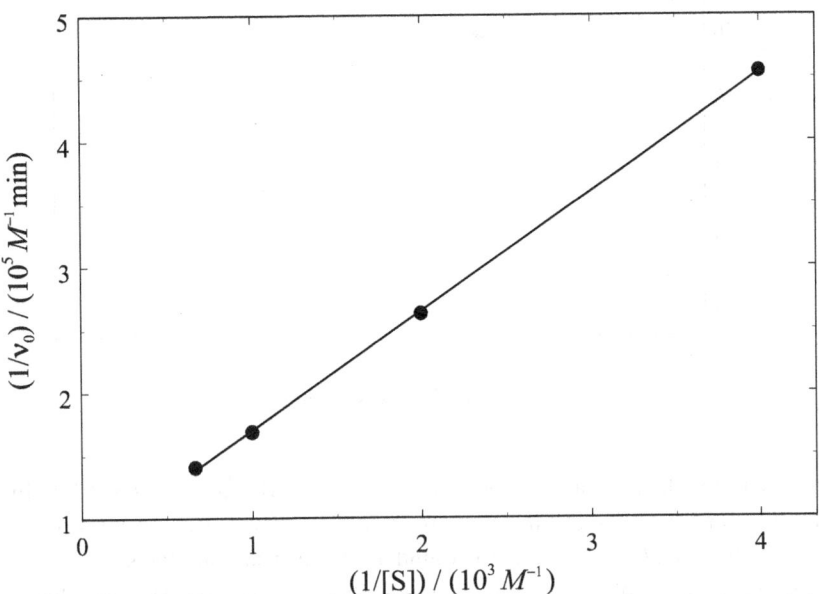

The best-fit line to the data has an equation of $y = 94.6x + 7.56 \times 10^4$. The intercept of a Lineweaver–Burk plot is $1/V_{max}$ giving

$$V_{max} = \frac{1}{7.56 \times 10^4\ M^{-1}\ \text{min}}$$

$$= 1.32 \times 10^{-5}\ M\ \text{min}^{-1}$$

$$= 1.3 \times 10^{-5}\ M\ \text{min}^{-1}$$

The slope is K_M/V_{max} so that

$$K_M = (94.6\ \text{min})(1.32 \times 10^{-5}\ M\ \text{min}^{-1})$$

$$= 1.2 \times 10^{-3}\ M$$

Finally,

$$k_2 = \frac{V_{max}}{[E]_0}$$

$$= \frac{1.32 \times 10^{-5}\ M\ \text{min}^{-1}}{4.0 \times 10^{-6}\ M}$$

$$= 3.3\ \text{min}^{-1}$$

The Eadie–Hofstee plot uses the following data,

$(v_0/[S])/10^{-3}\ \text{min}^{-1}$	8.80	7.60	5.90	4.73
$v_0/10^{-6}\ M\ \text{min}^{-1}$	2.2	3.8	5.9	7.1

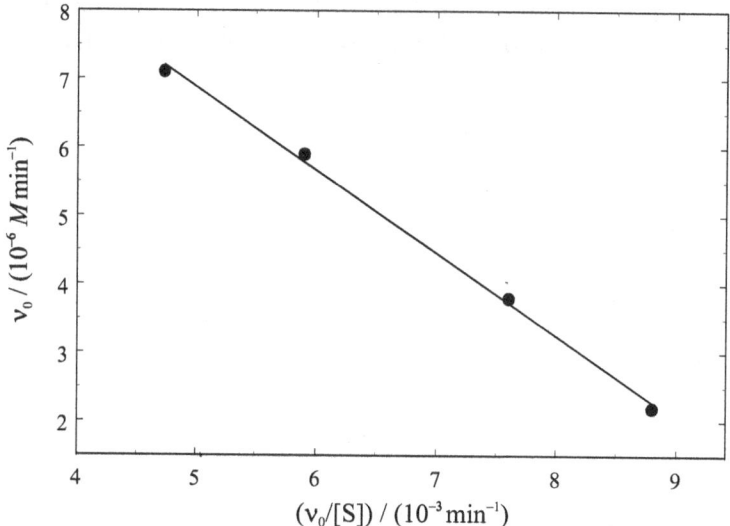

The best-fit line to the data has an equation of $y = -1.21 \times 10^{-3}x + 1.29 \times 10^{-5}$. In a Eadie–Hofstee plot the slope is $-K_M$ and the y-intercept is V_{max}. Thus, $V_{max} = 1.3 \times 10^{-5}\ M\ min^{-1}$ and $K_M = 1.2 \times 10^{-3}\ M$. $k_2 = 3.3\ min^{-1}$ is found as above. These are the same values as found from the Lineweaver–Burk plot, which given good data is as expected. The two plots weight the data differently, so that the values determined may be different depending on the quality of the data.

15.48 The hydrolysis of urea,

$$(NH_2)_2CO + H_2O \rightarrow 2NH_3 + CO_2$$

has been studied by many researchers. At 100°C, the (pseudo) first-order rate constant is $4.2 \times 10^{-5}\ s^{-1}$. The reaction is catalyzed by the enzyme urease, which at 21°C has a rate constant of $3 \times 10^4\ s^{-1}$. If the enthalpies of activation for the uncatalyzed and catalyzed reactions are 134 kJ mol^{-1} and 43.9 kJ mol^{-1}, respectively, **(a)** calculate the temperature at which the nonenzymatic hydrolysis of urea would proceed at the same rate as the enzymatic hydrolysis at 21°C; **(b)** calculate the lowering of $\Delta G^{\ddagger}$ due to urease; and **(c)** comment on the sign of $\Delta S^{\ddagger}$. Assume that $\Delta H^{\ddagger} = E_a$ and that $\Delta H^{\ddagger}$ and $\Delta S^{\ddagger}$ are independent of temperature.

(a) The Arrhenius equation relates reaction rate and activation energy via $k = Ae^{-E_a/RT}$. Requiring that the rates of the catalyzed and uncatalyzed reactions be equal at their respective temperatures then means (assuming A to be the same for both the catalyzed and uncatalyzed reactions),

$$k_{cat} = k_{uncat}$$

$$Ae^{-E_a^{cat}/RT_1} = Ae^{-E_a^{uncat}/RT_2}$$

$$\frac{E_a^{cat}}{T_1} = \frac{E_a^{uncat}}{T_2}$$

taking $E_a \approx \Delta H^{\ddagger}$,

$$\frac{\Delta H_{\text{cat}}^{\ddagger}}{T_1} = \frac{\Delta H_{\text{uncat}}^{\ddagger}}{T_2}$$

$$\frac{43.9 \times 10^3 \text{ J mol}^{-1}}{294.15 \text{ K}} = \frac{134 \times 10^3 \text{ J mol}^{-1}}{T_2}$$

$$T_2 = 898 \text{ K}$$

At this temperature the solvent would be vaporized and the urea thermally decomposed, so that it is in fact impossible to achieve the enzymatic rate without the catalyst.

(b) From Equation 15.48, $k = \frac{k_B T}{h} e^{-\Delta G^{\ddagger}/RT}$, or $\Delta G^{\ddagger} = -RT \ln \frac{kh}{k_B T}$.

For the uncatalyzed reaction at 100°C,

$$\Delta G^{\ddagger} = -\left(8.314 \text{ J K}^{-1} \text{mol}^{-1}\right) (373.15 \text{ K}) \ln \frac{\left(4.2 \times 10^{-5} \text{ s}^{-1}\right)\left(6.626 \times 10^{-34} \text{ J s}\right)}{\left(1.381 \times 10^{-23} \text{ J K}^{-1}\right)(373.15 \text{ K})}$$

$$= 1.234 \times 10^5 \text{ J mol}^{-1}$$

For the catalyzed reaction at 21°C,

$$\Delta G^{\ddagger} = -\left(8.314 \text{ J K}^{-1} \text{mol}^{-1}\right) (294.15 \text{ K}) \ln \frac{\left(3 \times 10^4 \text{ s}^{-1}\right)\left(6.626 \times 10^{-34} \text{ J s}\right)}{\left(1.381 \times 10^{-23} \text{ J K}^{-1}\right)(294.15 \text{ K})}$$

$$= 4.68 \times 10^4 \text{ J mol}^{-1}$$

Thus, $\Delta G^{\ddagger}$ is lowered by $1.234 \times 10^5 \text{ J mol}^{-1} - 4.68 \times 10^4 \text{ J mol}^{-1} = 7.7 \times 10^4 \text{ J mol}^{-1}$, although the comparison is being made at two different temperatures.

(c) Since $\Delta G^{\ddagger} = \Delta H^{\ddagger} - T\Delta S^{\ddagger}$, $\Delta S^{\ddagger} = \left(\Delta H^{\ddagger} - \Delta G^{\ddagger}\right)/T$.

For the uncatalyzed reaction,

$$\Delta S^{\ddagger} = \frac{134 \times 10^3 \text{ J mol}^{-1} - 1.234 \times 10^5 \text{ J mol}^{-1}}{373.15 \text{ K}} = 28.4 \text{ J K}^{-1} \text{mol}^{-1}$$

There is an increase in entropy upon approaching the transition state as would be expected in a case where a single molecule is breaking apart in two or more fragments in the transition state.

For the catalyzed reaction,

$$\Delta S^{\ddagger} = \frac{43.9 \times 10^3 \text{ J mol}^{-1} - 4.68 \times 10^4 \text{ J mol}^{-1}}{294.15 \text{ K}} = -9.9 \text{ J K}^{-1} \text{mol}^{-1}$$

Here there is a decrease in entropy upon approaching the transition state, since the rate determining step now involves the binding of two molecules, enzyme and substrate.

15.50 A flask contains a mixture of compounds A and B. Both compounds decompose by first-order kinetics. The half-lives are 50.0 min for A and 18.0 min for B. If the concentrations of A and B are equal initially, how long will it take for the concentration of A to be four times that of B?

For first-order kinetics,

$$\ln \frac{[X]}{[X]_0} = -kt$$

$$k = \frac{\ln 2}{t_{1/2}}$$

Let the initial concentrations of A and B be c_0. The concentrations of A and B at a later time, t, satisfy

$$\ln \frac{[A]}{c_0} = -k_A t$$

$$\ln \frac{[B]}{c_0} = -k_B t$$

Subtracting the second equation from the first,

$$\ln \left(\frac{[A]}{c_0} \frac{c_0}{[B]} \right) = \ln \frac{[A]}{[B]} = -\left(k_A - k_B\right) t = -\left(\frac{\ln 2}{t_{1/2,A}} - \frac{\ln 2}{t_{1/2,B}} \right) t$$

When [A] becomes four times [B], the above expression becomes

$$\ln 4 = -\left(\frac{\ln 2}{50.0 \text{ min}} - \frac{\ln 2}{18.0 \text{ min}} \right) t$$

$$t = 56.3 \text{ min}$$

15.52   The recombination of iodine atoms in an organic solvent, such as carbon tetrachloride, is a diffusion-controlled process:

$$I + I \rightarrow I_2$$

Given that the viscosity of CCl_4 is 9.69×10^{-4} N s m^{-2} at 20°C, calculate the rate constant of recombination at this temperature.

$$k_D = \frac{8}{3} \frac{RT}{\eta}$$

$$= \frac{8}{3} \frac{\left(8.314 \text{ J K}^{-1} \text{ mol}^{-1}\right) (293.15 \text{ K})}{9.69 \times 10^{-4} \text{ N s m}^{-2}}$$

$$= \left(6.71 \times 10^6 \text{ m}^3 \text{ mol}^{-1} \text{ s}^{-1}\right) \left(\frac{1000 \text{ L}}{1 \text{ m}^3} \right)$$

$$= 6.71 \times 10^9 \ M^{-1} \text{ s}^{-1}$$

15.54   Polyethylene is used in many items, including water pipes, bottles, electrical insulation, toys, and mailing envelopes. It is a *polymer*, a molecule with a very high molar mass made by joining many ethylene molecules (the basic unit is called a *monomer*) together. The initiation step is

$$R_2 \xrightarrow{k_i} 2R \cdot \qquad \text{initiation}$$

The R· species (called a radical) reacts with an ethylene molecule (M) to generate another radical

$$R \cdot + M \longrightarrow M_1 \cdot$$

Reaction of $M_1 \cdot$ with another monomer leads to the growth or propagation of the polymer chain:

$$M_1 \cdot + M \xrightarrow{k_p} M_2 \cdot \qquad \text{propagation}$$

This step can be repeated with hundreds of monomer units. The propagation terminates when two radicals combine

$$M' \cdot + M'' \cdot \xrightarrow{k_t} M' \!-\! M'' \qquad \text{termination}$$

The initiator in the polymerization of ethylene commonly is benzoyl peroxide [$(C_6H_5COO)_2$]:

$$[(C_6H_5COO)_2] \longrightarrow 2C_6H_5COO \cdot$$

This is a first-order reaction. The half-life of benzoyl peroxide at 100°C is 19.8 min. **(a)** Calculate the rate constant (in min^{-1}) of the reaction. **(b)** If the half-life of benzoyl peroxide is 7.30 h, or 438 min, at 70°C, what is the activation energy (in kJ/mol) for the decomposition of benzoyl peroxide? **(c)** Write the rate laws for the elementary steps in the above polymerization process and identify the reactant, product, and intermediates. **(d)** What condition would favor the growth of long high-molar-mass polyethylenes?

(a)
$$k = \frac{\ln 2}{t_{1/2}} = \frac{\ln 2}{19.8 \text{ min}} = 3.501 \times 10^{-2} \text{ min}^{-1} = 3.50 \times 10^{-2} \text{ min}^{-1}$$

(b) At 70°C,

$$k = \frac{\ln 2}{t_{1/2}} = \frac{\ln 2}{438 \text{ min}} = 1.583 \times 10^{-3} \text{ min}^{-1}$$

The activation energy can be calculated using the rate constants at 100°C and 70°C.

$$\ln \frac{k_2}{k_1} = -\frac{E_a}{R} \left(\frac{1}{T_2} - \frac{1}{T_1} \right)$$

$$\ln \frac{3.501 \times 10^{-2} \text{ min}^{-1}}{1.583 \times 10^{-3} \text{ min}^{-1}} = -\frac{E_a}{8.314 \text{ J K}^{-1} \text{ mol}^{-1}} \left(\frac{1}{373.15 \text{ K}} - \frac{1}{343.15 \text{ K}} \right)$$

$$E_a = 1.10 \times 10^5 \text{ J mol}^{-1} = 110 \text{ kJ mol}^{-1}$$

(c) Since all steps are elementary steps, we can deduce the rate laws simply from the equations representing the steps. The rate laws are

$$\text{Initiation} \quad \text{Rate} = k_i[R_2]$$

$$\text{Propagation} \quad \text{Rate} = k_p[M_1 \cdot][M]$$

$$\text{Termination} \quad \text{Rate} = k_t[M' \cdot][M'' \cdot]$$

The reactant molecules are ethylene monomers, the product is polyethylene, and the intermediates are the radicals $R \cdot$, $M_1 \cdot$, $M' \cdot$, $M'' \cdot$, etc.

(d) The growth of long polymers would be favored by a high rate of propagation and a slow rate of termination. Since the rate law of propagation depends on the concentration of monomer, an increase in the concentration of ethylene would increase the propagation (growth) rate. From the rate law for termination, a low concentration of the radical fragment M'· or M''· would lead to a slower rate of termination. This can be accomplished by using a low concentration of the initiator, R_2.

15.56 Explain why grain dust in grain elevators can be explosive.

The answer here is related to that of Problem 15.55. A finely dispersed dust presents a very large surface area to the atmosphere and combustion can occur with extreme rapidity. Just about any organic material can serve as a fuel in this manner, and if it is in a fine enough form, dry enough, and in contact with sufficient air in a confined area, it will be explosive. Wheat and corn starch dusts are among the most explosive grain dusts.

15.58 The *activity* of a radioactive sample is the number of nuclear disintegrations per second, which is equal to the first-order rate constant times the number of radioactive nuclei present. The fundamental unit of radioactivity is the *curie* (Ci), where 1 Ci corresponds to exactly 3.70×10^{10} disintegrations per second. This decay rate is equivalent to that of 1 g of radium-226. Calculate the rate constant and half-life for the radium decay. Starting with 1.0 g of the radium sample, what is the activity after 500 years? The molar mass of Ra-226 is 226.03 g mol^{-1}.

For nuclear decay,

$$\text{rate} = \lambda N$$

where λ is the first order rate constant and N is the number of nuclei. In 1.0 g of ^{226}Ra, there are

$$N = (1.0 \text{ g}) \left(\frac{1 \text{ mol}}{226.06 \text{ g}} \right) \left(6.022 \times 10^{23} \text{ mol}^{-1} \right) = 2.66 \times 10^{21} \text{ nuclei}$$

Since this sample has an activity of 1 Ci,

$$3.70 \times 10^{10} \text{ s}^{-1} = \lambda \left(2.66 \times 10^{21} \right)$$

$$\lambda = 1.39 \times 10^{-11} \text{ s}^{-1} = 1.4 \times 10^{-11} \text{ s}^{-1}$$

$$t_{1/2} = \frac{\ln 2}{\lambda} = \frac{\ln 2}{1.39 \times 10^{-11} \text{ s}^{-1}} = 5.0 \times 10^{10} \text{ s} = 1.6 \times 10^3 \text{ yr}$$

For first-order decay, $N = N_0 e^{-\lambda t}$, but because $N = \text{rate}/\lambda = R/\lambda$,

$$\frac{R}{\lambda} = \frac{R_0}{\lambda} e^{-\lambda t}$$

$$R = R_0 e^{-\lambda t}$$

$$= \left(3.70 \times 10^{10} \text{ s}^{-1} \right) e^{-\left(1.39 \times 10^{-11} \text{ s}^{-1} \right)(500 \text{ yr})\left(3.15 \times 10^7 \text{ s yr}^{-1} \right)}$$

$$= 3.0 \times 10^{10} \text{ s}^{-1}$$

15.60 Consider the following parallel first-order reactions:

$$A \begin{array}{c} \xrightarrow{k_1} B \\ \xrightarrow{k_2} C \end{array}$$

(a) Write the expression for $d[B]/dt$ at time t, given that $[A]_0$ is the concentration of A at $t = 0$.
(b) What is the ratio of $[B]/[C]$ upon the completion of the reactions?

(a) As long as the back reactions can be ignored, $\dfrac{d[B]}{dt} = k_1[A] = k_1\left([A]_0 - [B] - [C]\right)$. Likewise, $\dfrac{d[C]}{dt} = k_2[A] = k_2\left([A]_0 - [B] - [C]\right)$

(b) Regardless of the importance of the back reactions,

$$-\frac{d[A]}{dt} = \frac{d[B]}{dt} + \frac{d[C]}{dt}$$

but if the back reactions can be ignored,

$$-\frac{d[A]}{dt} = \left(k_1 + k_2\right)[A]$$

and

$$[A] = [A]_0 e^{-(k_1+k_2)t}$$

From part **(a)**,

$$\frac{d[B]}{dt} = k_1[A]$$

$$= k_1[A]_0 e^{-(k_1+k_2)t}$$

$$d[B] = k_1[A]_0 e^{-(k_1+k_2)t} dt$$

$$\int_0^t d[B] = \int_0^t k_1[A]_0 e^{-(k_1+k_2)t} dt$$

$$[B]\big|_0^t = -\frac{k_1[A]_0}{k_1 + k_2} e^{-(k_1+k_2)t}\bigg|_0^t$$

$$[B]_t - [B]_0 = [B] = \frac{k_1[A]_0}{k_1 + k_2}\left[1 - e^{-(k_1+k_2)t}\right]$$

because $[B]_0 = 0$. Similarly,

$$[C] = \frac{k_2[A]_0}{k_1 + k_2}\left[1 - e^{-(k_1+k_2)t}\right]$$

Thus, at any point during the reactions, as long as the back reactions are ignored, and initially $[B]_0 = [C]_0 = 0$, it holds that

$$\frac{[B]}{[C]} = \frac{k_1}{k_2}$$

At completion, when the reaction reaches equilibrium, the ratio of concentrations is determined by thermodynamic considerations, specifically the difference in Gibbs energy between B and C. At this point, the back reactions can no longer be ignored, as the principle of microscopic reversibility requires that they occur at the same rate as the respective forward reactions.

15.62   A certain protein molecule, P, of molar mass $\mathcal{M}$ dimerizes when it is allowed to stand in solution at room temperature. A plausible mechanism is that the protein molecule is first denatured before it dimerizes:

$$P \xrightarrow{k_1} P^* \text{ (denatured)} \quad \text{slow}$$

$$2P^* \xrightarrow{k_2} P_2 \qquad\qquad \text{fast}$$

The progress of this reaction can be followed by making viscosity measurements of the average molar mass $\overline{\mathcal{M}}$. Derive an expression for $\overline{\mathcal{M}}$ in terms of the initial concentration, $[P]_0$, and the concentration at time t, $[P]$, and $\mathcal{M}$. Write a rate equation consistent with this scheme.

The average molar mass is given by

$$\overline{\mathcal{M}} = \frac{[P]\mathcal{M} + 2[P_2]\mathcal{M}}{[P] + [P_2]}$$

The stoichiometry of the reaction requires $[P_2] = \left([P]_0 - [P]\right)/2$

$$\overline{\mathcal{M}} = \frac{[P]\mathcal{M} + [P]_0\mathcal{M} - [P]\mathcal{M}}{[P] + \frac{1}{2}[P]_0 - \frac{1}{2}[P]}$$

$$= \frac{2\mathcal{M}[P]_0}{[P]_0 + [P]}$$

In the proposed mechanism the denaturation step is rate determining. Thus,

$$-\frac{d[P]}{dt} = k_1[P]$$

This first order rate equation has solution

$$[P] = [P]_0 e^{-k_1 t}$$

Using this result for [P] in the expression for average molar mass,

$$\overline{\mathcal{M}} = \frac{2\mathcal{M}[P]_0}{[P]_0 + [P]_0 e^{-k_1 t}}$$

$$= \frac{2\mathcal{M}}{\cdot 1 + e^{-k_1 t}}$$

or

$$\frac{2\mathcal{M} - \overline{\mathcal{M}}}{\overline{\mathcal{M}}} = e^{-k_1 t}$$

$$\ln\left(\frac{2\mathcal{M} - \overline{\mathcal{M}}}{\overline{\mathcal{M}}}\right) = -k_1 t$$

Thus, a plot of $\ln\left(\frac{2\mathcal{M} - \overline{\mathcal{M}}}{\overline{\mathcal{M}}}\right)$ versus t will give a straight line with slope $-k_1$.

15.64 The rate law for the reaction $2NO_2(g) \rightarrow N_2O_4(g)$ is rate $= k\,[NO_2]^2$. Which of the following changes will alter the value of k? **(a)** The pressure of NO_2 is doubled. **(b)** The reaction is run in an organic solvent. **(c)** The volume of the container is doubled. **(d)** The temperature is decreased. **(e)** A catalyst is added to the container.

(a) Changing the concentration of a reactant has no effect on the value of k.

(b) If a reaction is run in a solvent other than in the gas phase, then the reaction mechanism will probably change and will thus change the value of k.

(c) Doubling the volume simply changes the concentration. There is no effect on the value of k, as in **(a)**.

(d) The value of k will change with temperature.

(e) A catalyst changes the reaction mechanism and therefore changes the value of k.

15.66 Oxygen for metabolism is taken up by hemoglobin (Hb) to form oxyhemoglobin (HbO$_2$) according to the simplified equation

$$Hb(aq) + O_2(aq) \xrightarrow{k} HbO_2(aq)$$

where the second-order rate constant is $2.1 \times 10^6\ M^{-1}s^{-1}$ at 37°C. For an average adult, the concentrations of Hb and O_2 in the blood in the lungs are $8.0 \times 10^{-6}\ M$ and $1.5 \times 10^{-6}\ M$, respectively. **(a)** Calculate the rate of formation of HbO$_2$. **(b)** Calculate the rate of consumption of O_2. **(c)** The rate of formation of HbO$_2$ increases to $1.4 \times 10^{-4}\ M\,s^{-1}$ during exercise to meet the demand of an increased metabolic rate. Assuming the Hb concentration remains the same, what oxygen concentration is necessary to sustain this rate of HbO$_2$ formation?

(a) $\text{Rate of formation of HbO}_2 = \dfrac{d\,[HbO_2]}{dt} = k[Hb][O_2]$

$$= \left(2.1 \times 10^6\ M^{-1}s^{-1}\right)\left(8.0 \times 10^{-6}\ M\right)\left(1.5 \times 10^{-6}\ M\right)$$

$$= 2.5 \times 10^{-5}\ M\,s^{-1}$$

(b) The rate of consumption of O_2 is

$$-\dfrac{d\,[O_2]}{dt} = \dfrac{d\,[HbO_2]}{dt} = 2.5 \times 10^{-5}\ M\,s^{-1}$$

(c) $\text{Rate of formation of HbO}_2 = k[Hb][O_2]$

$$[O_2] = \dfrac{\text{Rate of formation of HbO}_2}{k[Hb]}$$

$$= \dfrac{1.4 \times 10^{-4}\ M\,s^{-1}}{\left(2.1 \times 10^6\ M^{-1}s^{-1}\right)\left(8.0 \times 10^{-6}\ M\right)}$$

$$= 8.3 \times 10^{-6}\ M$$

Thus, 5.6 times as much oxygen is required to sustain the increase in metabolism.

15.68 Thallium(I) is oxidized by cerium(IV) in solution as follows:

$$TI^+ + 2Ce^{4+} \longrightarrow TI^{3+} + 2Ce^{3+}$$

The elementary steps, in the presence of Mn(II), are

$$Ce^{4+} + Mn^{2+} \longrightarrow Ce^{3+} + Mn^{3+}$$
$$Ce^{4+} + Mn^{3+} \longrightarrow Ce^{3+} + Mn^{4+}$$
$$TI^+ + Mn^{4+} \longrightarrow TI^{3+} + Mn^{2+}$$

(a) Identify the catalyst, intermediates, and the rate-determining step if the rate law is rate = $k\,[Ce^{4+}]\,[Mn^{2+}]$. **(b)** Explain why the reaction is slow without the catalyst. **(c)** Classify the type of catalysis (homogeneous or heterogeneous).

(a) The catalyst is Mn^{2+}. It participates in the reaction but is regenerated at the end. The intermediates are Mn^{3+} and Mn^{4+}.

The first step is the rate-determining step because the rate depends on the concentrations of the reactants for that step.

(b) Without the catalyst, the reaction would be a termolecular one involving 3 cations (TI^+ and two Ce^{4+}). The reaction would be slow.

(c) The catalyst is a homogeneous catalyst because it has the same phase (aqueous) as the reactants.

15.70   The rate constants for the reaction

$$CH_2{=}CH{-}CH{=}CH_2 \;+\; CH_2{=}CH{-}CHO \longrightarrow$$

have been measured at several temperatures:

$10^3 k/M^{-1}s^{-1}$	0.138	1.63	7.2	36.8	81
$t/°C$	155.3	208.3	246.5	295.8	330.8

Calculate the values of the pre-exponential factor, E_a, $\Delta S^{o\ddagger}$, and $\Delta H^{o\ddagger}$ for the reaction. Use 516 K as the mean temperature for your calculation. [Data taken from G. B. Kistiakowsky and J. R. Lacher, *J. Am. Chem. Soc.* **58,** 123 (1936).]

Since

$$k = Ae^{-E_a/RT}$$

$$\ln k = \ln A - \frac{E_a}{RT}$$

A plot of $\ln k$ vs $1/T$ gives a straight line with a slope of $-E_a/R$ and an intercept of $\ln A$.

10^3 K/T	2.3337	2.0768	1.9242	1.7575	1.6556
$\ln(k/M^{-1}s^{-1})$	−8.8883	−6.4192	−4.934	−3.3023	−2.513

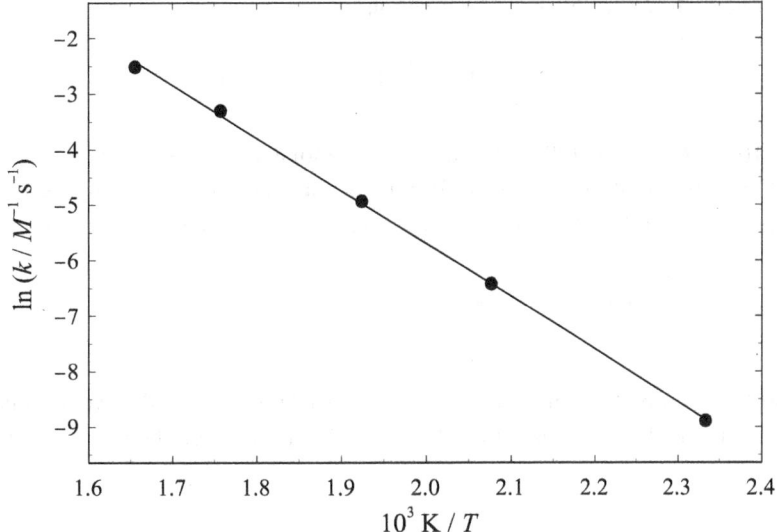

The best fit line has a formula of $y = -9499x + 13.31$. Therefore,

$$E_a = -(-9499 \text{ K}) \left(8.314 \text{ J K}^{-1} \text{mol}^{-1}\right) = 7.897 \times 10^4 \text{ J mol}^{-1} = 79.0 \text{ kJ mol}^{-1}$$

$$A = e^{13.31} = 6.032 \times 10^5 \text{ } M^{-1} \text{s}^{-1} = 6.03 \times 10^5 \text{ } M^{-1} \text{s}^{-1}$$

For this gas-phase, bimolecular reaction, (see Equation 15.56)

$$A = e^2 \frac{k_B T}{h} e^{\Delta S^{\text{o‡}}/R}$$

$$\Delta S^{\text{o‡}} = R \left(\ln \frac{hA}{k_B T} - 2 \right)$$

$$= \left(8.314 \text{ J K}^{-1} \text{mol}^{-1}\right) \left[\ln \frac{(1 \text{ } M) \left(6.626 \times 10^{-34} \text{ J s}\right) \left(6.032 \times 10^5 \text{ } M^{-1} \text{s}^{-1}\right)}{\left(1.381 \times 10^{-23} \text{ J K}^{-1}\right) (516 \text{ K})} - 2 \right]$$

$$= -155 \text{ J K}^{-1} \text{mol}^{-1}$$

and

$$\Delta H^{\text{o‡}} = E_a - 2RT$$

$$= 7.897 \times 10^4 \text{ J mol}^{-1} - 2 \left(8.314 \text{ J K}^{-1} \text{mol}^{-1}\right) (516 \text{ K})$$

$$= 7.04 \times 10^4 \text{ J mol}^{-1}$$

15.72 To prevent brain damage, a drastic medical procedure is to lower the body temperature of someone who has suffered cardiac arrest. What is the physiochemical basis for this treatment?

The low temperature slows down the rates of the chemical reactions of metabolism and decreases the body's need for oxygen. With oxygen being consumed less rapidly, its concentration in the blood takes longer to fall the point at which brain damage would occur.

15.74   The rate constant for the gaseous reaction

$$H_2(g) + I_2(g) \rightarrow 2HI(g)$$

is $2.4 \times 10^{-2}\, M^{-1}\, s^{-1}$ at 400°C. Initially an equimolar sample of H_2 and I_2 is placed in a vessel at 400°C and the total pressure is 1658 mmHg. **(a)** What is the initial rate ($M\, min^{-1}$) of the formation of HI? **(b)** What are the rate of formation of HI and the concentration of HI after 10.0 min?

(a)   With an equimolar sample of H_2 and I_2, the partial pressures of the two reactants are the same: $1658\ mmHg/2 = 829\ mmHg = 1.091\ atm$. Assuming ideal behavior, their concentrations are

$$[H_2] = \frac{n_{H_2}}{V} = \frac{P_{H_2}}{RT}$$

$$= \frac{1.091\ \text{atm}}{\left(0.08206\ \text{L atm K}^{-1}\,\text{mol}^{-1}\right)(673.15\ \text{K})}$$

$$= 1.975 \times 10^{-2}\ M$$

$$[I_2] = 1.975 \times 10^{-2}\ M$$

The units for the rate constant indicate that this reaction is second-order overall, and Problem 15.24 indicates that the reaction orders with respect to H_2 and I_2 are both 1. The initial rate is

$$\text{Rate} = k[H_2][I_2]$$

$$= \left(2.4 \times 10^{-2}\ M^{-1}\,s^{-1}\right)\left(1.975 \times 10^{-2}\ M\right)\left(1.975 \times 10^{-2}\ M\right)$$

$$= \left(9.36 \times 10^{-6}\ M\,s^{-1}\right)\left(\frac{60\ s}{1\ min}\right)$$

$$= 5.62 \times 10^{-4}\ M\ min^{-1} = 5.6 \times 10^{-4}\ M\ min^{-1}$$

Since $\text{Rate} = \frac{1}{2}\frac{d[HI]}{dt}$, the rate of formation of HI is $2(\text{Rate}) = 1.1 \times 10^{-3}\ M\ min^{-1}$

(b)   After 10 minutes, the amount of H_2 remaining is (see the discussion after Equation 15.14 in the text)

$$\frac{1}{[H_2]} = kt + \frac{1}{[H_2]_0}$$

$$= \left(2.4 \times 10^{-2}\ M^{-1}\,s^{-1}\right)(10.0\ min)\left(\frac{60\ s}{1\ min}\right) + \frac{1}{1.975 \times 10^{-2}\ M}$$

$$[H_2] = 1.54 \times 10^{-2}\ M$$

The amount of I_2 remaining is also $1.54 \times 10^{-2}\ M$. Therefore, the amount of H_2 consumed is $1.975 \times 10^{-2}\ M - 1.54 \times 10^{-2}\ M = 4.35 \times 10^{-3}\ M$. Consequently, the amount of HI formed is $2\left(4.35 \times 10^{-3}\ M\right) = 8.7 \times 10^{-3}\ M$.

The rate of reaction after 10 minutes is

Rate $= k[H_2][I_2]$

$$= \left(2.4 \times 10^{-2} M^{-1} s^{-1}\right) \left(1.54 \times 10^{-2} M\right) \left(1.54 \times 10^{-2} M\right)$$

$$= \left(5.69 \times 10^{-6} M\, s^{-1}\right) \left(\frac{60\, s}{1\, min}\right)$$

$$= 3.41 \times 10^{-4} M\, min^{-1} = 3.4 \times 10^{-4} M\, min^{-1}$$

Since Rate $= \frac{1}{2} \frac{d[HI]}{dt}$, the rate of formation of HI is 2(Rate) $= 5.69 \times 10^{-6} M\, s^{-1}$, or $6.8 \times 10^{-4} M\, min^{-1}$

15.76 The diameter of the methyl radical is 3.80 Å. Calculate the rate constant for the second-order gas phase reaction

$$2 \cdot CH_3 \rightarrow C_2H_6$$

at 50°C. Is this the maximum possible rate constant? Explain.

Radical recombination reactions, such as this one for the methyl radical, often require no activation energy. That is, they occur on every collision. The rate constant in this case is given by the collision rate for two methyl radicals divided by the square of their concentration (see Equation 15.39).

$$k = \frac{Z_{CH_3\text{-}CH_3}}{\left(N_{CH_3}/V\right)^2}$$

$$= 2d^2 \sqrt{\frac{\pi k_B T}{m_{CH_3}}}$$

$$= 2\left(3.80 \times 10^{-10}\, m\right)^2 \sqrt{\frac{\pi \left(1.381 \times 10^{-23}\, J\, K^{-1}\right)(323.15\, K)}{(15.03\, amu)\left(1.6605 \times 10^{-27}\, kg\, amu^{-1}\right)}}$$

$$= 2.165 \times 10^{-16}\, m^3\, molecule^{-1}\, s^{-1}$$

Converting to more commonly used chemical units,

$$k = 2.165 \times 10^{-16}\, m^3\, molecule^{-1}\, s^{-1} \left(\frac{1000\, L}{1\, m^3}\right)\left(6.022 \times 10^{23}\, molecules\, mol^{-1}\right)$$

$$= 1.30 \times 10^{11}\, M^{-1}\, s^{-1}$$

This is the maximum possible rate constant, since it assumes that a reaction takes place on every collision.

CHAPTER 16

Photochemistry

PROBLEMS AND SOLUTIONS

16.2 Convert 450 nm to kJ einstein^{-1}.

$$E = h\nu = \frac{hc}{\lambda}$$

$$= \frac{(6.626 \times 10^{-34}\,\text{J s})\,(2.998 \times 10^8\,\text{m s}^{-1})}{450 \times 10^{-9}\,\text{m}} \left(\frac{6.022 \times 10^{23}}{1\,\text{mol}}\right)\left(\frac{1\,\text{kJ}}{1000\,\text{J}}\right)$$

$$= 266\,\text{kJ einstein}^{-1}$$

16.4 An organic molecule absorbs light at 549.6 nm. If 0.031 mole of the molecule is excited by 1.43 einsteins of light, what is the quantum efficiency for this process? Also, calculate the total energy taken up in the process.

The quantum efficiency is

$$\Phi = \frac{0.031\,\text{mol}}{1.43\,\text{einsteins}} = 0.022$$

The energy of each photon is

$$E = h\nu = \frac{hc}{\lambda} = \frac{(6.626 \times 10^{-34}\,\text{J s})\,(2.998 \times 10^8\,\text{m s}^{-1})}{549.6 \times 10^{-9}\,\text{m}} = 3.617 \times 10^{-19}\,\text{J}$$

The energy taken up in the process is

$$\left(3.617 \times 10^{-19}\,\text{J}\right)(1.43\,\text{einsteins})\left(\frac{6.022 \times 10^{23}}{1\,\text{einstein}}\right) = 3.11 \times 10^5\,\text{J}$$

16.6 The first-order rate constants for the fluorescence and phosphorescence of naphthalene ($C_{10}H_8$) are 4.5×10^7 s^{-1} and 0.50 s^{-1}, respectively. Calculate how long it takes for 1.0% of fluorescence and phosphorescence to occur following termination of excitation.

For a first-order reaction, the number of photoexcited molecules at time t, N_t, is related to that at time 0, N_0, by

$$N_t = N_0 e^{-kt}$$

Therefore,

$$t = -\frac{1}{k} \ln \frac{N_t}{N_0}$$

The time for 1.0% of fluorescence to occur is when 1.0% of the photoexcited molecules have decayed, leaving $N_t/N_0 = 0.990$,

$$t = -\frac{1}{4.5 \times 10^7 \text{ s}^{-1}} \ln 0.990 = 2.2 \times 10^{-10} \text{ s} = 2.2 \times 10^2 \text{ ps}$$

Similarly, the time for 1.0% of phosphorescence to occur is

$$t = -\frac{1}{0.50 \text{ s}^{-1}} \ln 0.990 = 2.0 \times 10^{-2} \text{ s} = 20. \text{ ms}$$

16.8 Name the major source of heat that originates from Earth.

The major source of heat that is of terrestrial origin is radioactive decay.

16.10 The hydroxyl radical in the atmosphere is most effectively removed by hydrocarbons such as methane according to the second-order reaction

$$\cdot OH + CH_4 \rightarrow H_2O + CH_3 \cdot$$

Given that the second-order rate constant is 4.6×10^6 L mol^{-1} s^{-1}, calculate the lifetime of the radical at 25°C if the concentration of CH_4 is 1.7×10^3 ppb by volume. (*Hint*: The lifetime of the radical is given by $1/k[CH_4]$.)

The volume ratio between CH_4 and the atmosphere is the same as their mole ratio. That is,

$$\frac{V_{CH_4}}{V_{total}} = \frac{n_{CH_4}}{n_{total}} = \frac{1.7 \times 10^3}{1 \times 10^9} = 1.7 \times 10^{-6}$$

The molar concentration of CH_4 is

$$[CH_4] = \frac{n_{CH_4}}{V_{total}} = \frac{P_{CH_4}}{RT} = \frac{P_{total}\left(\frac{n_{CH_4}}{n_{total}}\right)}{RT}$$

$$= \frac{(1 \text{ atm})\left(1.7 \times 10^{-6}\right)}{\left(0.08206 \text{ L atm K}^{-1} \text{ mol}^{-1}\right)(298.15 \text{ K})}$$

$$= 6.95 \times 10^{-8} \text{ mol L}^{-1}$$

The lifetime of the radical is

$$\frac{1}{k[CH_4]} = \frac{1}{\left(4.6 \times 10^6 \, L \, mol^{-1} \, s^{-1}\right)\left(6.95 \times 10^{-8} \, mol \, L^{-1}\right)} = 3.1 \, s$$

16.12 Deforestation contributes to the greenhouse effect in two ways. What are they?

Combustion of the forest products directly releases CO_2 to the atmosphere, and the reduction in photosynthetic activity decreases the rate of removal of CO_2 from the atmosphere.

16.14 Is ozone a greenhouse gas? Sketch three ways an ozone molecule can vibrate.

The vibrational modes of O_3, as shown below, are similar to those of water and absorb strongly in the infrared. Thus, ozone is a greenhouse gas.

asymmetric stretch   symmetric stretch   bend

16.16 Which of the following settings is the most suitable for photochemical smog formation? **(a)** Gobi desert at noon in June, **(b)** New York city at 1 p.m. in July, **(c)** Boston at noon in January. Explain your choice.

Photochemical smog formation requires both heavy traffic and intense sunlight. These two conditions are met in **(b)**, an urban setting during a summer midday. (Note that daylight savings time is in effect in New York in July, so the sun is directly overhead at 1:00 p.m., not noon.)

16.18 The gas-phase decomposition of peroxyacetyl nitrate (PAN) obeys first-order kinetics:

$$CH_3COOONO_2 \rightarrow CH_3COOO + NO_2$$

with a rate constant of $4.9 \times 10^{-4} \, s^{-1}$. Calculate the rate of decomposition in $M \, s^{-1}$ if the concentration of PAN is 0.55 ppm by volume. Assume STP conditions.

The volume ratio between PAN and the atmosphere is the same as their mole ratio. That is,

$$\frac{V_{PAN}}{V_{total}} = \frac{n_{PAN}}{n_{total}} = \frac{0.55}{1 \times 10^6} = 5.5 \times 10^{-7}$$

The molar concentration of PAN is

$$[\text{PAN}] = \frac{n_{\text{PAN}}}{V_{\text{total}}} = \frac{P_{\text{PAN}}}{RT} = \frac{P_{\text{total}}\left(\frac{n_{\text{PAN}}}{n_{\text{total}}}\right)}{RT}$$

$$= \frac{(1\ \text{atm})\left(5.5 \times 10^{-7}\right)}{\left(0.08206\ \text{L atm K}^{-1}\ \text{mol}^{-1}\right)(273.15\ \text{K})}$$

$$= 2.45 \times 10^{-8}\ \text{mol L}^{-1}$$

Thus, the rate of decomposition of PAN is

$$\text{Rate} = k[\text{PAN}] = \left(4.9 \times 10^{-4}\ \text{s}^{-1}\right)\left(2.45 \times 10^{-8}\ \text{mol L}^{-1}\right) = 1.2 \times 10^{-11}\ M\ \text{s}^{-1}$$

16.20 The safety limits of ozone and carbon monoxide are 120 ppb by volume and 9 ppm by volume, respectively. Why does ozone have a lower limit?

Ozone is much more damaging to lung tissues, and hence more toxic than carbon monoxide. Although carbon monoxide is quite dangerous, the human body can tolerate a fairly large amount of CO and still survive. For example, a person can still function even when 10% or so of the body's hemoglobin is complexed with CO.

16.22 Given that the quantity of ozone in the stratosphere is equivalent to a 3.0-mm-thick layer of ozone on Earth at 1 atm and 25°C, calculate the number of O_3 molecules in the stratosphere and their mass in kilograms. See Problem 16.7 for other information.

The volume of ozone can be calculated from the Earth's average radius, r, and the thickness of the ozone layer, h.

$$V = 4\pi r^2 h = 4\pi \left(6371 \times 10^3\ \text{m}\right)^2 \left(3.0 \times 10^{-3}\ \text{m}\right)\left(\frac{1000\ \text{L}}{1\ \text{m}^3}\right) = 1.53 \times 10^{15}\ \text{L}$$

The number of moles of ozone is

$$n = \frac{PV}{RT} = \frac{(1\ \text{atm})\left(1.53 \times 10^{15}\ \text{L}\right)}{\left(0.08206\ \text{L atm K}^{-1}\ \text{mol}^{-1}\right)(298.15\ \text{K})} = 6.25 \times 10^{13}\ \text{mol}$$

The number of O_3 molecules and the mass are

$$\text{Number of } O_3 \text{ molecules} = \left(6.25 \times 10^{13}\ \text{mol}\right)\left(\frac{6.022 \times 10^{23}\ \text{molecules}}{1\ \text{mol}}\right)$$

$$= 3.8 \times 10^{37}\ \text{molecules}$$

$$\text{Mass} = \left(6.25 \times 10^{13}\ \text{mol}\right)\left(\frac{48.00\ \text{g}}{1\ \text{mol}}\right)\left(\frac{1\ \text{kg}}{1000\ \text{g}}\right)$$

$$= 3.00 \times 10^{12}\ \text{kg} = 3.0 \times 10^{12}\ \text{kg}$$

16.24 Why are CFCs not decomposed by UV radiation in the troposphere?

The short-wavelength UV radiation with sufficient photon energy to decompose CFCs is absorbed by species in the upper atmosphere, and only UV radiation of longer wavelengths and lower photon energy reaches the troposphere.

16.26 Like CFCs, certain bromine-containing compounds, such as CF_3Br, can participate in the destruction of ozone by a similar mechanism starting with the Br atom:

$$CF_3Br \xrightarrow{h\nu} CF_3 + Br\cdot$$

Given that the average C–Br bond enthalpy is 276 kJ mol^{-1}, estimate the longest wavelength required to break this bond. Will the decomposition of CF_3Br occur in the troposphere or in both the troposphere and stratosphere?

The energy of the radiation with the longest wavelength must be the same as the average C–Br bond enthalpy.

$$\lambda = \frac{c}{\nu} = \frac{ch}{E}$$

$$= \frac{(2.998 \times 10^8 \text{ m s}^{-1})(6.626 \times 10^{-34} \text{ J s})}{(276 \times 10^3 \text{ J mol}^{-1})\left(\frac{1 \text{ mol}}{6.022 \times 10^{23}}\right)}$$

$$= 4.34 \times 10^{-7} \text{ m} = 433 \text{ nm}$$

This is light in the visible region of the spectrum, and the compound will be decomposed in both the troposphere and the stratosphere.

16.28 Why are CFCs more effective greenhouse gases than methane and carbon dioxide?

The greater polarity of the C–F and C–Cl bonds compared to the C–H and C=O bonds results in greater dipole derivatives for CFCs. This makes them stronger absorbers of IR radiation (see Section 11.3).

16.30 Calculate the standard enthalpy of formation $(\Delta_f \overline{H}^\circ)$ of ClO from the following bond dissociation enthalpies: Cl_2: 242.7 kJ mol^{-1}; O_2: 498.8 kJ mol^{-1}; ClO: 206 kJ mol^{-1}.

First we estimate the standard enthalpy of reaction for $Cl_2(g) + O_2(g) \rightarrow 2ClO(g)$ using bond dissociation enthalpies.

$$\Delta_r H^\circ = \sum BE(\text{reactants}) - \sum BE(\text{products})$$

$$= 242.7 \text{ kJ mol}^{-1} + 498.8 \text{ kJ mol}^{-1} - 2\left(206 \text{ kJ mol}^{-1}\right)$$

$$= 329.5 \text{ kJ mol}^{-1}$$

The standard enthalpy of reaction is also related to the standard enthalpies of formation of the reactants and products.

$$\Delta_r H^\circ = 2\Delta_f \overline{H}^\circ[\mathrm{ClO}(g)] - \Delta_f \overline{H}^\circ[\mathrm{Cl}_2(g)] - \Delta_f \overline{H}^\circ[\mathrm{O}_2(g)]$$

$$329.5\ \mathrm{kJ\ mol^{-1}} = 2\Delta_f \overline{H}^\circ[\mathrm{ClO}(g)] - 0\ \mathrm{kJ\ mol^{-1}} - 0\ \mathrm{kJ\ mol^{-1}}$$

$$\Delta_f \overline{H}^\circ[\mathrm{ClO}(g)] = \frac{329.5\ \mathrm{kJ\ mol^{-1}}}{2} = 165\ \mathrm{kJ\ mol^{-1}}$$

16.32 The transparency of a certain type of sunglasses to light depends on the intensity of light in the environment. The lenses are clear in dimly lit rooms but darken when the wearer goes outdoors. The material responsible for this change is the very tiny AgCl crystals incorporated in the glass. Suggest a photochemical mechanism that would account for this change.

When irradiated with the appropriate wavelength (which must be in the visible region of the spectrum), AgCl decomposes according to

$$\mathrm{AgCl} \longrightarrow \mathrm{Ag} + \mathrm{Cl}$$

The Ag and Cl atoms formed are trapped in the glass matrix. The small Ag particles diminish the amount of light transmitted. In the absence of intense light, the reactive atoms recombine to form transparent AgCl.

16.34 Consider the photochemical isomerization $A \rightleftharpoons B$. At 650 nm, the quantum yields for the forward and reverse reactions are 0.73 and 0.44, respectively. If the molar absorptivities at 650 nm of A and B are $1.3 \times 10^3\ \mathrm{L\ mol^{-1}\ cm^{-1}}$ and $0.47 \times 10^3\ \mathrm{L\ mol^{-1}\ cm^{-1}}$, respectively, what is the ratio [B]/[A] in the photostationary state?

The rate of formation of B from A is given by $I_A \Phi_A$, where I_A is the intensity of light *absorbed* by A. Likewise, the rate of formation of A from B is $I_B \Phi_B$. At the photostationary state, the two rates are equal, or

$$I_A \Phi_A = I_B \Phi_B$$

The intensity of light absorbed by a sample is given by $I_0 - I$, where I is the transmitted light intensity. Since according to the Beer–Lambert law, $I = I_0 10^{-A} = I_0 10^{-\varepsilon bc}$, then intensity of light absorbed is

$$I_0 - I = I_0 \left(1 - 10^{-\varepsilon bc}\right)$$

Thus, at the photostationary state,

$$I_0 \left(1 - 10^{-\varepsilon_A b c_A}\right) \Phi_A = I_0 \left(1 - 10^{-\varepsilon_B b c_B}\right) \Phi_B$$

This is difficult to solve in the general case, but in the limit of small absorption of light ($\varepsilon bc \ll 1$), the approximation $10^{-\varepsilon bc} \approx 1 - 2.303 \varepsilon bc$ may be made, leading to

$$I_0 \left(2.303 \varepsilon_A b c_A\right) \Phi_A = I_0 \left(2.303 \varepsilon_B b c_B\right) \Phi_B$$

$$\frac{c_B}{c_A} = \frac{[B]}{[A]} = \frac{\varepsilon_A \Phi_A}{\varepsilon_B \Phi_B} = \frac{\left(1.3 \times 10^3\ \mathrm{L\ mol^{-1}\ cm^{-1}}\right)(0.73)}{\left(0.47 \times 10^3\ \mathrm{L\ mol^{-1}\ cm^{-1}}\right)(0.44)} = 4.6$$

16.36 In 1991, it was discovered that nitrous oxide (N_2O) is produced in the synthesis of nylon. This compound, which is released into the atmosphere, contributes *both* to the depletion of ozone in the stratosphere and to the greenhouse effect. **(a)** Write equations representing the reactions between N_2O and oxygen atoms in the stratosphere to produce nitric oxide, which is then oxidized by ozone to form nitrogen dioxide. **(b)** Is N_2O a more effective greenhouse gas than carbon dioxide? Explain. **(c)** One of the intermediates in nylon manufacture is adipic acid [$HOOC(CH_2)_4COOH$]. About 2.2×10^9 kg of adipic acid are consumed every year. Estimates are that for every mole of adipic acid produced, 1 mole of N_2O is generated. What is the maximum number of moles of O_3 that can be destroyed as a result of this process per year?

(a) The individual reactions and the overall reaction are

$$N_2O + O \rightleftharpoons 2NO$$

$$2NO + 2O_3 \rightleftharpoons 2O_2 + 2NO_2$$

$$\overline{\text{Overall: }\quad N_2O + O + 2O_3 \rightleftharpoons 2O_2 + 2NO_2}$$

(b) As a polar molecule, the dipole derivatives associated with the vibrational motions of N_2O are larger than those for the nonpolar CO_2. This makes the IR absorptions of N_2O stronger than those of CO_2 (see Section 11.3), making N_2O a more effective greenhouse gas.

(c) The number of moles of N_2O generated is the same as the number of moles of adipic acid consumed.

$$n_{N_2O} = \frac{\text{mass of adipic acid consumed}}{\text{molar mass of adipic acid}} = \frac{2.2 \times 10^9 \text{ kg}}{146.1 \times 10^{-3} \text{ kg mol}^{-1}} = 1.51 \times 10^{10} \text{ mol}$$

For each mole of N_2O generated, 2 mol of O_3 can be destroyed. Thus, the maximum amount of O_3 destroyed is

$$n_{O_3} = 2n_{N_2O} = 2\left(1.51 \times 10^{10} \text{ mol}\right) = 3.0 \times 10^{10} \text{ mol}$$

16.38 Given that the collision diameter of ozone is about 4.2 Å, calculate the mean free path of ozone at sea level (1 atm and 25°C) and in the stratosphere (3×10^{-3} atm and −23°C).

The mean free path can be calculated using Equation 2.19:

$$\lambda = \frac{RT}{\sqrt{2}\pi d^2 P N_A}$$

At sea level,

$$\lambda = \frac{\left(8.314 \text{ J K}^{-1} \text{ mol}^{-1}\right)(298.15 \text{ K})}{\sqrt{2}\pi \left(4.2 \times 10^{-10} \text{ m}\right)^2 \left[(1 \text{ atm})\left(\frac{101.3 \times 10^3 \text{ Pa}}{1 \text{ atm}}\right)\right]\left(6.022 \times 10^{23} \text{ mol}^{-1}\right)}$$

$$= 5.2 \times 10^{-8} \text{ m} = 52 \text{ nm}$$

In the stratosphere,

$$\lambda = \frac{\left(8.314\,\text{J K}^{-1}\,\text{mol}^{-1}\right)(250.15\,\text{K})}{\sqrt{2}\pi\,(4.2 \times 10^{-10}\,\text{m})^2\left[(3 \times 10^{-3}\,\text{atm})\left(\frac{101.3 \times 10^3\,\text{Pa}}{1\,\text{atm}}\right)\right](6.022 \times 10^{23}\,\text{mol}^{-1})}$$

$$= 1.5 \times 10^{-5}\,\text{m} = 14\,\mu\text{m}$$

16.40   Account for the oscillation in atmospheric concentration CO_2 shown in Figure 16.6.

This is a seasonal oscillation caused by increased photosynthesis in the summer that reduces atmospheric CO_2. During the winter, the CO_2 level increases due to reduced photosynthesis.

16.42   A light source of power 2×10^{-16} W is sufficient to be detected by the human eye. Assuming the wavelength of light is at 550 nm, calculate the number of photons that must be absorbed by rhodopsin per second. (*Hint*: Vision persists for only 1/30 of a second.)

The energy of 1 photon at 550 nm is

$$E = h\nu = \frac{hc}{\lambda} = \frac{(6.626 \times 10^{-34}\,\text{J s})(2.998 \times 10^8\,\text{m s}^{-1})}{550 \times 10^{-9}\,\text{m}} = 3.614 \times 10^{-19}\,\text{J}$$

$1\,\text{W} = 1\,\text{J s}^{-1}$, so the light source produces $2 \times 10^{-16}\,\text{J s}^{-1}$. Therefore, The number of photons that must be absorbed by rhodopsin each second is

$$\frac{2 \times 10^{-16}\,\text{J}}{3.614 \times 10^{-19}\,\text{J}} = 5.53 \times 10^2 = 5.5 \times 10^2$$

Since vision persists for only 1/30 of a second, the number of photons detected is

$$\left(5.53 \times 10^2\right)\left(\frac{1}{30}\right) = 18$$

CHAPTER 17

Intermolecular Forces

PROBLEMS AND SOLUTIONS

17.2 Arrange the following species in order of decreasing melting points: Ne, KF, C_2H_6, MgO, H_2S.

$MgO > KF > H_2S > C_2H_6 > Ne$

17.4 If you lived in Alaska, which of the following natural gases would you keep in an outdoor storage tank in winter? Methane (CH_4), propane (C_3H_8), or butane (C_4H_{10}). Explain.

CH_4 has the weakest intermolecular forces and, as a result, the lowest boiling point, making it the best choice for a cold climate.

17.6 The boiling points of the three different structural isomers of pentane (C_5H_{12}) are 9.5°C, 27.9°C, and 36.1°C. Draw their structures, and arrange them in order of decreasing boiling points. Justify your arrangement.

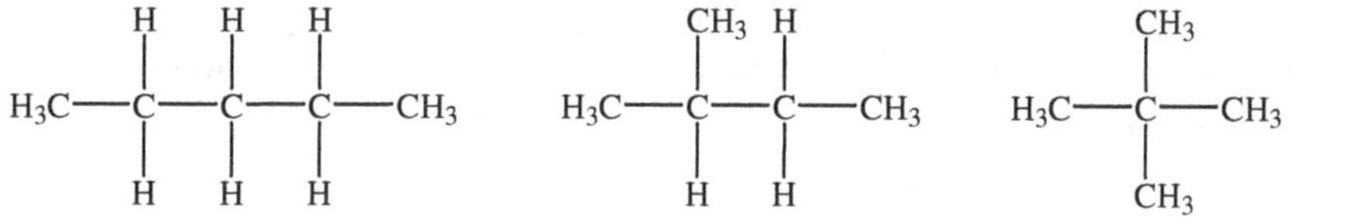

n-pentane
36.1 °C

2-methylbutane
27.9 °C

2,2-dimethylpropane
9.5 °C

The boiling points depend on the ease of packing the molecules together. The *n*-pentane packs together most easily, and it has the highest boiling point. The packing is least favorable for 2,2-dimethylpropane, which has the lowest boiling point.

17.8 Coulombic forces are usually referred to as long-range forces (they depend on $1/r^2$) whereas van der Waals forces are called short-range forces (they depend on $1/r^7$). **(a)** Assuming that the forces (F) depend only on distances, plot F as a function of r at $r = 1$ Å, 2 Å, 3 Å, 4 Å, and 5 Å.

(b) Based on your results, explain the fact that although a 0.2 *M* nonelectrolyte solution usually behaves ideally, nonideal behavior is quite noticeable in a 0.02 *M* electrolyte solution.

(a) A plot with graphs of $1/r^2$ vs r and $1/r^7$ vs r is presented below. The forces will be proportional to these functions.

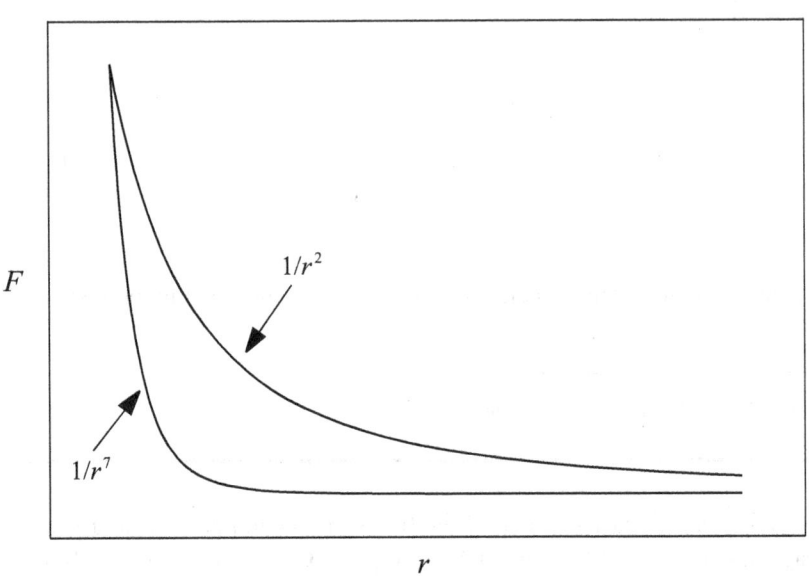

(b) In a nonelectrolyte solution, the attractive forces have a $1/r^7$ dependence, and as the graph shows, they fall off very rapidly with distance. In an electrolyte solution, the ionic (Coulombic) forces have a $1/r^2$ dependence that extends to large distances. These "long-range" forces are responsible for nonideal behavior, even at low concentrations.

17.10   Differentiate Equation 17.21 with respect to r to obtain an expression for σ and ε. Express the equilibrium distance, r_e, in terms of σ and show that $V = -\varepsilon$.

Starting with Equation 17.21,

$$V = 4\varepsilon \left[\left(\frac{\sigma}{r} \right)^{12} - \left(\frac{\sigma}{r} \right)^{6} \right]$$

and differentiating gives

$$\frac{dV}{dr} = 4\varepsilon \left[-\frac{12\sigma^{12}}{r^{13}} + \frac{6\sigma^{6}}{r^{7}} \right]$$

The minimum of the potential energy occurs when $r = r_e$ and $\dfrac{dV}{dr} = 0$.

$$4\varepsilon\left[-\frac{12\sigma^{12}}{r_e^{13}}+\frac{6\sigma^6}{r_e^7}\right]=0$$

$$-\frac{12\sigma^{12}}{r_e^{13}}+\frac{6\sigma^6}{r_e^7}=0$$

$$-\frac{2\sigma^6}{r_e^6}+1=0$$

$$r_e=2^{1/6}\sigma$$

To calculate the potential energy at the equilibrium distance, substitute the expression for r_e into that for the potential energy.

$$V=4\varepsilon\left[\left(\frac{\sigma}{2^{1/6}\sigma}\right)^{12}-\left(\frac{\sigma}{2^{1/6}\sigma}\right)^6\right]$$

$$=4\varepsilon\left[\frac{1}{4}-\frac{1}{2}\right]$$

$$=-\varepsilon$$

17.12 **(a)** From the data in Table 17.2, determine the van der Waals radius for argon. **(b)** Use this radius to determine the fraction of the volume occupied by 1 mole of argon at 25°C and 1 atm.

(a) Since σ gives the distance of closest approach of two argon atoms, the van der Waals radius for argon is

$$r=\frac{\sigma}{2}=\frac{3.40\text{ Å}}{2}=1.70\text{ Å}$$

(b) The volume of 1 mole of Ar atoms is

$$\frac{4}{3}\pi r^3 N_A=\frac{4}{3}\pi\left(1.70\times10^{-10}\text{ m}\right)^3\left(\frac{6.022\times10^{23}}{1\text{ mol}}\right)$$

$$=1.239\times10^{-5}\text{ m}^3\text{ mol}^{-1}$$

$$=1.239\times10^{-2}\text{ L mol}^{-1}$$

Assuming ideal behavior, the volume occupied by one mole of argon gas is

$$\frac{V}{n}=\frac{RT}{P}=\frac{(0.08206\text{ L atm K}^{-1}\text{ mol}^{-1})(298.15\text{ K})}{1\text{ atm}}=24.47\text{ L mol}^{-1}$$

The fraction of this volume occupied by the one mole of argon atoms is

$$\frac{1.239\times10^{-2}\text{ L mol}^{-1}}{24.47\text{ L mol}^{-1}}=5.1\times10^{-4}$$

17.14 If water were a linear molecule, **(a)** would it still be polar and **(b)** would the water molecules still be able to form hydrogen bonds with one another?

(a) A "linear" water molecule would not be polar.

(b) Such a molecule could still form hydrogen bonds, although it would assume two-dimensional hydrogen bond structures.

17.16 Explain why ammonia is soluble in water but nitrogen trichloride is not.

Ammonia, NH_3, can form hydrogen bonds with water, but NCl_3 cannot.

17.18 Which of the following molecules has a higher melting point? Explain your answer.

The *para* isomer can form intermolecular hydrogen bonds, while the *ortho* isomer preferentially forms only intramolecular hydrogen bonds as shown below. Thus, the *para* form with stronger intermolecular forces will have the higher melting point.

17.20 Assume the energy of hydrogen bonds per base pair to be 10 kJ mol^{-1}. Given two complementary strands of DNA containing 100 base pairs each, calculate the ratio of two separate strands to hydrogen-bonded double helix in solution at 300 K.

For one pair of bases, the ratio of the two separate strands to hydrogen-bonded double helix is

$$\exp\left(-\frac{\Delta E}{RT}\right) = \exp\left[-\frac{10 \times 10^3 \text{ J mol}^{-1}}{\left(8.314 \text{ J K}^{-1} \text{mol}^{-1}\right)(300 \text{ K})}\right] = 1.8 \times 10^{-2}$$

For 100 base pairs, the ratio of the two separate strands to hydrogen-bonded double helix is

$$\exp\left[-\frac{(100)\left(10 \times 10^3 \text{ J mol}^{-1}\right)}{\left(8.314 \text{ J K}^{-1} \text{mol}^{-1}\right)(300 \text{ K})}\right] = 7.6 \times 10^{-175} \approx 0$$

17.22   List all the intra- and intermolecular forces that could exist between hemoglobin molecules in water.

All of the intermolecular interactions discussed in the chapter (dispersion, dipole–dipole, hydrogen bonding, ionic) exist between hemoglobin molecules in water.

17.24   Which of the following properties indicates very strong intermolecular forces in a liquid? **(a)** A very low surface tension, **(b)** a very low critical temperature, **(c)** a very low boiling point, **(d)** a very low vapor pressure.

Only **(d)** indicates very strong intermolecular forces in a liquid. The others indicate weak intermolecular forces.

17.26   Using values listed in Table 17.1 and a handbook of chemistry, plot the polarizabilities of the noble gases versus their boiling points. On the same graph, also plot their molar masses versus boiling points. Comment on the trends.

The necessary data are in the table. The polarizability of Rn is not accurately known; this noble gas is not included.

Noble Gas	Molar Mass/g mol^{-1}	$\alpha/10^{-30}$ m^3	b.p./K
He	4.00	0.20	4.2
Ne	20.18	0.40	27.1
Ar	39.95	1.66	87.3
Kr	83.80	2.54	120.0
Xe	131.29	4.15	165.2

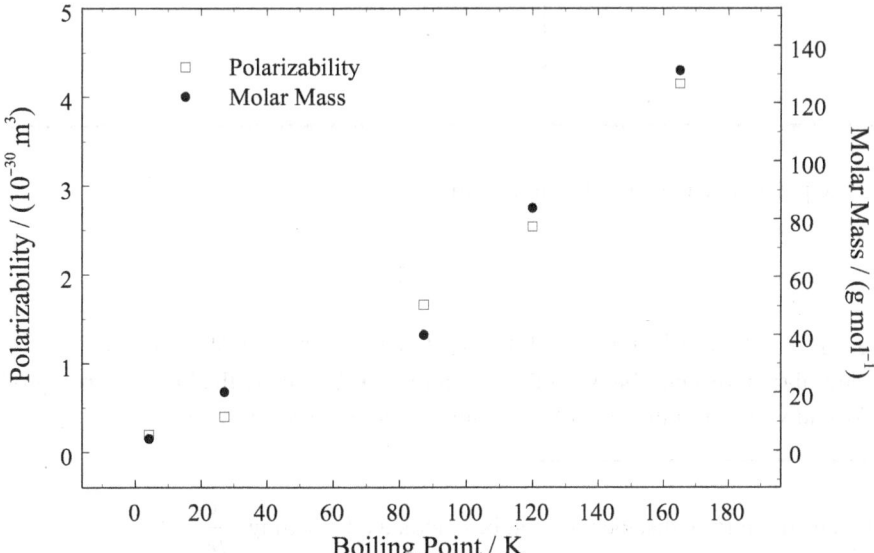

Both polarizability and molar mass seem to track the boiling point of the noble gas.

17.28   The HF_2^- ion exists as

$$\left[:\ddot{F}-H\cdots:\ddot{F}:\right]^{-}$$

The fact that both HF bonds are the same length suggests that proton tunneling occurs. **(a)** Draw resonance structures for the ion. **(b)** Give a molecular orbital description (with an energy-level diagram) of hydrogen bonding in the ion.

(a)

$$:\ddot{F}-H\cdots:\ddot{F}:^{-} \longleftrightarrow \quad^{-}:\ddot{F}:\cdots H-\ddot{F}:$$

(b) The $1s$ orbital on the H atom and a $2p$ orbital on each of the F atoms (the ones along the internuclear axis) combine to form 3 σ molecular orbitals: one bonding, one nonbonding, and one antibonding. There are four electrons to be accommodated in these molecular orbitals, and they are placed, paired, in the lowest two. Thus, there is a delocalized σ bond extending over the entire ion and a delocalized "lone pair" as a result of the nonbonding molecular orbital that has significant electron density at the fluorines, but a node at the hydrogen. (The other 12 valence electrons in the ion are in localized orbitals on the two fluorines.)

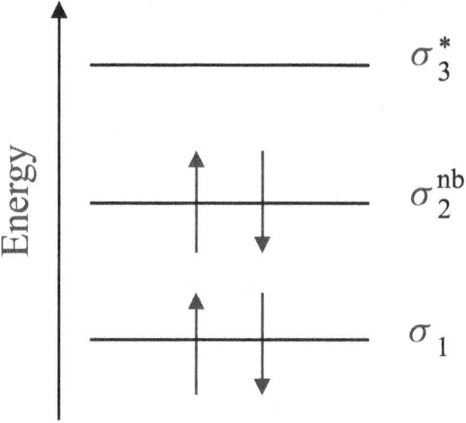

17.30 The potential energy of the helium dimer (He_2) is given by

$$V = \frac{B}{r^{13}} - \frac{C}{r^6}$$

where $B = 9.29 \times 10^4$ kJ Å^{13} (mol dimer)$^{-1}$ and $C = 97.7$ kJ Å^6 (mol dimer)$^{-1}$. **(a)** Calculate the equilibrium distance between the He atoms. **(b)** Calculate the binding energy of the dimer. **(c)** Would you expect the dimer to be stable at room temperature (300 K)?

(a) The equilibrium distance, r_e, can be calculated by setting $\dfrac{dV}{dr} = 0$.

$$\frac{dV}{dr} = -\frac{13B}{r^{14}} + \frac{6C}{r^7}$$

$$-\frac{13B}{r_e^{14}} + \frac{6C}{r_e^7} = 0$$

$$r_e = \left(\frac{13B}{6C}\right)^{1/7} = \left\{\frac{13\left[9.29 \times 10^4 \text{ kJ Å}^{13} \text{ (mol dimer)}^{-1}\right]}{6\left[97.7 \text{ kJ Å}^6 \text{ (mol dimer)}^{-1}\right]}\right\}^{1/7} = 2.975 \text{ Å} = 2.98 \text{ Å}$$

(b) The binding energy, $V(r_e)$, is

$$V(r_e) = \frac{B}{r_e^{13}} - \frac{C}{r_e^6}$$

$$= \frac{9.29 \times 10^4 \text{ kJ Å}^{13} \text{ (mol dimer)}^{-1}}{\left(2.975 \text{ Å}\right)^{13}} - \frac{97.7 \text{ kJ Å}^6 \text{ (mol dimer)}^{-1}}{\left(2.975 \text{ Å}\right)^6}$$

$$= -7.60 \times 10^{-2} \text{ kJ (mol dimer)}^{-1}$$

(c) The thermal energy at 300 K is

$$RT = \left(8.314 \text{ J K}^{-1} \text{ mol}^{-1}\right)(300 \text{ K}) = 2.49 \times 10^3 \text{ J mol}^{-1} = 2.49 \text{ kJ mol}^{-1}$$

which is much larger than 7.60×10^{-2} kJ (mol dimer)$^{-1}$. Thus the dimer would not be stable at room temperature. This species has been observed and studied at low temperature.

17.32   The energy of a hydrogen bond between two water molecules is about 10 times that of van der Waals interaction between two xenon atoms. Yet water molecules dimerizes in air only about 30 percent more frequently than xenon atoms. Explain.

The equilibrium between the dimer and the two monomers that form it is governed by the difference in Gibbs energy, $\Delta_r G^\circ = -RT \ln K$. Both enthalpy and entropy contribute to the difference in Gibbs energy. The stronger intermolecular hydrogen bonds between two water molecules do indeed gives the formation of a water dimer an enthalpic advantage over the formation of xenon dimer as a result of the weaker dispersion interactions between the two xenon atoms. However, the water molecules must approach each other in one of a few specific orientations to allow hydrogen bonding. (See Figure 17.13) Xenon atoms, on the other hand are spherical, and one can form a van der Waals bond with another in any direction. Thus, the formation of the xenon dimer has an entropic advantage over the formation of the water dimer. The two effects work against each other, and the increased entropy cost of forming the water dimer partially cancels the favorable enthalpy difference to reduce the frequency of dimerization.

CHAPTER 18

The Solid State

PROBLEMS AND SOLUTIONS

18.2 When X rays with a wavelength of 0.85 Å are diffracted by a metallic crystal, the angle of first-order diffraction ($n = 1$) is measured to be 14.8°. What is the distance between the layers of atoms responsible for the diffraction?

$$2d \sin \theta = n\lambda$$

$$d = \frac{n\lambda}{2 \sin \theta} = \frac{(1)\left(0.85 \text{ Å}\right)}{2 \sin 14.8°} = 1.7 \text{ Å}$$

18.4 The distance between layers in a NaCl crystal is 282 pm. X rays are diffracted from these layers at an angle of 23.0°. Assuming that $n = 1$, calculate the wavelength of the X rays in nm.

$$2d \sin \theta = n\lambda$$

$$\lambda = \frac{2d \sin \theta}{n} = \frac{(2)\left(282 \text{ pm}\right) \sin 23.0°}{1} = 220 \text{ pm} = 0.220 \text{ nm}$$

18.6 Aluminum has a face-centered cubic lattice. The cell dimension is 4.05 Å. Calculate the closest interatomic distance and the density of the metal.

The face of the unit cell is shown below. The atomic radius is denoted by r. Thus, the diagonal of the face has a length of $4r$.

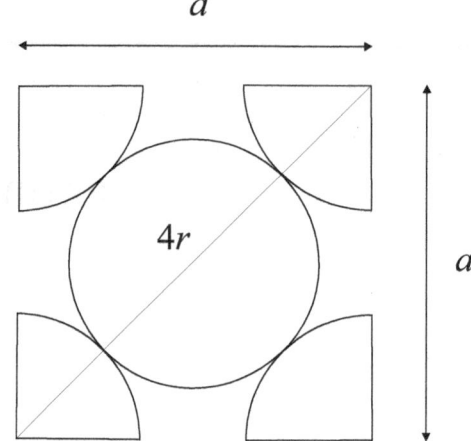

According to the figure and Pythagoras' theorem,

$$(4r)^2 = a^2 + a^2$$

$$r = \frac{\sqrt{2}a}{4} = \frac{\sqrt{2}\left(4.05 \text{ Å}\right)}{4} = 1.432 \text{ Å}$$

The interatomic distance is $2r = 2.86$ Å.

The density of the metal is

$$d = \frac{\text{Mass of metal in one unit cell}}{\text{Volume of a unit cell}}$$

$$= \frac{4 \text{ (mass of 1 Al atom)}}{a^3}$$

$$= \frac{(4)\,(26.98 \text{ amu})\,\left(1.6605 \times 10^{-24} \text{ g amu}^{-1}\right)}{\left(4.05 \times 10^{-8} \text{ cm}\right)^3}$$

$$= 2.70 \text{ g cm}^{-3}$$

18.8 Explain why diamond is harder than graphite. Why is graphite an electrical conductor but diamond is not?

In the diamond lattice, each carbon atom is covalently bonded to four other carbon atoms in a three-dimensional array. The bonds are strong and uniform, leading to a very hard substance. In graphite, the carbon atoms in each layer are covalently bonded to each other, but the individual layers are held together only by weak dispersion forces. Thus, graphite is easily cleaved between layers and is not hard.

In diamond, the bonding electrons are localized in covalent σ bonds so that the material is not an electrical conductor. Graphite contains delocalized π bonds in the planes of the layers, and the electrons in these delocalized orbitals are relatively free to move throughout the layer, rendering the material electrically conductive.

18.10   Metallic iron crystallizes in a cubic lattice. The unit cell edge length is 287 pm. The density of iron is 7.87 g cm^{-3}. How many iron atoms are there within a unit cell?

The mass of iron atoms in a unit cell is

$$m = dV = \left(7.87 \text{ g cm}^{-3}\right)\left(287 \times 10^{-10} \text{ cm}\right)^3 = 1.860 \times 10^{-22} \text{ g}$$

The number of iron atoms in a unit cell is

$$\left(\frac{1.860 \times 10^{-22} \text{ g}}{55.85 \text{ g mol}^{-1}}\right)\left(\frac{6.022 \times 10^{23} \text{ atoms}}{1 \text{ mol}}\right) = 2.006 \text{ atoms} = 2 \text{ atoms}$$

Thus, iron crystallizes in a body-centered cubic lattice. (The simple cubic lattice has 1 atom per unit cell and the face-centered cubic lattice has 4 atoms per unit cell.)

18.12   Barium metal crystallizes in a body-centered cubic lattice (the Ba atoms are at the lattice points only). The unit cell edge length is 501.5 pm, and the density of the metal is 3.50 g cm^{-3}. Using this information, calculate the Avogadro constant.

The mass of Ba atoms in one unit cell is

$$m = dV = \left(3.50 \text{ g cm}^{-3}\right)\left(501.5 \times 10^{-10} \text{ cm}\right)^3 = 4.414 \times 10^{-22} \text{ g}$$

Because there are 2 Ba atoms in a bcc unit cell, the mass of one Ba atom is

$$\frac{4.414 \times 10^{-22} \text{ g}}{2} = 2.207 \times 10^{-22} \text{ g}$$

The Avogadro constant is

$$\frac{\text{Molar mass of Ba}}{\text{Mass of 1 Ba atom}} = \frac{137.3 \text{ g mol}^{-1}}{2.207 \times 10^{-22} \text{ g}} = 6.22 \times 10^{23} \text{ mol}^{-1}$$

This is 3.3% larger than the accepted value.

18.14   Europium crystallizes in a body-centered cubic lattice (the Eu atoms occupy only the lattice points). The density of Eu is 5.26 g cm^{-3}. Calculate the unit cell edge length in pm.

There are 2 Eu atoms in one unit cell. Thus, the mass of Eu in a unit cell is

$$(2 \text{ atoms})\left(152.0 \text{ g mol}^{-1}\right)\left(\frac{1 \text{ mol}}{6.022 \times 10^{23} \text{ atoms}}\right) = 5.048 \times 10^{-22} \text{ g}$$

The volume of the unit cell is

$$V = \frac{m}{d} = \frac{5.048 \times 10^{-22} \text{ g}}{5.26 \text{ g cm}^{-3}} = 9.597 \times 10^{-23} \text{ cm}^3$$

Thus, the unit cell edge length is

$$a = V^{1/3} = \left(9.597 \times 10^{-23} \text{ cm}^3\right)^{1/3} = 4.58 \times 10^{-8} \text{ cm} = 458 \text{ pm}$$

18.16 A face-centered cubic cell contains 8 X atoms at the corners of the cell and 6 Y atoms at the faces. What is the empirical formula of the solid?

Each corner atom is shared by 8 unit cells and each atom at a face is shared by 2 unit cells. Thus, in a unit cell, there are $8 \times 1/8 = 1$ X atom and $6 \times 1/2 = 3$ Y atoms. The empirical formula is XY_3.

18.18 Argon crystallizes in the face-centered cubic arrangement. Given that the atomic radius of argon is 191 pm, calculate the density of solid argon.

There are 4 Ar atoms in one unit cell. Thus, the mass of Ar in a unit cell is

$$(4 \text{ atoms}) \left(39.95 \text{ g mol}^{-1}\right) \left(\frac{1 \text{ mol}}{6.022 \times 10^{23} \text{ atoms}}\right) = 2.654 \times 10^{-22} \text{ g}$$

The volume of the unit cell is

$$V = a^3 = \left(\sqrt{8}r\right)^3 = \left[\sqrt{8}\left(191 \times 10^{-10} \text{ cm}\right)\right]^3 = 1.577 \times 10^{-22} \text{ cm}^3$$

Thus, the density of solid Ar is

$$d = \frac{m}{V} = \frac{2.654 \times 10^{-22} \text{ g}}{1.577 \times 10^{-22} \text{ cm}^3} = 1.68 \text{ g cm}^{-3}$$

18.20 Use the Born–Haber cycle (see Section 17.2) to calculate the lattice energy of LiF. [The heat of sublimation of Li is 155.2 kJ mol^{-1} and $\Delta_f\overline{H}^\circ$(LiF) $= -594.1$ kJ mol^{-1}. Bond enthalpy for F_2 is 158.8 kJ mol^{-1}. Other data may be found in Tables 12.7 and 12.8.]

The Born–Haber cycle is

The lattice energy is $-\Delta H_5^\circ$. Because

$$\Delta_f\overline{H}^\circ = \Delta H_1^\circ + \Delta H_2^\circ + \Delta H_3^\circ + \Delta H_4^\circ + \Delta H_5^\circ$$

The lattice energy is

$$-\Delta H_5^\circ = \Delta H_1^\circ + \Delta H_2^\circ + \Delta H_3^\circ + \Delta H_4^\circ - \Delta_f\overline{H}^\circ$$

The individual enthalpy changes are

$$\Delta H_1^\circ = 155.2 \text{ kJ mol}^{-1} \qquad \text{[enthalpy of sublimation of Li(s), given]}$$

$$\Delta H_2^\circ = \frac{158.8 \text{ kJ mol}^{-1}}{2} = 79.4 \text{ kJ mol}^{-1} \qquad \text{[bond enthalpy of } F_2(g) \text{, given]}$$

$$\Delta H_3^\circ = 520 \text{ kJ mol}^{-1} \qquad \text{[first ionization energy of Li(g), Table 12.7]}$$

$$\Delta H_4^\circ = -328 \text{ kJ mol}^{-1} \qquad \text{[negative electron affinity of F(g), Table 12.8]}$$

$$\Delta_f \overline{H}^\circ = -594.1 \text{ kJ mol}^{-1} \qquad \text{[enthalpy of formation, given]}$$

Thus, the lattice energy is

$$(155.2 + 79.4 + 520 - 328 + 594.1) \text{ kJ mol}^{-1} = 1021 \text{ kJ mol}^{-1}$$

18.22 From the following data, explain why magnesium chloride in the solid state is $MgCl_2$ and not MgCl, whereas sodium chloride is NaCl and not $NaCl_2$.

	Mg	Na
First ionization energy	738 kJ mol^{-1}	496 kJ mol^{-1}
Second ionization energy	1450 kJ mol^{-1}	4560 kJ mol^{-1}

The lattice energy of $MgCl_2$ is 2527 kJ mol^{-1}.

The lattice energy for $MgCl_2$ is 2527 kJ mol^{-1}, which is more than enough to compensate for the energy needed, $\left(738 \text{ kJ mol}^{-1} + 1450 \text{ kJ mol}^{-1}\right) = 2188 \text{ kJ mol}^{-1}$, to remove the first two electrons from the Mg atom to form the Mg^{2+} cation.

For NaCl, even though a lattice with the Na^{2+} ion, such as would occur in $NaCl_2$, would have a lattice energy greater than that of NaCl, it is unable to compensate for the large second ionization energy of Na. The lattice energy of a hypothetical $NaCl_2$ crystal could be estimated as being similar to that for $MgCl_2$, namely 2527 kJ mol^{-1}, but the sum of the first two ionization energies for Na is $\left(496 \text{ kJ mol}^{-1} + 4560 \text{ kJ mol}^{-1}\right) = 5056 \text{ kJ mol}^{-1}$.

18.24 Without referring to a handbook of chemistry, decide which of the following has a greater density: diamond or graphite.

Referring to Figure 18.27, it is apparent that with the relatively large spacing between layers in graphite, there is more "empty space" in this structure than in diamond. Thus, diamond has the greater density.

18.26 Compare the temperature dependence of electrical conduction in an aqueous solution and in a metal.

In a solution, the electrical conductance depends on the motion of the ions so that increasing the temperature increases the conductance. In a metal, however, electrical conductance is a result of electron delocalization, which is disrupted by lattice vibrations. Thus, in a metal, increasing temperature results in a decrease in conductance.

18.28 Classify the solid state of the following substances as ionic crystals, covalent crystals, molecular crystals, or metallic crystals: **(a)** SiO_2, **(b)** SiC, **(c)** S_8, **(d)** KBr, **(e)** Mg, **(f)** LiCl, **(g)** Cr.

(a) Covalent crystal

(b) Covalent crystal

(c) Molecular crystal

(d) Ionic crystal

(e) Metallic crystal

(f) Ionic crystal

(g) Metallic crystal

18.30 Zinc selenide (ZnSe) crystallizes in the zinc blende structure (see Figure 18.24) and has a density of 5.42 g cm^{-3}. **(a)** How many Zn^{2+} and Se^{2-} ions are in each unit cell? **(b)** What is the mass of a unit cell? **(c)** What is the edge length of a unit cell?

(a) For Se^{2-}, there is one ion at each of the eight corners and one ion at each of the six faces, giving $8 \times 1/8 + 6 \times 1/2 = 4$ ions per cell.

For Zn^{2+}, there are 4 ions (not shared) in the unit cell.

(b) The mass of a unit cell is

$$m = 4\,(65.38 + 78.96) \text{ g mol}^{-1}$$

$$= \left(577.36 \text{ g mol}^{-1}\right)\left(\frac{1 \text{ mol}}{6.022 \times 10^{23}}\right)$$

$$= 9.5875 \times 10^{-22} \text{ g}$$

(c) The length of each cell is

$$a = V^{1/3} = \left(\frac{m}{d}\right)^{1/3}$$

$$= \left(\frac{9.5875 \times 10^{-22} \text{ g}}{5.42 \text{ g cm}^{-3}}\right)^{1/3}$$

$$= 5.61 \times 10^{-8} \text{ cm} = 561 \text{ pm}$$

18.32 The isolated O^{2-} ion is unstable, so it is not possible to measure the electron affinity of the O^- ion directly. Show how you can calculate the value by using the lattice energy of MgO

(3890 kJ mol^{-1}) and the Born–Haber cycle. [Useful information: Mg(s) → Mg(g) $\Delta H^\circ = 148$ kJ mol^{-1}.]

The enthalpy of formation of MgO(s) can be represented by the following equation:

$$\text{Mg}(s) + 1/2\, \text{O}_2(g) \rightarrow \text{MgO}(s) \quad \Delta_f \overline{H}^\circ = -601.8 \text{ kJ mol}^{-1} \quad \text{[Appendix B]}$$

This is equivalent to the sum of the following processes:

Mg(s) → Mg(g)	$\Delta H_1^\circ = 148 \text{ kJ mol}^{-1}$	[Given]
1/2 O$_2$(g) → O(g)	$\Delta H_2^\circ = 498.8/2 \text{ kJ mol}^{-1} = 249.4 \text{kJ mol}^{-1}$	[Table 3.4]
Mg(g) → Mg^{2+}(g) + 2e^-	$\Delta H_3^\circ = (738.1 + 1450) \text{ kJ mol}^{-1} = 2188.1 \text{ kJ mol}^{-1}$	[Table 12.7]
O(g) + e^- → O$^-$(g)	$\Delta H_4^\circ = -141 \text{ kJ mol}^{-1}$	[Table 12.8]
O$^-$(g) + e^- → O^{2-}(g)	$\Delta H_5^\circ = ?$	
Mg^{2+}(g) + O^{2-}(g) → MgO(s)	$\Delta H_6^\circ = -3890 \text{ kJ mol}^{-1}$	[Given]

Consequently,

$$\Delta_f \overline{H}^\circ = \Delta H_1^\circ + \Delta H_2^\circ + \Delta H_3^\circ + \Delta H_4^\circ + \Delta H_5^\circ + \Delta H_6^\circ$$

$$\Delta H_5^\circ = \Delta_f \overline{H}^\circ - \Delta H_1^\circ - \Delta H_2^\circ - \Delta H_3^\circ - \Delta H_4^\circ - \Delta H_6^\circ$$

$$= (-601.8 - 148 - 249.4 - 2188.1 + 141 + 3890) \text{ kJ mol}^{-1}$$

$$= 844 \text{ kJ mol}^{-1}$$

Thus, the electron affinity of the O$^-$ ion is $-\Delta H_5^\circ$, or -844 kJ mol^{-1}. As with the noble gases (see Table 12.8), the electron affinity of the O$^-$ ion is negative.

18.34 Crystals of a pure compound and the same compound with deuterium atoms substituted for hydrogen atoms are separately grown. Compare and contrast how the neutron and X-ray diffraction patterns will look for these two isotopically-substituted crystals.

Because X-rays are scattered primarily by electrons, they are scattered weakly by both hydrogen and deuterium atoms. Furthermore, the scattering intensity (or lack of it) would be the same for both isotopes. Thus, the X-ray diffraction patterns for the two isotopically-substituted crystals would look identical, or at least extremely similar. Neutrons, on the other hand are scattered by nuclei, and are scattered particularly strongly by ^{1}H nuclei compared to ^{2}H nuclei. Consequently, the neutron diffraction pattern of the normal isotopologue will show many strong lines that do not appear in the X-ray pattern. These lines will be weaker in the neutron diffraction pattern of the deuterated version. Additionally, the reduced zero-point motion of the C–D bonds relative to the C–H bonds may be significant enough to cause a slight shift in the positions of the diffraction lines for the two crystals.

The Liquid State

PROBLEMS AND SOLUTIONS

19.2 At 293 K, the time of flow for water through an Ostwald viscometer is 342.5 s; for the same volume of an organic solvent, the time of flow is 271.4 s. Calculate the viscosity of the organic liquid relative to that of water. The density of the organic solvent is 0.984 g cm^{-3}.

Rearrange Equation 19.7,

$$\frac{\eta_{\text{organic}}}{\eta_{\text{water}}} = \frac{(\rho t)_{\text{organic}}}{(\rho t)_{\text{water}}}$$

to give

$$\eta_{\text{organic}} = \frac{(\rho t)_{\text{organic}}}{(\rho t)_{\text{water}}} \eta_{\text{water}}$$

$$= \frac{\left(0.984 \text{ g cm}^{-3}\right)(271.4 \text{ s})}{\left(1.00 \text{ g cm}^{-3}\right)(342.5 \text{ s})} \left(0.00101 \text{ N s m}^{-2}\right)$$

$$= 7.88 \times 10^{-4} \text{ N s m}^{-2}$$

The viscosity of the organic liquid is 78.0% of that of water.

19.4 An arteriole has a diameter of 2.4×10^{-5} m and a blood flow rate of 2.6×10^{-3} m s^{-1}. Calculate the pressure drop, ΔP, from one end to the other if the length of the arteriole is 5.0×10^{-3} m.

Rearranging Equation 19.9, and using $\eta = 0.004$ N s m^{-2} from Table 19.1,

$$\Delta P = \frac{8\eta L Q}{\pi R^4} = \frac{8\eta L \left(\pi R^2 v\right)}{\pi R^4} = \frac{8\eta L v}{R^2}$$

$$= \frac{8\left(0.004 \text{ N s m}^{-2}\right)\left(5.0 \times 10^{-3} \text{ m}\right)\left(2.6 \times 10^{-3} \text{ m s}^{-1}\right)}{\left(\frac{2.4 \times 10^{-5} \text{ m}}{2}\right)^2}$$

$$= 3 \times 10^3 \text{ N m}^{-2}$$

The pressure drop is 3% of an atmosphere, or about 20 torr.

19.6   Show that the Reynolds number (see Equation 19.4) is dimensionless.

The Reynolds number is given by

$$\frac{2Rv\rho}{\eta}$$

The units are

$$\frac{(m)\left(m\,s^{-1}\right)\left(kg\,m^{-3}\right)}{N\,s\,m^{-2}} = \frac{(m)\left(m\,s^{-1}\right)\left(kg\,m^{-3}\right)}{\left(kg\,m\,s^{-2}\right)\,s\,m^{-2}} = \text{dimensionless}$$

19.8   The rate of flow of a liquid through a cylindrical tube that has an inner radius of 0.12 cm and a length of 26 cm is 364 cm³ in 88 s. The pressure drop between the ends of the tube is 57 torr. Calculate the liquid's viscosity. Is the flow laminar? The density of the liquid is 0.98 g cm⁻³.

The viscosity can be calculated by rearranging the pressure-drop equation from p. 850:

$$\Delta P = \frac{8\eta L Q}{\pi R^4}$$

to give

$$\eta = \frac{\Delta P \pi R^4}{8LQ}$$

$$= \frac{(57\ \text{torr})\left(\frac{1\ \text{atm}}{760\ \text{torr}}\right)\left(\frac{1.01325 \times 10^5\ \text{Pa}}{1\ \text{atm}}\right)\pi\left(0.12 \times 10^{-2}\ \text{m}\right)^4}{8\left(26 \times 10^{-2}\ \text{m}\right)\left(\frac{364 \times 10^{-6}\ \text{m}^3}{88\ \text{s}}\right)}$$

$$= 5.75 \times 10^{-3}\ \text{N}\,\text{s}\,\text{m}^{-2}$$

$$= 5.8 \times 10^{-3}\ \text{N}\,\text{s}\,\text{m}^{-2}$$

Before calculating the Reynolds number, the velocity of the liquid needs to be determined. In time t, a volume $\pi R^2 L$ flows through the tube. The velocity is

$$v = \frac{L}{t} = \frac{\frac{V}{\pi R^2}}{t} = \frac{V}{\pi R^2 t} = \frac{Qt}{\pi R^2 t}$$

$$= \frac{Q}{\pi R^2}$$

$$= \frac{\frac{364 \times 10^{-6}\ \text{m}^3}{88\ \text{s}}}{\pi\left(0.12 \times 10^{-2}\ \text{m}\right)^2} = 0.914\ \text{m}\,\text{s}^{-1}$$

The Reynolds number is

$$\frac{2Rv\rho}{\eta} = \frac{(2)\left(0.12 \times 10^{-2}\ \text{m}\right)\left(0.914\ \text{m}\,\text{s}^{-1}\right)\left(0.98 \times 10^3\ \text{kg}\,\text{m}^3\right)}{5.75 \times 10^{-3}\ \text{N}\,\text{s}\,\text{m}^{-2}} = 374$$

Since the Reynolds number is less than 2000, the flow is laminar.

19.10   Give a molecular interpretation for the decrease in surface tension of a liquid with increasing temperature.

The decrease in surface tension with temperature is analogous to the decrease in viscosity with temperature (see Problem 19.1). At higher temperatures, molecules have a greater average kinetic energy and are able to overcome the attractive intermolecular forces responsible for surface tension. Consequently, it is easier to stretch the surface of the liquid, since the molecules are not held together as tightly.

19.12   Both ethanol and mercury are used in thermometers. Explain the difference between the menisci of the liquids in these two types of thermometers.

In ethanol, adhesion is stronger than cohesion giving rise to an upward-curving meniscus. The reverse holds true for mercury. (See Figure 19.7.)

19.14   The surface tension of quinoline is twice that of acetone at 20°C. If the capillary rise is 2.5 cm for quinoline, what is the rise for acetone in the same capillary? Assume that the angles of contact are zero. The densities of quinoline and acetone at 20°C are 1.09 g cm^{-3} and 0.79 g cm^{-3}, respectively.

Because the contact angles are assumed to be 0, Equation 19.12 gives

$$h_{acetone} = \frac{2\gamma_{acetone}}{rg\rho_{acetone}}$$

$$h_{quinoline} = \frac{2\gamma_{quinoline}}{rg\rho_{quinoline}}$$

Thus,

$$\frac{h_{acetone}}{h_{quinoline}} = \frac{\gamma_{acetone}\rho_{quinoline}}{\rho_{acetone}\gamma_{quinoline}}$$

$$h_{acetone} = \left(\frac{\gamma_{acetone}}{\gamma_{quinoline}}\right)\left(\frac{\rho_{quinoline}}{\rho_{acetone}}\right) h_{quinoline}$$

$$= \left(\frac{1}{2}\right)\left(\frac{1.09\ \text{g cm}^{-3}}{0.79\ \text{g cm}^{-3}}\right)(2.5\ \text{cm})$$

$$= 1.7\ \text{cm}$$

19.16   Two capillary tubes with inside diameters of 1.4 mm and 1.0 mm, respectively, are inserted into a liquid of density 0.95 g cm^{-3}. Calculate the surface tension of the liquid if the difference between the capillary rises in the tubes is 1.2 cm. Assume the contact angle is zero.

Let h_1 be the liquid rise for the capillary with diameter $r_1(= 1.4 \text{ mm})$ and h_2 be the liquid rise for the capillary with diameter $r_2(= 1.0 \text{ mm})$. Then, with contact angles of 0,

$$h_1 = \frac{2\gamma}{r_1 g\rho}$$

$$h_2 = \frac{2\gamma}{r_2 g\rho}$$

$$h_2 - h_1 = \frac{2\gamma}{g\rho}\left(\frac{1}{r_2} - \frac{1}{r_1}\right)$$

Therefore, the surface tension is

$$\gamma = \frac{g\rho\left(h_2 - h_1\right)}{2\left(\frac{1}{r_2} - \frac{1}{r_1}\right)}$$

$$= \frac{\left(9.81 \text{ m s}^{-2}\right)\left(0.95 \times 10^3 \text{ kg m}^{-3}\right)\left(1.2 \times 10^{-2} \text{ m}\right)}{(2)\left(\frac{1}{1.0\times 10^{-3}\text{ m}} - \frac{1}{1.4\times 10^{-3}\text{ m}}\right)}$$

$$= 0.20 \text{ N m}^{-1}$$

Note that this is an unusually large surface tension, an order of magnitude larger than most common organic solvents.

19.18   The diffusion coefficient of sucrose in water at 298 K is $0.46 \times 10^{-5} \text{ cm}^2 \text{ s}^{-1}$, and the viscosity of water at the same temperature is $0.0010 \text{ N s m}^{-2}$. From these data, estimate the effective radius of a sucrose molecule.

Rearrange Equation 19.29 to give

$$r = \frac{k_B T}{6\pi \eta D}$$

$$= \frac{\left(1.381 \times 10^{-23} \text{ J K}^{-1}\right)(298 \text{ K})}{6\pi\left(0.0010 \text{ N s m}^{-2}\right)\left(0.46 \times 10^{-9} \text{ m}^2 \text{ s}^{-1}\right)}$$

$$= 4.7 \times 10^{-10} \text{ m} = 4.7 \text{ Å}$$

19.20   Diffusion coefficients have been measured for many solid systems. If the diffusion coefficient of bismuth in lead is $1.1 \times 10^{-16} \text{ cm}^2 \text{ s}^{-1}$ at 20°C, calculate how long it will take (in years) for a bismuth atom to travel 1.0 cm.

Rearrange Equation 19.26 to give

$$t = \frac{\overline{x^2}}{2D}$$

$$t = \frac{\left(1.0 \times 10^{-2} \text{ m}\right)^2}{(2)\left(1.1 \times 10^{-20} \text{ m}^2 \text{ s}^{-1}\right)}$$

$$= \left(4.55 \times 10^{15} \, \text{s}\right) \left(\frac{1 \, \text{h}}{3600 \, \text{s}}\right) \left(\frac{1 \, \text{day}}{24 \, \text{h}}\right) \left(\frac{1 \, \text{year}}{365 \, \text{days}}\right)$$

$$= 1.4 \times 10^8 \, \text{years}$$

This is about 3% of the age of the Earth.

19.22 Two soap bubbles of radii r_1 and r_2 ($r_2 > r_1$) are connected by a piece of tubing with a stopcock. Predict how the size of the bubbles will change when the stopcock is opened.

The smaller bubble has the larger internal pressure (see Appendix 19.1). Thus, the smaller bubble will get smaller and the larger bubble gets larger.

19.24 The carbon monoxide–hemoglobin complex has a diffusion coefficient of 0.062×10^{-9} $\text{m}^2 \, \text{s}^{-1}$ in water at 298 K. In the more viscous cytoplasm, the diffusion coefficient is only $0.013 \times 10^{-9} \, \text{m}^2 \, \text{s}^{-1}$. How long would it take for such a complex to travel the 3.0-μm length of a bacterial cell?

Rearrange Equation 19.26 to calculate travel time.

$$t = \frac{\overline{x^2}}{2D} = \frac{\left(3.0 \times 10^{-6} \, \text{m}\right)^2}{(2) \left(0.013 \times 10^{-9} \, \text{m}^2 \, \text{s}^{-1}\right)} = 0.35 \, \text{s}$$

19.26 A hypodermic syringe is filled with a solution of viscosity $1.6 \times 10^{-3} \, \text{N s m}^{-2}$. The plunger area of the syringe is $7.5 \times 10^{-5} \, \text{m}^2$, and the length of the needle is 0.026 m. The internal radius of the needle is 4.0×10^{-4} m. The gauge pressure in a vein is 1850 Pa (14 mmHg). Calculate the force in newtons that must be applied to the plunger so that $1.2 \times 10^{-6} \, \text{m}^3$ of the solution can be injected in 4.0 s.

First calculate ΔP by rearranging Equation 19.9.

$$\Delta P = \frac{8 \eta L Q}{\pi R^4}$$

$$= \frac{(8) \left(1.6 \times 10^{-3} \, \text{N s m}^2\right) (0.026 \, \text{m}) \left(\frac{1.2 \times 10^{-6} \, \text{m}^3}{4.0 \, \text{s}}\right)}{\pi \left(4.0 \times 10^{-4} \, \text{m}\right)^4}$$

$$= 1.24 \times 10^3 \, \text{N m}^{-2}$$

The pressure that must be applied to the plunger must be ΔP above the gauge pressure in a vein; that is

$$1.24 \times 10^3 \, \text{N m}^{-2} + 1850 \, \text{N m}^{-2} = 3.09 \times 10^3 \, \text{N m}^{-2}$$

which corresponds to a force of

$$\left(3.09 \times 10^3 \, \text{N m}^{-2}\right) \left(7.5 \times 10^{-5} \, \text{m}^2\right) = 0.23 \, \text{N}$$

19.28   How much work is required to break up 1 mole of water at 20°C into spherical droplets that have a radius of 4.16×10^{-3} m? [*Hint*: The volume of a sphere is $(4/3)\pi r^3$, and the surface area of a sphere is $4\pi r^2$, where r is the radius of the sphere.] The density of water is 1.0 g cm^{-3}.

Work must be done to increase the surface area in forming the droplets. A spherical molar volume of water has a mass of 18.02 g, and a volume of

$$V = \frac{m}{\rho} = \frac{18.02 \text{ g}}{1.0 \text{ g cm}^{-3}} \left(\frac{1 \text{ m}}{100 \text{ cm}}\right)^3 = 1.802 \times 10^{-5} \text{ m}^3$$

Since the volume of a sphere is given by $V = \frac{4}{3}\pi r^3$, the radius of the spherical molar volume of water is

$$r = \left(\frac{3V}{4\pi}\right)^{1/3} = \left[\frac{3\left(1.802 \times 10^{-5} \text{ m}^3\right)}{4\pi}\right]^{1/3} = 1.6264 \times 10^{-2} \text{ m}$$

and the surface area of this sphere is

$$A_{\text{i}} = 4\pi r^2 = 4\pi \left(1.6264 \times 10^{-2} \text{ m}\right)^2 = 3.3240 \times 10^{-3} \text{ m}^2$$

The volume of each small droplet is

$$V_{\text{drop}} = \frac{4}{3}\pi r^3 = \frac{4}{3}\pi \left(4.16 \times 10^{-3} \text{ m}\right)^3 = 3.016 \times 10^{-7} \text{ m}^3$$

The number of these droplets required to make a mole of water is

$$\frac{1.802 \times 10^{-5} \text{ m}^3}{3.016 \times 10^{-7} \text{ m}^3} = 59.75 \approx 60$$

The total surface area of 60 of these drops is

$$A_{\text{f}} = 60 \left(4\pi r^2\right) = 60 \left(4\pi\right) \left(4.16 \times 10^{-3} \text{ m}\right)^2 = 1.30 \times 10^{-2} \text{ m}^2$$

The increase in surface is opposed by the surface tension, and the work that must be done to overcome this resistance is

$$dw = \gamma \, dA \quad \text{(Table 3.1)}$$

$$w = \gamma \Delta A = \gamma \left(A_{\text{f}} - A_{\text{i}}\right)$$

$$= \left(0.07275 \text{ N m}^{-1}\right) \left(1.30 \times 10^{-2} \text{ m}^2 - 3.3240 \times 10^{-3} \text{ m}^2\right) = 7.0 \times 10^{-4} \text{ J}$$

19.30   The diffusion coefficient of oxygen in air is 0.20 cm^2 s^{-1}; the diffusion coefficient of the same gas in water is about 10^4 times smaller. **(a)** Explain the huge difference in the magnitude of these diffusion coefficients. **(b)** Most animal cells are bathed in fluids, so that a hemoglobinlike molecule and a circulatory system are necessary for the purpose of transporting O_2 to the cells and carrying CO_2 away. (The diffusion coefficients of CO_2 in air and in water are comparable in magnitude to those of oxygen.) Because plants do not have a circulatory system, explain how the O_2 and CO_2 gases are transported efficiently in these systems. **(c)** Insects do possess a circulating system but lack a hemoglobinlike molecule. Considering the diffusion coefficients of CO_2 and

O_2 in water, do you think it likely that ants, bees, and cockroaches can grow to human size, as they sometimes do in horror movies?

(a) Since diffusion involves molecular motion, it will depend greatly upon the medium through which the molecules in question must move. Because the density of a liquid is greater than that of a gas, an oxygen molecule will encounter many more collisions when traveling through a liquid. Thus, diffusion through a liquid is much slower.

(b) Plants have conspicuous intercellular air spaces by which they can take advantage of the large diffusion coefficients of O_2 and CO_2 in the gas phase.

(c) Without a means of efficient transport for O_2 and CO_2, insects are limited to small sizes so that their metabolic processes can obtain enough O_2 via diffusion through their circulating system. Because of this limitation, they can not grow to horror movie size.

19.32 A capillary tube of constant inner diameter d is inserted into an aqueous solution of concentration c at temperature T and pressure P. The solution rises within the tube to a height h. Describe at least five different ways to change this experiment to produce a larger rise h.

Rearranging Equation 19.12 gives

$$h = \frac{2\gamma \cos\theta}{r\rho g}$$

Changing any of the quantities on the right hand side of this equation in the appropriate manner will lead to a larger rise.

1. A tube with a smaller diameter will have a smaller radius, r, and increase h.

2. An increase in temperature will lead to an increase in the volume of the solution and a decrease in the density. This will produce a larger rise.

3. If the solute is a surfactant, lowering the concentration will increase the surface tension and give a greater rise.

4. Reducing the pressure over the capillary, but not over the surface of the liquid sample, will reduce the downward force of the column of liquid in the capillary, resulting in a larger rise.

5. The experiment can be performed on the international space station, where the force of gravity is significantly reduced.

19.34 Explain how a water strider is able to "walk" on water.

The relatively large surface tension of water is an important contributor to the ability of the water strider to walk on water. The insect has long, flexible, hydrophobic legs that distribute its small weight over a large area. As a result, the downward force of the strider is not sufficient to break the surface tension. Since the legs are hydrophobic, they resist wetting, which would draw them into the bulk regions of the water. Recently, however, it has been found that the water strider's legs possess microscopic hairs that trap air. These air bubbles act like miniature flotation devices that keep the insect from sinking. [See X. Gao and L. Jiang, *Nature* **432**, 36 (2004).]

CHAPTER 20

Statistical Thermodynamics

PROBLEMS AND SOLUTIONS

20.2   What is the high temperature limit (i.e., as $T \rightarrow \infty$) of Equation 20.23?

$$\lim_{T \rightarrow \infty} \frac{n_2}{n_1} = \lim_{T \rightarrow \infty} \frac{g_2}{g_1} e^{-\Delta\varepsilon/k_B T} = \frac{g_2}{g_1} e^0 = \frac{g_2}{g_1}$$

At the limit of infinite temperature, the probability that any state is occupied becomes equal, regardless of their energy. Thus each state has the same population, and the relative population of any two energy levels is just given by the ratio of the number of states with the respective energies, which is the ratio of the degeneracies.

20.4   The fundamental vibrational wavenumber for N_2 is 2360 cm^{-1}. For 1 mole of the molecules, calculate the number of N_2 molecules in the $v = 0$ and $v = 1$ levels at (a) 298 K and (b) 1000 K.

The ratio between the number of molecules in the $v = i$ level and the total number of molecules is

$$\frac{n_i}{N} = \frac{e^{-\Delta\varepsilon_i/k_B T}}{\sum e^{-\Delta\varepsilon_i/k_B T}} = \frac{e^{-\Delta\varepsilon_i/k_B T}}{q_{\text{vib}}}$$

Thus, the number of molecules in the $v = i$ level is

$$n_i = N \left(\frac{e^{-\Delta\varepsilon_i/k_B T}}{q_{\text{vib}}} \right)$$

$$= N \left(\frac{e^{-\Delta\varepsilon_i/k_B T}}{\frac{1}{1-e^{-h\nu/k_B T}}} \right)$$

$$= N \exp\left(-\frac{\Delta\varepsilon_i}{k_B T}\right) \left[1 - \exp\left(-\frac{h\nu}{k_B T}\right)\right]$$

ν is related to the vibrational wavenumber:

$$\nu = \tilde{\nu} c = \left(2360 \text{ cm}^{-1}\right)\left(2.998 \times 10^{10} \text{ cm s}^{-1}\right) = 7.0753 \times 10^{13} \text{ s}^{-1}$$

The energy of the v level relative to the $v = 0$ level is $vh\nu$. Thus,

$$E_0 = 0$$

$$E_1 = hv = \left(6.626 \times 10^{-34} \text{ J s}\right) \left(7.0753 \times 10^{13} \text{ s}^{-1}\right) = 4.688 \times 10^{-20} \text{ J}$$

(a) At 298 K,

$$1 - \exp\left(-\frac{hv}{k_B T}\right) = 1 - \exp\left[-\frac{\left(6.626 \times 10^{-34} \text{ J s}\right) \left(7.0753 \times 10^{13} \text{ s}^{-1}\right)}{\left(1.381 \times 10^{-23} \text{ J K}^{-1}\right) (298 \text{ K})}\right] = 1.000$$

Thus,

$$n_0 = \left(6.022 \times 10^{23}\right) \left[\exp(0)\right] (1.000) = 6.02 \times 10^{23}$$

$$n_1 = \left(6.022 \times 10^{23}\right) \left[\exp\left(-\frac{\left(4.688 \times 10^{-20} \text{ J}\right)}{\left(1.381 \times 10^{-23} \text{ J K}^{-1}\right) (298 \text{ K})}\right)\right] (1.000) = 6.80 \times 10^{18}$$

(b) At 1000 K,

$$1 - \exp\left(-\frac{hv}{k_B T}\right) = 1 - \exp\left[-\frac{\left(6.626 \times 10^{-34} \text{ J s}\right) \left(7.0753 \times 10^{13} \text{ s}^{-1}\right)}{\left(1.381 \times 10^{-23} \text{ J K}^{-1}\right) (1000 \text{ K})}\right] = 0.9665$$

Thus,

$$n_0 = \left(6.022 \times 10^{23}\right) \left[\exp(0)\right] (0.9665) = 5.82 \times 10^{23}$$

$$n_1 = \left(6.022 \times 10^{23}\right) \left[\exp\left(-\frac{\left(4.600 \times 10^{-20} \text{ J}\right)}{\left(1.381 \times 10^{-23} \text{ J K}^{-1}\right) (1000 \text{ K})}\right)\right] (0.9665)$$

$$= 1.95 \times 10^{22}$$

20.6 Explain why q_{trans} increases with (a) m and (b) T.

(a) As m increases, the spacing between translational energy levels decreases, so that there are more levels at lower energies, and more levels available to the system. Thus, the partition function increases.

(b) As T increases, more levels with energies above the ground state energy will be populated, and the partition function will increase.

20.8 Calculate the entropy of HCl at 298 K and 1 bar, given that the bond length is 1.275 Å and the masses of ^{1}H and ^{35}Cl are 1.008 amu and 34.97 amu, respectively. The vibrational wavenumber is 2886 cm^{-1}.

The contributions to entropy are translational, rotational, and vibrational.

Translational contribution

$$\overline{S}_{trans} = R \ln \left[\frac{(2\pi m k_B T)^{3/2}}{h^3} \frac{k_B T}{P} e^{5/2} \right]$$

$$= \left(8.314 \text{ J K}^{-1} \text{ mol}^{-1} \right)$$

$$\times \ln \left\{ \frac{\left[2\pi \left[(34.97 + 1.008) \text{ amu} \right] \left(1.6605 \times 10^{-27} \text{ kg amu}^{-1} \right) \left(1.381 \times 10^{-23} \text{ J K}^{-1} \right) (298 \text{ K}) \right]^{3/2}}{\left(6.626 \times 10^{-34} \text{ J s} \right)^3} \right.$$

$$\left. \times \frac{\left(1.381 \times 10^{-23} \text{ J K}^{-1} \right) (298 \text{ K})}{(1 \text{ bar}) \left(\frac{10^5 \text{ N m}^{-2}}{1 \text{ bar}} \right)} e^{5/2} \right\}$$

$$= 153.5 \text{ J K}^{-1} \text{ mol}^{-1}$$

Rotational contribution

To calculate the rotational contribution to the entropy, the moment of inertia of HCl must be found, which in turn requires the reduced mass of the molecule.

$$I = \mu r^2$$

$$= \frac{m_H m_{Cl}}{m_H + m_{Cl}}$$

$$= \frac{(1.008 \text{ amu}) (34.97 \text{ amu})}{1.008 \text{ amu} + 34.97 \text{ amu}} \left(1.6605 \times 10^{-27} \text{ kg amu}^{-1} \right) \left(1.275 \times 10^{-10} \text{ m} \right)^2$$

$$= 2.6447 \times 10^{-47} \text{ kg m}^2$$

This is used in the expression for the rotational contribution to the entropy

$$\overline{S}_{rot} = R \ln q_{rot} + R = R \ln \frac{8\pi^2 I k_B T}{\sigma h^2} + R$$

$$= \left(8.314 \text{ J K}^{-1} \text{ mol}^{-1} \right) \ln \frac{8\pi^2 \left(2.6447 \times 10^{-47} \text{ kg m}^2 \right) \left(1.381 \times 10^{-23} \text{ J K}^{-1} \right) (298 \text{ K})}{(1) \left(6.626 \times 10^{-34} \text{ J s} \right)^2}$$

$$+ 8.314 \text{ J K}^{-1} \text{ mol}^{-1}$$

$$= 33.04 \text{ J K}^{-1} \text{ mol}^{-1}$$

Vibrational contribution

First calculate $h\nu / k_B T$:

$$\frac{h\nu}{k_B T} = \frac{h\tilde{\nu}c}{k_B T} = \frac{\left(6.626 \times 10^{-34} \text{ J s} \right) \left(2886 \text{ cm}^{-1} \right) \left(2.998 \times 10^{10} \text{ cm s}^{-1} \right)}{\left(1.381 \times 10^{-23} \text{ J K}^{-1} \right) (298 \text{ K})} = 13.93$$

The vibrational contribution to entropy is

$$\overline{S}_{\text{vib}} = -R \ln \left(1 - e^{-h\nu/k_B T}\right) + R \frac{h\nu}{k_B T} \frac{1}{e^{h\nu/k_B T} - 1}$$

$$= -\left(8.314\,\text{J K}^{-1}\,\text{mol}^{-1}\right) \ln \left(1 - e^{-13.93}\right) + \left(8.314\,\text{J K}^{-1}\,\text{mol}^{-1}\right) (13.93) \left(\frac{1}{e^{13.93} - 1}\right)$$

$$= 1.107 \times 10^{-4}\,\text{J K}^{-1}\,\text{mol}^{-1}$$

As expected, vibrational motion makes a negligible contribution to entropy at 298 K.

The entropy of HCl is

$$\overline{S} = \left(153.5 + 33.04 + 1.107 \times 10^{-4}\right)\,\text{J K}^{-1}\,\text{mol}^{-1} = 186.5\,\text{J K}^{-1}\,\text{mol}^{-1}$$

20.10 Calculate the translational partition function of helium at 1 bar in a 1.00-m^3 container. The large value of q_{trans} means that this motion can be treated classically. When $q_{\text{trans}} \leq 10$, however, the motion must be treated quantum mechanically. Calculate the temperature at which this change occurs.

The translational partition function at 298 K is

$$q_{\text{trans}} = \frac{\left(2\pi m k_B T\right)^{3/2} V}{h^3}$$

$$= \frac{\left[2\pi\,(4.003\,\text{amu})\left(1.6605 \times 10^{-27}\,\text{kg amu}^{-1}\right)\left(1.381 \times 10^{-23}\,\text{J K}^{-1}\right)(298\,\text{K})\right]^{3/2}\left(1.00\,\text{m}^3\right)}{\left(6.626 \times 10^{-34}\,\text{J s}\right)^3}$$

$$= 7.75 \times 10^{30}$$

When $q_{\text{trans}} = 10$,

$$q_{\text{trans}} = \frac{\left(2\pi m k_B T\right)^{3/2} V}{h^3} = 10$$

$$T^{3/2} = \frac{10 h^3}{\left(2\pi m k_B\right)^{3/2} V}$$

$$= \frac{10\left(6.626 \times 10^{-34}\,\text{J s}\right)^3}{\left[2\pi\,(4.003\,\text{amu})\left(1.6605 \times 10^{-27}\,\text{kg amu}^{-1}\right)\left(1.381 \times 10^{-23}\,\text{J K}^{-1}\right)\right]^{3/2}\left(1.00\,\text{m}^3\right)}$$

$$= 6.641 \times 10^{-27}\,\text{K}^{3/2}$$

$$T = 3.53 \times 10^{-18}\,\text{K}$$

This is several orders of magnitude lower than the lowest temperature recorded. This system may certainly be treated classically at any temperature for which helium is still a gas.

20.12 List the symmetry number (σ) for each of the following molecules: Cl_2, N_2O (NNO), H_2O, HDO, BF_3, CH_4, CH_3Cl. (*Hint*: For CH_4, note that each of the four C–H bonds represents a three-fold symmetry axis about which three successive 120° indistinguishable rotations are possible.)

Cl_2: 2; N_2O: 1; H_2O: 2; HDO: 1; BF_3: 6; CH_4: $4 \times 3 = 12$; CH_3Cl: 3.

20.14 Calculate the approximate value of $\Delta_r S^\circ$ for

$$^{16}O_2(g) + {}^{18}O_2(g) \rightarrow 2\,{}^{16}O^{18}O(g)$$

Assume that differences in molar masses, moments of inertia, and vibrational frequencies are negligible.

The major contribution to $\Delta_r S^\circ$ is the symmetry factor. Assuming that any differences in molar masses, moments of inertia, and vibrational frequencies are negligible, only the symmetry factor in the rotational contribution to entropy need to be considered.

$$\overline{S}_{rot} = R \ln q_{rot} + R = R \ln \frac{8\pi^2 I k_B T}{\sigma h^2} + R$$

Let $A = \frac{8\pi^2 I k_B T}{h^2}$. A is the same for the reactants and product in the reaction considered. Then the entropy expression becomes

$$\overline{S}_{rot} = R \ln A - R \ln \sigma + R$$

The entropy of reaction is

$$\Delta \overline{S}^\circ = 2\overline{S}_{rot}(^{16}O^{18}O) - \overline{S}_{rot}(^{16}O_2) - \overline{S}_{rot}(^{18}O_2)$$

$$= 2\left[R \ln A - R \ln \sigma(^{16}O^{18}O) + R\right] - \left[R \ln A - R \ln \sigma(^{16}O_2) + R\right]$$

$$- \left[R \ln A - R \ln \sigma(^{18}O_2) + R\right]$$

$$= (-R)\left\{2\left[\ln \sigma(^{16}O^{18}O)\right] - \ln \sigma(^{16}O_2) - \ln \sigma(^{18}O_2)\right\}$$

$$= \left(-8.314 \text{ J K}^{-1}\text{ mol}^{-1}\right)[2 \ln 1 - \ln 2 - \ln 2]$$

$$= 11.53 \text{ J K}^{-1}\text{ mol}^{-1}$$

20.16 Provide a simple physical interpretation for the partition function.

Physically, the partition function represents the number of states that are thermally accessible to a molecule at the temperature of interest.

20.18 Normally we expect the molar entropy to increase from N_2 to F_2 across the second row of the periodic table. According to Appendix B, however, we find the molar entropy of O_2 is actually greater than that of F_2. Explain.

Typically, there is no electronic contribution to the entropy of a diatomic molecule at room temperature because the electronic partition function is adequately represented by $q_{elec} = g_0 = 1$ for most molecules. Additionally, $E_{elec} = 0$, so that normally

$$\bar{S}_{elec} = k_B \ln q_{elec}^{N_A} + \frac{E_{elec}}{T}$$

$$= k_B \ln 1^{N_A} + \frac{0}{T}$$

$$= 0$$

O_2 is an exception because the degeneracy of its ground electronic state is $g_0 = 3$, and this causes a non-zero contribution to the entropy.

$$\bar{S}_{elec} = k_B \ln q_{elec}^{N_A} + \frac{E_{elec}}{T}$$

$$= k_B \ln 3^{N_A} + \frac{0}{T}$$

$$= (k_B N_A) \ln 3$$

$$= R \ln 3$$

$$= \left(8.314 \, \mathrm{J\,K^{-1}\,mol^{-1}}\right)(1.099)$$

$$= 9.134 \ \mathrm{J\,K^{-1}\,mol^{-1}}$$

Thus, if O_2 were not a ground state triplet, its molar entropy would be $(205.0 - 9.134)$ $\mathrm{J\,K^{-1}}$ $\mathrm{mol^{-1}}$ or 195.9 $\mathrm{J\,K^{-1}\,mol^{-1}}$, which would fall between that of N_2 (191.6 $\mathrm{J\,K^{-1}\,mol^{-1}}$) and that of F_2 (202.8 $\mathrm{J\,K^{-1}\,mol^{-1}}$).

Useful Conversion Factors

$1\ \text{Å} = 10^{-8}\ \text{cm} = 10^{-10}\ \text{m} = 0.1\ \text{nm} = 100\ \text{pm}$

$1\ \text{atm} = 760\ \text{torr} = 1.01325 \times 10^5\ \text{Pa} = 101.325\ \text{kPa} = 1.01325\ \text{bar}$

$1\ \text{bar} = 1 \times 10^5\ \text{Pa} = 100\ \text{kPa} = 0.9\ 923\ \text{atm}$

$1\ \text{cal} = 4.184\ \text{J}$ (defined)

$1\ e\text{V} = 1.60218 \times 10^{-19}\ \text{J} = 96.4853\ \text{kJ mol}^{-1}$

$1\ \text{L atm} = 101.325\ \text{J}$

At 298 K, $k_{\text{B}}T = 207.1\ \text{cm}^{-1} = 2.478\ \text{kJ mol}^{-1}$

$R = 8.314\ \text{J K}^{-1}\ \text{mol}^{-1} = 0.08206\ \text{L atm} \ldots ^{-1} = 0.0831\ \text{L bar K}^{-1}\ \text{mol}^{-1}$

Some Commonly Used Non-SI Units

Unit	Quantity	Symbol	Conversion Factor
Angstrom	length	Å	$1\ \text{Å} = 10^{-10}\ \text{m} = 100\ \text{pm}$
Calorie	energy	cal	$1\ \text{cal} = 4.184\ \text{J}$
Debye	dipole moment	D	$1\ \text{D} = 3.3356 \times 10^{-30}\ \text{C m}$
Gauss	magnetic field	G	$1\ \text{G} = 10^{-4}\ \text{T}$
Liter	volume	L	$1\ \text{L} = 10^{-3}\ \text{m}^3 = 10^3\ \text{cm}^3$
Torr	pressure	torr	$1\ \text{torr} = 1.3332 \times 10^{-3}\ \text{bar}$ $= \dfrac{1}{760}\ \text{atm}$

Index of Important Figures and Tables

Topic	Text Page
Bond enthalpies	112
Covalent radii	585
Critical constants	20
Dielectric constants	273
Electron affinities of the elements	531
Electron configurations of the elements	526
Electronegativities of the elements	577
Intermolecular interactions	792
Ionic radii	829
Ionization energies of the elements	529
IR frequencies	480
Molecular geometries	580
NMR chemical shifts	643
Observables/operators	410
Real hydrogen atomic wave functions	512

The Greek Alphabet

A	α	alpha
B	β	beta
Γ	γ	gamma
Δ	δ	delta
E	ε	epsilon
Z	ζ	zeta
H	η	eta
Θ	θ	theta
I	ι	iota
K	κ	kappa
Λ	λ	lambda
M	μ	mu
N	ν	nu
Ξ	ξ	xi
O	o	omicron
Π	π	pi
P	ρ	rho
Σ	σ	sigma
T	τ	tau
Y	υ	upsilon
Φ	φ	phi
X	χ	chi
Ψ	ψ	psi
Ω	ω	omega

Values of Some Fundamental Constants

Constant	Value
Atomic mass unit (amu)	$1.660\ 538\ 921 \times 10^{-27}$ kg
Avogadro's constant (N_A)	$6.022\ 141\ 29 \times 10^{23}$ mol^{-1}
Bohr radius (a_0)	$5.291\ 772\ 1092 \times 10^{-11}$ m
Boltzmann constant (k_B)	$1.380\ 6488 \times 10^{-23}$ J K^{-1}
Electron charge (e)	$1.602\ 176\ 565 \times 10^{-19}$ C
Electron mass (m_e)	$9.109\ 382\ 91 \times 10^{-31}$ kg
Faraday constant (F)	96485.3365 C mol^{-1}
Gas constant (R)	$8.314\ 4621$ J K^{-1} mol^{-1}
Neutron mass (m_n)	$1.674\ 927\ 351 \times 10^{-27}$ kg
Permittivity of vacuum (ε_0)	$8.854\ 187\ 817 \times 10^{-12}$ C^2 N^{-1} m^{-2}
Planck constant (h)	$6.626\ 069\ 57 \times 10^{-34}$ J s
Proton mass (m_p)	$1.672\ 621\ 777 \times 10^{-27}$ kg
Rydberg constant $(\tilde{R}_H)$	$109737.315\ 685\ 39$ cm^{-1}
Speed of light in vacuum (c)	$299\ 792\ 458$ m s^{-1} (exactly)

Pressure of Water Vapor at Various Temperatures

Temperature/°C	Water Vapor Pressure/mmHg
0	4.58
5	6.54
10	9.21
15	12.79
20	17.54
25	23.77
30	31.84
35	42.26
40	55.36
45	71.88
50	92.59
55	118.15
60	149.51
65	187.69
70	233.85
75	289.26
80	355.34
85	433.66
90	525.94
95	634.04
100	760.00

Publisher contact:
The MIT Press
Massachusetts Institute of Technology
77 Massachusetts Avenue, Cambridge, MA 02139
mitpress.mit.edu

EU Authorised Representative:
Easy Access System Europe, Mustamäe tee 50,
10621 Tallinn, Estonia
gpsr.requests@easproject.com

Printed by Integrated Books International,
United States of America